EXERCISES FOR
FOURIER ANALYSIS

EXERCISES FOR *FOURIER ANALYSIS*

T.W. KÖRNER

Trinity Hall, Cambridge

Published by the Press Syndicate of the University of Cambridge
The Pitt Building, Trumpington Street, Cambridge CB2 1RP
40 West 20th Street, New York, NY 10011–4211, USA
10 Stamford Road, Oakleigh, Melbourne 3166, Australia

First published 1993
Reprinted 1996

A catalogue record for this book is available from the British Library

Library of Congress cataloguing in publication data

Körner, T. W. (Thomas William), 1946–
Exercises for Fourier analysis / T.W. Körner.
p. cm.
ISBN 0-521-43276-6 (hardback). – ISBN 0-521-43849-7 (pbk.)
1. Fourier analysis–Problems, exercises, etc. I. Körner, T. W.
(Thomas William), 1946– Fourier analysis. II. Title.
QA403.5.K66 1993
515′.723–dc20 92-30757 CIP

ISBN 0 521 43276 6 hardback
ISBN 0 521 43849 7 paperback

Transferred to digital printing 2002

KT

CONTENTS

PREFACE

My book *Fourier Analysis* has no exercises and, in my view, is complete without them. However, exercises are useful both to the teacher of a course and to the student who wishes to learn by doing. This supplement provides such exercises (the exercises are grouped by chapter, although not all chapters have exercises).

The two remarks that follow are addressed to students using this book by themselves.

(1) I have tried to produce exercises and not problems. You should find them more in the nature of a hill top walk than a rock climbing expedition. I have marked some of the easier questions with a minus sign to prevent you searching for non-existent subtleties. Very occasionally part of a question is marked by a plus sign to indicate that further reflection may be required.

(2) Unless you intend to do all the questions, you should browse until you find a question or sequence of questions that interest you. You are more likely to pick up knowledge or technique from an exercise that interests you than from one that does not.

The references to other books and papers which occur from time to time are intended to encourage further reading, and not as a complete record of any indebtedness to other sources. The Cambridge Tripos examinations of various years have been the largest single source of questions, but experts will recognise the influence of the texts of Helson, Katznelson, Rogosinski, Dym and McKean and many others. Experts will also recall the verses of Kipling.

When Homer smote his blooming lyre,
He'd heard men sing by land and sea:
And what he thought he might require,
He went and took – the same as me!

The market-girls and fishermen,
The shepherds and the sailors, too,
They heard old songs turn up again,
But kept it quiet – same as you.

I should like to thank Mrs L. Clark for typing these exercises with unfailing cheerfulness and competence. Mr T. Wilkins has kindly worked through them all. I owe him a debt of gratitude not only for uncovering numerous errors but for showing that the exercises can be done. Since it is, in the nature of things, harder to find errors in exercises than in plain text I have no doubt that quite a few remain and I would be grateful to be told of them. My e-mail address is **twk@pmms.cam.ac.uk.**

At the suggestion of my wife I dedicate this book to two charming, though non mathematical, problems – our children Michael and Katy.

1

INTRODUCTION

1.1 (i) Let $s_0 = 1/2$, $s_n = 1/2 + \sum_{j-1}^{n-1} \cos jx$ for $n \geqslant 1$. By writing $s_n = (\sum_{j=-n}^{n} e^{ijx})/2$ and summing geometric series show that $(n+1)^{-1}\sum_{j=0}^{n} s_j \to 0$ as $n \to \infty$ for all $x \not\equiv 0 \bmod 2\pi$, and so

$$0 = 1/2 + \sum_{j=1}^{\infty} \cos jx \text{ in the Cesàro sense.}$$

(ii) Show similarly that, if $x \not\equiv 0 \bmod 2\pi$, then

$$\cot(x/2) = 2\sum_{j=1}^{\infty} \sin jx \text{ in the Cesàro sense.}$$

1.2$^{(-)}$ (i) Suppose $s_r = (-1)^r(2r+1)$ for $r = 0, 1, 2, \ldots$. Show that $t_n = (n+1)^{-1}\sum_{j=0}^{n} s_j$ does not tend to a limit but that $(n+1)^{-1}\sum_{j=0}^{n} t_j$ does. In other words, applying the Cesàro procedure once does not produce a limit, but applying it twice does.

(ii) Give an example of a sequence where applying the Cesàro procedure twice does not produce a limit, but applying the Cesàro procedure three times does.

1.3 Can we improve on Cesàro? In particular can we find $a_{nj} \in \mathbb{C}$ such that

(A) if $s_n \to s$ then $\sum_{j=0}^{n} a_{nj}s_j \to s$,

(B) $\sum_{j=0}^{n} a_{nj}s_j$ converges for every sequence (s_n)?

A little thought suggests that this is too much to hope, but may leave open the question if we replace condition (B) by

(B)′ $\sum_{j=0}^{n} a_{nj}s_j$ converges for every bounded sequence (s_n).

To tackle this question, we first try to find necessary and sufficient conditions on the a_{nj} to make (A) hold. Two conditions are obviously necessary.

(i) By taking $s_m = 1$ and $s_n = 0$ for $n \neq m$, show that if (A) holds then $a_{nm} \to 0$ as $n \to \infty$ for each fixed m.

(ii) By taking $s_n = 1$ for all n show that, if (A) holds, then $\sum_{j=0}^{n} a_{nj} \to 1$ as $n \to \infty$.

1.4 Suppose now that condition (A) of Question 1.3 holds.

(i) Using part (i) of Question 1.3 show that, given any $M \geqslant 0$ and any $\varepsilon > 0$, we can find an N such that $\sum_{j=0}^{n} |a_{nj}| \leqslant \varepsilon$ for all $n \geqslant N$.

(ii) Hence show, using induction and part (ii) of Question 1.3, that we can find integers $0 = N(0) < N(1) < N(2) < \ldots$ such that

$$\sum_{j=0}^{N(r-1)} |a_{N(r)j}| \leqslant 2^{-r},$$

$$\left| \sum_{j=0}^{N(r)} a_{N(r)j} - 1 \right| \leqslant 2^{-r}$$

for all $r \geqslant 1$.

(iii) Now define $s_j = (-1)^r$ for $N(r-1) \leqslant j < N(r)$ $[r \geqslant 1]$ and observe that the s_j form a bounded sequence with $\sum_{j=0}^{N(2r-1)} a_{N(2r-1)j} s_j \to -1$, $\sum_{j=0}^{N(2r)} a_{N(2r)j} s_j \to 1$. Conclude that (A) and (B)′ can not be simultaneously satisfied.

(iv)* Find necessary and sufficient conditions on $a_{nj} \in \mathbb{C}$ for condition (B)′ of Question 1.3 to be satisfied.

1.5 In Question 1.3 we saw that the conditions

(α) $a_{nj} \to 0$ as $n \to \infty$ for each fixed j,

(β) $\sum_{j=0}^{n} a_{nj} \to 1$ as $n \to \infty$,

are necessary for condition (A) to be satisfied. There is a third condition,

(γ) there exists a K with $\sum_{j=0}^{n} |a_{nj}| < K$ for all n,

which is also necessary, as we prove in this question.

(i) Suppose the a_{nj} satisfy condition (A), but the sequence $K(n) = \sum_{j=0}^{n} |a_{nj}|$ is unbounded. By imitating parts (i) and (ii) of Question 1.4, or otherwise, show that we can find integers $0 = N(0) < N(1) < N(2) < \ldots$ such that

$$\sum_{j=0}^{N(r-1)} |a_{N(r)j}| \leqslant 2^{-r},$$

$$\sum_{j=0}^{N(r)} |a_{N(r)j}| \geq 2^{2r}.$$

(ii) Now recall the definition $\operatorname{sgn} \lambda = \lambda^*/|\lambda|$ if $\lambda \neq 0$, $\operatorname{sgn} 0 = 0$ and take $s_j = 2^{-r} \operatorname{sgn}(a_{N(r)j})$ for $N(r-1) \leq j < N(r)$ $[r \geq 1]$. Show that $s_j \to 0$ as $j \to \infty$ but $\sum_{j=0}^{N(r)} a_{N(r)j} s_j \to \infty$ as $r \to \infty$.
Conclude that condition (γ) is necessary for condition (A) to be satisfied.

(iii) By imitating the proof of Lemma 1.4(i), or otherwise, show that the three conditions (α), (β), (γ) are together also sufficient for condition (A) to hold.

(iv) Find necessary and sufficient conditions on $a_{nj} \in \mathbb{C}$ for condition (B) of Question 1.3 to hold.

1.6 (Generalisations)

(i) Suppose $0 = N(0) < N(1) < N(2) < \ldots$ and $a_{nj} \in \mathbb{C}$ $[0 \leq j \leq N(r),$ $r \geq 1]$. Show that the condition,

(A) if $s_n \to s$ then $\sum_{j=0}^{N(n)} a_{nj} s_j \to s$,

holds if and only if

(α) $a_{nj} \to 0$ as $n \to \infty$ for each fixed j,

(β) $\sum_{j=0}^{N(n)} a_{nj} \to 1$,

(γ) there exists a K such that $\sum_{j=0}^{N(n)} |a_{nj}| < K$.

(We could repeat the earlier proofs with trivial changes. Another method is to define

$$b_{nj} = a_{N(r-1)j} \text{ for } 0 \leq j \leq N(r-1) \leq n < N(r)$$
$$b_{nj} = 0 \text{ for } N(r-1) < j \leq n < N(r)$$

and apply the result of Question 1.5 to the b_{nj}.)

(ii) In fact many classical techniques involve replacing the sum $\sum_{j=0}^{N(n)} a_{nj} s_j$ in (A) by an infinite sum $\sum_{j=0}^{\infty} a_{nj} s_j$. (We shall see an important instance of such a technique in Chapter 27.) Suppose now that $a_{nj} \in \mathbb{C}$ for all $j, n \geq 0$. Show using the ideas of Question 1.5, or otherwise, that, if $\sum_{j=0}^{\infty} a_{nj} s_j$ converges whenever $s_j \to 0$, then $\sum_{j=0}^{\infty} |a_{nj}|$ converges.

(iii) Now show that, if $a_{nj} \in \mathbb{C}$ for all $j, n \geq 0$, then the condition

(A)′ if $s_n \to s$ as $n \to \infty$, then $\sum_{j=0}^{\infty} a_{nj} s_j$ converges for each fixed n and $\sum_{j=0}^{\infty} a_{nj} s_j \to s$ as $n \to \infty$,

holds if and only if

(α)′ $a_{nj} \to 0$ as $n \to \infty$ for each fixed j,

$(\beta)'$ $\sum_{j=0}^{\infty} a_{nj} \to 1$ as $n \to \infty$,

$(\gamma)'$ there exists a K such that $\sum_{j=0}^{\infty} |a_{nj}| \leqslant K$ for each n.

(Again we could repeat earlier proofs with minor changes. Alternatively we could construct $0 = N(0) < N(1) < \ldots$ in such a way that $\sum_{j=N(n)+1}^{\infty} |a_{nj}| \leqslant 2^{-n}$, and apply part (i).)

This result is due to Toeplitz. From now on we shall refer to an array (a_{nj}) as *regular* if it satisfies condition (A)$'$.

(iv) Show that if (a_{nj}) is regular, then there exists a bounded sequence s_j such that $\sum_{j=0}^{\infty} a_{nj}s_j$ does not tend to a limit as $n \to \infty$.

(v)* Find necessary and sufficient conditions on $a_{nj} \in \mathbb{C}$ for the following to hold:

(A)$''$ if $s_n \to s$ as $n \to \infty$, then $\sum_{j=0}^{\infty} a_{nj}s_j$ converges for each fixed n and $\sum_{j=0}^{\infty} a_{nj}s_j$ tends to a limit as $n \to \infty$.

1.7 (i) Suppose s_n is a sequence in $\mathbb{C}$ which fails to converge as $n \to \infty$ and let $\gamma \in \mathbb{C}$. Show that either

(I) We can find $n(1) < n(2) < \ldots$ such that $|s_{n(j)}| \to \infty$, or

(II) we can find $\alpha, \beta \in \mathbb{C}$ and $n(1) < n(2) < \ldots$ such that $\alpha \neq \beta$ and $s_{n(2r+1)} \to \alpha$, $s_{n(2r)} \to \beta$.

(ii) In case (II) show that we can find $\lambda, \mu \in \mathbb{C}$ such that $\lambda + \mu = 1$ and $\lambda\alpha + \mu\beta = \gamma$. By setting $a_{rn(2r)} = \mu$, $a_{rn(2r+1)} = \lambda$, $\alpha_{rj} = 0$ otherwise, show that there is a regular array (a_{rj}) such that $\sum_{j=0}^{\infty} a_{rj}s_j \to \gamma$ as $r \to \infty$.

(iii) Show that the conclusion of (ii) holds in case (I) also. Conclude that, given any non convergent sequence s_n and any $\gamma \in \mathbb{C}$, we can find a regular array (a_{rj}) such that $\sum_{j=0}^{\infty} a_{rj}s_j \to \gamma$ as $r \to \infty$.

(iv) For which sequences s_n can we find a regular array (a_{rj}) such that $|\sum_{j=0}^{\infty} a_{rj}s_j| \to \infty$?

1.8 Among the arrays (a_{nj}) which are regular in the sense of Question 1.6 there are some which are more natural than others.

(i) Show that the following two conditions on a regular array (a_{nj}) are equivalent.

(A) If s_j is a sequence with $M \geqslant s_j \geqslant 0$ for all j and $\lim_{n\to\infty} \sum_{j=0}^{\infty} a_{nj}s_j$ exists then $\lim_{n\to\infty} \sum_{j=0}^{\infty} a_{nj}s_j \geqslant 0$.

(B) There exists a regular array (b_{nj}) such that $b_{nj} \geqslant 0$ for all $n, j \geqslant 0$ and $\sum_{j=0}^{\infty} |a_{nj} - b_{nj}| \to 0$ as $n \to \infty$.

(ii) Let us call a regular array (a_{nj}) *positive* if $a_{nj} \geqslant 0$ for all $n, j \geqslant 0$. Show that for any positive regular array (a_{nj}) and any real sequence s:

$$\liminf_{j\to\infty} s_j \leqslant \liminf_{n\to\infty} \sum_{j=1}^{\infty} a_{nj}s_j \leqslant \limsup_{n\to\infty} \sum_{j=1}^{\infty} a_{nj}s_j \leqslant \limsup_{j\to\infty} s_j.$$

Show also that for any real sequence s_j and any s with

$$\liminf_{j\to\infty} s_j \leqslant s \leqslant \limsup_{j\to\infty} s_j$$

there exists a positive regular array (a_{nj}) such that $\sum_{j=0}^{\infty} a_{nj}s_j \to s$ as $n \to \infty$.

1.9 There is another natural condition on regular arrays. To introduce it, we have recourse to the ideas of linear algebra. Let us write $\boldsymbol{v}$ for the sequence of complex numbers who jth term is v_j and, if $\boldsymbol{u}$, $\boldsymbol{v}$ are sequences, let us write $\lambda\boldsymbol{u} + \mu\boldsymbol{v}$ for the sequence whose jth term is $\lambda u_j + \mu v_j$. Let V be the collection of sequences such that $\lim_{n\to\infty}\sum_{j=0}^{\infty} a_{nj}v_j$ exists. If $\boldsymbol{v} \in V$, we write $T\boldsymbol{v} = \lim_{n\to\infty}\sum_{j=0}^{\infty} a_{nj}v_j$.

(i) Show that, if $\boldsymbol{u}, \boldsymbol{v} \in V$ and $\lambda, \mu \in \mathbb{C}$, then $\lambda\boldsymbol{u} + \mu\boldsymbol{v} \in V$ and $T(\lambda\boldsymbol{u} + \mu\boldsymbol{v}) = \lambda T\boldsymbol{u} + \mu T\boldsymbol{v}$.

(ii) If $\boldsymbol{v}$ is a sequence, let us write $S\boldsymbol{v}$ for the sequence whose jth term is v_{j+1}. Show that $S(\lambda\boldsymbol{u} + \mu\boldsymbol{v}) = \lambda S\boldsymbol{u} + \mu S\boldsymbol{v}$. Show that, if $u_j \to u$, then $u_{j+1} \to u$ as $j \to \infty$.

(iii) In view of (ii), it is natural to demand that the regular array (a_{nj}) should be *translation consistent* in the sense that, if $\boldsymbol{v} \in V$, then $S\boldsymbol{v} \in V$ and $TS\boldsymbol{v} = T\boldsymbol{v}$.

Now suppose that (a_{nj}) is indeed regular and translation consistent and that $v_j = z^j$ for some $z \in \mathbb{C}$. Show that, if $\boldsymbol{v} \in V$, then $T\boldsymbol{v} = TS\boldsymbol{v} = zT\boldsymbol{v}$, and deduce that, if $z \neq 1$, then $T\boldsymbol{v} = 0$. What happens if $z = 1$?

(iv) With the same hypotheses on (a_{nj}) and with the same definition of V show that, if $u_j = \sum_{r=0}^{j} v_r$, then

(a) $\boldsymbol{u} \in V$ if and only if $\boldsymbol{v} \in V$ and $z \neq 1$,
(b) if $\boldsymbol{u} \in V$, then $T\boldsymbol{u} = (1 - z)^{-1}$.

Taking $z = -1$ we obtain a formal version of the argument, which goes back at least as far as Leibnitz, that if the sum S of $1 - 1 + 1 - \dots$ has any meaning then we must have $S = 1/2$.

(v) Suppose now that (a_{nj}) is also positive. By taking $z = e^{i\theta}$ and taking real and imaginary parts (explain carefully why this is possible), show that if the formula $u_j = \sum_{r=0}^{j} e^{ir\theta}$ defines a sequence $\boldsymbol{u} \in V$ for some $\theta \in \mathbb{R}$, then writing $b_j = 1/2 + \sum_{r=0}^{j} \cos r\theta$, $c_j = \sum_{r=0}^{j} \sin r\theta$ we have $\boldsymbol{b}, \boldsymbol{c} \in V$ and

$$Tb = 0, \quad 2Tc = \cot\theta/2.$$

Comment on the relation of these results to those of Question 1.1.

(vi) Suppose (a_{nj}) is a regular, translation consistent array. Let $u_j = (-1)^j j$. Show that, if $u \in V$, then $Tu = -1/4$. Let $w_j = (-1)^j j^2$. Assuming that $w \in V$, compute Tw.

(vii) Let (a_{nj}) be a regular array. Show that (a_{nj}) is translation consistent if and only if $\sum_{j=0}^{\infty} a_{njk+l} \to k^{-1}$ for each integer $k \geq 1$ and each integer $l \geq 0$. (You may wish to consider the sequence s_r given be $s_{jk} = 1$ for $j \geq 0$, $s_r = 0$ otherwise.)

(viii) Show that the Cesàro array given by $a_{nj} = (n+1)^{-1}$ for $0 \leq j \leq n$, $a_{nj} = 0$ otherwise, is translation consistent.

(ix) Let $s_j = (-1)^r$ for $2^r - 1 \leq j < 2^{r+1} - 1$ $[r \geq 0]$. Show that given any λ with $-1 \leq \lambda \leq 1$ we can find a regular, positive, translation consistent array (a_{nj}) such that $\sum_{j=0}^{\infty} a_{nj} s_j \to \lambda$ as $n \to \infty$.

1.10 (i) We start with a preliminary calculation. Observe that, if $f(n) \leq f(n+1) \leq \ldots \leq f(n+m)$, $0 \leq l < k$ and $l + rk \leq m$, then

$$\sum_{j=0}^{r-1} f(n+jk) \leq \sum_{j=0}^{r-1} f(n+jk+l) \leq \sum_{j=1}^{r} f(n+jk).$$

Now suppose $g\colon \mathbb{N} \to \mathbb{R}$ satisfies $g(r) \geq 0$ for all r,

$$g(0) \leq g(1) \leq \ldots \leq g(M), \quad g(M) \geq g(M+1) \geq \ldots,$$

and that $\sum_{r=0}^{\infty} g(r)$ converges. By using the idea of the first sentence show that

$$\left| \sum_{j=0}^{\infty} g(kj) - \sum_{j=0}^{\infty} g(kj+l) \right| \leq 4g(M)$$

for all $0 \leq l < k$.

(ii) Next we have a trivial but, to my mind, interesting observation. Let $X_0, X_1, \ldots$ be random variables taking non negative integer values and such that $P(X_n = r) \to 0$ as $n \to \infty$ for each fixed r. Show that, if $a_{nr} = P(X_n = r)$, then (a_{nr}) is a regular positive array. It is thus possible that interesting random variables such as binomial (coin tossing) and Poisson will give rise to interesting arrays.

(iii) Let $0 < p < 1$ and let

$$a_{nr} = \binom{n}{r} p^r (1-p)^{n-r}$$

for $0 \leq r \leq n$, $a_{nr} = 0$ otherwise. Show that a_{nr} form a regular positive translation consistent array. (You may find Question 1.9 (vii) and part (i) of this question useful.) Show that $\sum_{r=0}^{n} a_{nr} z^r$ tends to a limit as $n \to \infty$ if and only if $|1 + (z-1)p| < 1$ or $z = 1$. Find a value of p such that $\sum_{r=0}^{n} a_{nr}(-2)^r$ converges. Returning to the ideas of Question 1.9(iv), we may say that, if the sum $1-2+4-\ldots$ has any meaning, it must be 1/3 and we have found a method for giving meaning to that sum.

(iv) Let $\lambda(n)$ be a sequence of real positive numbers with $\lambda(n)\to\infty$. Let $a_{nr}=\mathrm{e}^{-\lambda(n)}\lambda(n)^r/r!$ for $n, r\geqslant 0$. Show that the a_{nr} form a regular positive translation consistent array. Show that $\sum_{r=0}^{\infty}a_{nr}z^r$ tends to a limit if and only if $\operatorname{Re} z<1$ or $z=1$.

(v) If $p\in\mathbb{C}$ and $p\neq 0, 1$ set

$$a_{nr}=\binom{n}{r}p^r(1-p)^{n-r}$$

for $0\leqslant r\leqslant n$, $a_{nr}=0$ otherwise. For what values of p is the array (a_{nr}) regular?

(vi) Let $\lambda(n)$ be a sequence of real positive numbers with $\lambda(n)\to\infty$. Let $\alpha\in\mathbb{C}$, $\alpha\neq 0$ and set $a_{nr}=\mathrm{e}^{-\alpha\lambda(n)}(\alpha\lambda(n))^r/r!$ for $n, r\geqslant 0$. For what values of α is (a_{nr}) regular? (Part (iii) goes back to Euler, part (iv) to Borel.)

1.11 Question 1.7 to 1.9 and Question 1.10(iv) suggest that our attempt to generalise the notion of a limit via regular arrays is at once too broad and too narrow. The attempt is too broad because many regular arrays do not give 'natural' results, and too narrow because there may be processes which give 'natural' results, but do not depend on arrays. Let us try another approach using the ideas of Question 1.9.

Let U be the space of all sequences of complex numbers and write $\boldsymbol{u}$ for the sequence of complex numbers $u_0, u_1\ldots$. Let B be the set of all bounded sequences and C the set of all convergent sequences. Let $S\colon U\to U$ be given by $S\boldsymbol{u}=\boldsymbol{u}$ where $w_i=u_{i+1}$ $[i\geqslant 0]$.

(i) Show that U is a vector space. Show that B is a subspace of U and C subspace of B. Show that $S\colon U\to U$ is linear and that S is surjective. Is S injective? Show that $S(B)=B$, $S(C)=C$.

(ii) Let V be a subspace of U and $L\colon V\to\mathbb{C}$ a linear map. We say that L is *a generalised limit* if

(A) $C\subseteq V$ and $L\boldsymbol{u}=\lim_{n\to\infty}u_n$ for all $\boldsymbol{u}\in C$,
(B) $S(V)=V$ and $LS\boldsymbol{u}=L\boldsymbol{u}$ for all $\boldsymbol{u}\in V$.

If we say that L preserves positivity if, in addition,

(C) if $\boldsymbol{u}\in V$ and u_n is real and non negative for all n, then $L\boldsymbol{u}$ is real and non negative.

Show that, if (a_{nj}) is regular array, and we write

$$V=\left\{\boldsymbol{u}\colon \lim_{n\to\infty}\sum_{j=0}^{\infty}a_{nj}u_j \text{ exists}\right\},\quad L\boldsymbol{u}=\lim_{n\to\infty}\sum_{j=0}^{\infty}a_{nj}u_j,$$

then L is a generalised limit if and only if (a_{nj}) is translation consistent. Show that L is a generalised limit which preserves positivity if and only if there is a positive translation consistent array (b_{nj}) such that $\sum_{j=1}^{\infty}|a_{nj}-b_{nj}|\to 0$.

(iii) Let V be the set of $\mathbf{u}$ such that $\sum_{j=0}^{\infty}u_jx^j/j!$ converges for all $x\geqslant 0$ and $e^{-x}\sum_{j=0}^{\infty}u_jx^j/j!$ tends to a limit $L(\mathbf{u})$, say, as $x\to\infty$. Using Question 1.10(iv), or otherwise, show that L is a generalised limit which preserves positivity.

(iv) Let $f_j\colon[0,\infty)\to\mathbb{C}$ be a collection of functions such that $\sum_{j=0}^{\infty}|f_j(x)|$ converges for all $x\geqslant 0$. Let V be the set of u such that $\sum_{j=0}^{\infty}u_jf_j(x)$ converges for all $x\geqslant 0$ and $\sum_{j=0}^{\infty}u_jf_j(x)$ tends to a limit $L(\mathbf{u})$, say, as $x\to\infty$. Find necessary and sufficient conditions on the f_j for L to be a generalised limit. When does L preserve positivity?

(v) By using the fact that power series can be multiplied term by term within their circles of convergence, show that there exist $b_n\in\mathbb{R}$ such that $\sum_{n=0}^{\infty}b_nx^n=e^x\sin\pi x$ for all $x\in\mathbb{R}$. Let $u_n=n!b_n$. Show that $\mathbf{u}\notin V$ where V is the set of part (iii) of this question but that, if, following Question 1.10(iv), we set $a_{nr}=e^{-n}n^r/r!$ then $\sum_{r=0}^{\infty}a_{nr}u_r$ tends to a limit as $n\to\infty$.

(vi) Let L be a generalised limit and V its associated vector space. Let $v_j=z^j$ and $w_j=j$. By recalling the proof in Question 1.9(iii), or otherwise, show that, if $\mathbf{v}\in V$, then $L\mathbf{v}=0$ or $z=1$. Show also that $\mathbf{w}\notin V$. Show that, if L preserves positivity and u_j is a sequence of real numbers with $u_j\to\infty$, then $\mathbf{u}\notin V$.

(Thus $V\neq U$. This does not exclude the possibility that we can find a generalised positivity preserving limit L with $V\supseteq B$. The reader is invited to try to decide whether such an L and V can exist, but warned that no answer is possible within the mind set of this book. What we have here is a seventeenth-century problem whose resolution depends on twentieth century ideas from the foundations of mathematics.)

2

PROOF OF FEJÉR'S THEOREM

2.1$^{(-)}$ By considering the function $f(t) = 0$ for $t \neq 0$, $f(0) = 1$ show that the condition f continuous cannot be dropped in Theorems 2.3 and 2.4.

2.2 If $f: \mathbb{T} \to \mathbb{C}$ is continuous and $g = \operatorname{Re} f$ show that $\hat{g}(r) = (\hat{f}(r) + \hat{f}(-r)^*)/2$. Find the Fourier coefficients of $\operatorname{Im} f$.

2.3 (An alternative proof of Theorem 2.4)

(i) Suppose $f: \mathbb{T} \to \mathbb{R}$ is continuous but $f \neq 0$ and so there exists an $c \in \mathbb{T}$ with $f(c) \neq 0$. By considering $-f$ if necessary, we may suppose $f(c) > 0$. Explain why we can find a $\pi/2 > \delta > 0$ such that $f(t) > f(c)/2$ for $|t - c| < \delta$ and a $K > 0$ such that $|f(t)| \leq K$ for all $t \in \mathbb{T}$.

(ii) Show that we can find an $\varepsilon > 0$ so small that $h(t) = \varepsilon + \cos(t - c)$ satisfies $|h(t)| \leq 1 - \varepsilon/2$ for $|t - c| > \delta$. Show that there exists an η with $\delta > \eta > 0$ such that $h(t) \geq 1 + \varepsilon/2$ for $|t - c| < \eta$. Show that

$$\left| \frac{1}{2\pi} \int_{|t-c| \geq \delta} h(t)^n f(t)\, dt \right| \leq K(1 - \varepsilon/2)^n \to 0 \quad \text{as } n \to \infty,$$

$$\frac{1}{2\pi} \int_{\delta > |t-c| \geq \eta} h(t)^n f(t)\, dt \geq 0 \quad \text{for all } n \geq 1,$$

and

$$\frac{1}{2\pi} \int_{\eta > |t-c|} h(t)^n f(t)\, dt \geq \frac{\eta f(c)}{4\pi} (1 + \varepsilon/2)^n \to \infty \quad \text{as } n \to \infty.$$

Deduce that

$$\frac{1}{2\pi} \int_{\mathbb{T}} h(t)^n f(t)\, dt \to \infty \quad \text{as } n \to \infty.$$

(iii) By first considering the case $n = 1$, show that we can write $h(t)^n = \sum_{j=-n}^{n} a_{nj} e^{ijt}$ for some $a_{nj} \in \mathbb{C}$. Show that

$$\frac{1}{2\pi}\int_{\mathbb{T}} h(t)^n f(t)\,dt = \sum_{j=-n}^{n} a_{nj}\hat{f}(-j)$$

and deduce that, if $\hat{f}(j) = 0$ for all j, then

$$\frac{1}{2\pi}\int_{\mathbb{T}} h(t)^n f(t)\,dt = 0 \quad \text{for all } n.$$

Conclude that, if $f\colon \mathbb{T} \to \mathbb{R}$ is continuous and $\hat{f}(r) = 0$ for all r, then $f = 0$.

(iv) By considering the real and imaginary parts of f, show that, if $f\colon \mathbb{T} \to \mathbb{C}$ is continuous and $\hat{f}(r) = 0$ for all r, then $f = 0$.

2.4 (i) Show that, if P and Q are trigonometric polynomials, then

$$\frac{1}{2\pi}\int_{\mathbb{T}} P(t)Q(mt)\,dt = \frac{1}{2\pi}\int_{\mathbb{T}} P(t)\,dt\,\frac{1}{2\pi}\int_{\mathbb{T}} Q(t)\,dt$$

whenever m is a sufficiently large integer.

(ii) By using Theroem 2.5 to produce a sequence Q_n of trigonometric polynomials with $Q_n \to g$ uniformly, show that, if P is a trigonometric polynomial and g a continuous function, then

$$\frac{1}{2\pi}\int_{\mathbb{T}} P(t)g(mt)\,dt \to \frac{1}{2\pi}\int_{\mathbb{T}} P(t)\,dt\,\frac{1}{2\pi}\int_{\mathbb{T}} g(t)\,dt \quad \text{as } m \to \infty.$$

(iii) Deduce that, if f and g are continuous,

$$\frac{1}{2\pi}\int_{\mathbb{T}} f(t)g(mt)\,dt \to \frac{1}{2\pi}\int_{\mathbb{T}} f(t)\,dt\,\frac{1}{2\pi}\int_{\mathbb{T}} g(t)\,dt \quad \text{as } m \to \infty.$$

(iv) Hence show that, if $f\colon\mathbb{T} \to \mathbb{C}$ is continuous, $\hat{f}(m) \to 0$ and $\hat{f}(-m) \to 0$ as $m \to \infty$. (This is the Riemann Lebesgue lemma proved later as Theorem 52.4.)

2.5 (A Bernstein inequality) If P is a trigonometric polynomial of degree n or less (i.e. $P(t) = \sum_{r=-n}^{n} a_r \exp(irt)$) show that

$$P'(t) = \frac{1}{2\pi}\int_{\mathbb{T}} P(t-y)L_n(y)\,dy$$

where $L_n(y) = -2nK_{n-1}(y)\sin ny$. (Hint: it suffices to verify the simple cases $P(t) = \exp(irt)$ with $-n \leqslant r \leqslant n$.) Hence deduce that

$$|P'(t)| \leqslant 2n \sup_{x\in\mathbb{T}} |P(x)|.$$

(This result can be improved by replacing $2n$ by n. See e.g. Question 43.8.)

3

WEYL'S EQUIDISTRIBUTION THEOREM

3.1 (Weyl's criterion) Let $g\colon \mathbb{Z}^+ \to \mathbb{R}$ be given. Let us say that the sequence $g(n)$ is equidistributed if, for every $0 \leqslant a \leqslant b \leqslant 1$,

$$n^{-1}\,\mathrm{card}\,\{1 \leqslant r \leqslant n\colon a \leqslant \langle g(r)\rangle \leqslant b\} \to b - a \quad \text{as } n \to \infty.$$

Modify the proof of Theorem 3.1′ to show that the following four conditions are equivalent.

(a) g is equidistributed.

(b) $n^{-1}\sum_{r=1}^{n} f(2\pi g(r)) \to \frac{1}{2\pi}\int_{\mathbb{T}} f(t)\,\mathrm{d}t$ as $n \to \infty$ for every continuous function $f\colon \mathbb{T} \to \mathbb{C}$.

(c) $n^{-1}\sum_{r=1}^{n} \exp(2\pi i m g(r)) \to 0$ as $n \to \infty$ for every integer $m \neq 0$.

(d) $n^{-1}\sum_{r=1}^{n} f(g(r)) \to \int_0^1 f(t)\,\mathrm{d}t$ as $n \to \infty$ for every continuous function $f\colon [0, 1] \to \mathbb{C}$.

3.2 Let $f\colon (0, \infty) \to \mathbb{R}$ be continuous. Question 3.1 suggests that the study of the behaviour $\sum_{r=1}^{n} \exp(2\pi i f(r))$ as $n \to \infty$ will be of great interest for number theory. This is the case, but we shall limit ourselves to a very simple example where f is slowly varying. (The arguments are due to Fejér.)

(i) Let $\phi(x) = x - [x] - 1/2$. Sketch the graph of ϕ.

(ii) Suppose f is differentiable. By integrating by parts, show that

$$\int_r^{r+1} \exp(2\pi i f(x))\,dx = 2^{-1}[\exp(2\pi i f(r+1)) + \exp(2\pi i f(r))] - 2\pi i \int_r^{r+1} \phi(x) f'(x) \exp(2\pi i f(x))\,dx.$$

(iii) By summing the previous result, show that

$$\sum_{r=1}^{n} \exp(2\pi i f(r)) = 2^{-1}[\exp(2\pi i f(n)) + \exp(2\pi i f(1))]$$
$$+ \int_1^n \exp(2\pi i f(x))\,dx$$
$$+ 2\pi i \int_1^n \phi(x) f'(x) \exp(2\pi i f(x))\,dx.$$

(iv) Conclude that, if in addition $f'(x) \to 0$ as $x \to \infty$, then

$$n^{-1} \sum_{r=1}^{n} \exp(2\pi i f(r)) - n^{-1} \int_1^n \exp(2\pi i f(x))\,dx \to 0 \quad \text{as } n \to \infty.$$

(v) Now suppose that f satisfies the previous hypotheses and, in addition, $xf'(x) \to A$ as $x \to \infty$. Show that

$$\left| \frac{1}{n} \int_1^n xf'(x) \exp(2\pi i f(x))\,dx - \frac{A}{n} \int_1^n \exp(2\pi i f(x))\,dx \right| \to 0$$

as $n \to \infty$. By integrating $\int_1^n xf'(x) \exp(2\pi i f(x))\,dx$ by parts, show using (iv) that

$$n^{-1} \sum_{r=1}^{n} \exp(2\pi i f(r)) - \frac{\exp(2\pi i f(n))}{(1 + 2\pi i A)} \to 0.$$

Conclude, using Question 3.1, that $f(n)$ is not equidistributed.

(vi) Suppose on the other hand that f satisfies the hypotheses of parts (ii) and (iv) but f' is increasing and $xf'(x) \to \infty$. Choose M such that $f'(x) > 0$ for $x > M - 1$. Explain why f has an inverse g on $(M-1, \infty)$ and why, if m is real,

$$\int_M^N \exp(2\pi i m f(x))\,dx = \int_{f(M)}^{f(N)} \exp(2\pi i mu) h(u)\,du$$

where $h(u) = 1/(f'(g(u)))$. Show that h is a decreasing continuous function and deduce, using an integral mean value theorem (see Question 52.7), that

$$\left| \int_M^N \exp(2\pi i m f(x))\,dx \right| \leqslant \sup_{f(M) \leqslant u \leqslant f(N)} |h(u)| \left| \int_{f(M)}^{f(N)} \exp(2\pi i mu)\,du \right|$$
$$\leqslant \frac{1}{f'(M)} \frac{1}{\pi |m|} \quad [m \neq 0].$$

Conclude, using Question 3.1, that $f(n)$ is equidistributed.

(vii) What can you say if $f'(x) \to 0$, $xf'(x) \to -\infty$?

(viii) For which values of α, β, γ with $\alpha, \gamma \in \mathbb{R}$, $0 \leqslant \beta < 1$ is $\alpha n^\beta (\log n)^\gamma$ equidistributed?

3.3 The results of Question 3.2 are very nice, but not very profound.

(i) Identify $\{r \in \mathbb{Z}: n \leqslant \log_4 r < n + 1/2\}$. (Here $\log_4$ means log to the base 4.)

(ii) Show that $\log_4 n$ is not equidistributed.

(iii) By estimating

$$n^{-1} \operatorname{card} \{1 \leqslant r \leqslant n: 0 \leqslant \langle \log r \rangle < 1/2\}$$

for suitable values of n, show, without using the methods of Chapter 3, that $\log n$ is not equidistributed.

(iv) Show also, without using the methods of Chapter 3, that $n^{1/2}$ is equidistributed.

(v)* Recover the results of part (viii) of Question 3.2 without using the methods of Chapter 3.

It appears to be much harder to deal with more rapidly varying functions with $f'(x) \nrightarrow 0$ as $x \to \infty$.

(vi) Let $f_1(x) = x + \log x$ and $f_2(x) = x + x^{-1/2}$. Is $f_1(n)$ equidistributed? Is $f_2(n)$ equidistributed? What happens to $f_1^{(r)}(x) - f_2^{(r)}(x)$ as $x \to \infty$? What happens to $f_1^{(r)}(x)/f_2^{(r)}(x)$ as $x \to \infty$?

In questions 32.11 to 32.13 we develop a method powerful enough to deal with $f(x) = x \log x$. (You should perhaps try this problem by yourself for a little while.)

3.4 (i) Using Theorem 3.1′, show that, if f is continuous,

$$n^{-1} \sum_{r=1}^{n} f(\sin r) \to \frac{1}{2\pi} \int_{\mathbb{T}} f(\sin t)\, dt \quad \text{as } n \to \infty.$$

(ii) Deduce as in Step 6 of the proof of Theorem 3.1′ that, if $0 \leqslant a \leqslant b \leqslant 1$,

$$n^{-1} \operatorname{card} \{n \geqslant r \geqslant 1: \langle \sin r \rangle \in [a, b]\} \to \pi^{-1}((\sin^{-1} b - \sin^{-1} a) + (\sin^{-1}(1-a) - \sin^{-1}(1-b)))$$

and conclude that $\sin n$ is not equidistributed.

(iii) Let $h: \mathbb{R} \to [0, 1]$ be a continuous periodic function with irrational period. By imitating the arguments above, find a necessary and sufficient condition for $h(n)$ to be equidistributed.

(iv) Can you find any h satisfying the conditions of (iii) which is equidistributed?

(v) Can you find any everywhere differentiable h satisfying the conditions of (iii) which is equidistributed?

3.5 (How to solve a difference equation) In this question we consider complex valued sequences u_n with $n \in \mathbb{Z}$. If P is a polynominal $P(z) = \sum_{r=0}^{m} a_r z^r$, we write $P(E)u_n = \sum_{r=0}^{m} a_r u_{n+r}$. If $P(z) = (z - \alpha)^k$, we write $P(E) = (E - \alpha I)^k$.

(i) Show that, if P and Q are polynomials with $P(z)Q(r) = R(z)$ for all z, then $P(E)(Q(E)u_n) = R(E)u_n$. [You may find it helpful to consider the special case $P(z) = z^k$.]

(ii) Show that, if $\alpha \in \mathbb{C}$ and S is a polynomial of degree k, then $(E - \alpha I)(S(n)\alpha^n) = \alpha^n T(n)$, where T is polynomial of degree $k - 1$. Deduce that $(E - \alpha I)^{k+1}(S(n)\alpha^n) = 0$.

(iii) Conclude that, if $P(z) = \prod_{j=1}^{r}(z - \alpha_j)^{n(j)}$ with $\alpha_1, \alpha_2, \ldots, \alpha_r$ distinct and $n(1), n(2), \ldots, n(r)$ strictly positive integers, and if S_j is a polynomial of degree at most $n(j) - 1$, then $P(E)(\sum_{r=1}^{n} S_j(n)\alpha_j^n) = 0$. In other words, if $P(z) = \sum_{r=0}^{m} a_j z^j$, then $u_n = \sum_{r=1}^{r} S_j(n)\alpha_j^n$ is a solution of

$$\sum_{j=0}^{m} a_j u_{n+j} = 0 \quad \text{for all } n \in \mathbb{Z}. \qquad (*)$$

(iv) We have found some solutions for $(*)$. Have we found all? To resolve this problem, we use ideas from a first course in vector spaces. Observe first that the space of complex valued sequences form an infinite-dimensional vector space under the usual operations. Let us write U for the collection of all sequences which satisfy $(*)$ and V for the collection of all sequences $u_n = \sum_{r=1}^{r} S_j(n)\alpha_j^n$ where the S_j and α_j are as specified in (iii). Show that U and V are vector subspaces and observe that $U \supseteq V$.

(v) Suppose $S_1, S_2, \ldots, S_r$ are polynomials of degree at most $n(1) - 1$, $n(2) - 1, \ldots, n(r) - 1$ respectively such that $\sum_{r=1}^{r} S_j(n)\alpha_j^n = 0$ for all n. Let $Q(z) = (\prod_{j=1}^{r-1}(z - \alpha_j)^{n(j)})(z - \alpha_r)^{n(r)-1}$. By considering $Q(E)(\sum_{r=1}^{r} S_j(n)\alpha_j^n)$, or otherwise, show that S_r has degree at most $n(r) - 2$ (we take the statement S_r has degree -1 to mean $S_r = 0$). By repeating this argument in a suitable form, show that in fact $S_1 = S_2 = \ldots = S_r = 0$. Conclude that V has dimension exactly $m = n(1) + n(2) + \ldots + n(r)$. (You should exhibit an explicit basis.)

(vi) Suppose $u_1, u_2, \ldots, u_m$ are given. Show that there exists one and only one u_{m+1} and one and only one u_0 such that

$$\sum_{j=0}^{m} a_j u_{1+j} = 0 \quad \text{and} \sum_{j=0}^{m} a_j u_j = 0.$$

Conclude that, if $v_1, v_2, \ldots, v_m$ are given, there exists one and only one $\boldsymbol{u} \in U$ with $u_j = v_j$ $[1 \leq j \leq m]$. If, when $1 \leq l \leq m$, $\boldsymbol{u}(l)$ is the unique sequence in U with $u_j(l) = 1$ if $j = l$, $u_j(l) = 0$ if $j \neq l$, $1 \leq j \leq m$, show that $\boldsymbol{u}(1), \boldsymbol{u}(2), \ldots, \boldsymbol{u}(m)$ is a basis for U and deduce that U has dimension m.

(vii) Conclude that $U = V$ and so the solutions of $(*)$ are precisely those given in part (iii).

3.6 (How to solve a linear differential equation) In this question we consider infinitely differentiable functions $f: \mathbb{R} \to \mathbb{C}$. If P is a polynomial $P(z) = \sum_{r=0}^{m} a_r z^r$, we write $P(D)f = \sum_{r=0}^{m} a_r f^{(r)}$. If $P(z) = (z - \alpha)^k$, we write $P(D) = (D - \alpha I)^k$.

(i) If $(D - \alpha I)g = 0$ and $g(0) = 0$, show, by applying the mean value theorem to $e^{-af} g(t)$, or otherwise, that $g = 0$.

(ii) Suppose Q is a polynomial of degree $m - 1$ and $P(z) = (z - \alpha)Q(z)$. Show that if f is a solution of

$$P(D)f = 0, \quad f(0) = f'(0) = f''(0) = \ldots = f^{(m-1)}(0) = 0,$$

then $g = Q(D)f$ is a solution of

$$(D - \alpha I)g = 0; \quad g(0) = 0.$$

Use induction to show that $f = 0$.

(iii) By an argument along the lines of the previous question, show that, if $P(z) = \prod_{j=1}^{r} (z - \alpha_j)^{n(j)}$ with $\alpha_1, \alpha_2, \ldots, \alpha_r$ distinct and $n(1), n(2), \ldots, n(r)$ strictly positive integers, then the general solution of $P(D)f = 0$ has the form $f(t) = \sum_{j=1}^{r} S_j(t) e^{\alpha_j t}$ where S_j is a polynomial of degree at most $n(j) - 1$. Show also that, if P has degree exactly m and $z_0, z_1, \ldots, z_{m-1} \in \mathbb{C}$, then the equations

$$P(D)f = 0; \quad f^{(k)}(0) = z_k \quad [0 \leqslant k \leqslant m - 1]$$

have exactly one solution.

3.7 Suppose P is a monic polynomial with integer coefficients and (not necessarily distinct) roots $\theta_1, \theta_2, \ldots, \theta_m$. Suppose $\theta_1 > 1$ and $|\theta_j| < 1$ for $2 \leqslant j \leqslant m$. (Such a θ_1 is called a *Pisot number*.) By considering $u(1), u(2), \ldots, u(m)$ defined as in Question 3.5(vi), or otherwise, show that there exists a sequence $u = (u_n)$ such that $u_1, u_2, \ldots, u_m$ are all integers $P(E)u_n = 0$ for all n and $u_n = \lambda\theta_1^n + \sum_{j=2}^{m} S_j(n)\theta_j^n$ where the S_j are polynomials and $\lambda \neq 0$. Arguing as in Lemma 3.2, show that, given $\varepsilon > 0$, we can find an N such that $\langle \lambda\theta_1^n \rangle \notin [\varepsilon, 1 - \varepsilon]$ for all $n \geqslant N$. (In the next question we show that this result can be improved substantially.)

3.8 (i) Let P be a polynomial $P(z) = \sum_{r=0}^{m} a_r z^r$, say, with (not necessarily distinct) roots $\theta_1, \theta_2, \ldots, \theta_m$. Let us write $s_k = \sum_{j=1}^{m} \theta_j^k$ for $k \geqslant 1$. Show (by writing $P(z) = a_m \prod_{j=1}^{m} (z - \theta_j)$, or otherwise) that

$$\frac{P'(z)}{P(z)} = \sum_{j=1}^{m} \frac{1}{(z - \theta_j)}.$$

By expanding in powers of z^{-1}, show that

$$\frac{P'(z)}{P(z)} = \sum_{k=0}^{\infty} \frac{s_k}{z^{k+1}} \quad \text{for } |z| > \max_{m \geq j \geq 1} |\theta_j|.$$

Multiplying both sides of $P(z)$ and equating coefficients, deduce that

$$-ja_j + \sum_{r=0}^{m-j} a_{j+r}s_r = 0 \quad \text{for all } 1 \leq j \leq m$$

and

$$\sum_{r=0}^{m} a_r s_{r+k} = 0 \quad \text{for } k \geq 1.$$

This result was first obtained by Newton. (More generally, it can be shown that any symmetric rational function of the roots of an equation can be expressed in terms of the coefficients. See e.g. Stewart and Tall, *Algebraic Number Theory*.)

(ii) Suppose now that P is monic (i.e. $a_m = 1$) and has integer coefficients. Show that s_k is an integer for each $k \geq 1$. If P, $\theta_1, \theta_2, \ldots, \theta_m$ are as in Question 3.7, deduce, as in Lemma 3.2, that, given $\varepsilon > 0$, we can find an N such that $\langle \theta_1^n \rangle \notin [\varepsilon, 1-\varepsilon]$ for all $n \geq N$. [It has been an open problem for more than half a century whether, if $\theta_1 > 1$ is such that it satisfies the condition just given, it must be a Pisot number. For background see Salem, *Algebraic Numbers and Fourier Analysis*.]

3.9 (The wrong envelope problem)

(i) Let e_n be the number of ways that I can place n letters in n envelopes (one letter to each envelope) so that each letter goes in the wrong envelope. By observing that, if Dr A's letter goes to Professor B, then either Professor B's letter goes to Dr A or it does not, show that, if $n \geq 3$,

$$e_n = (n-1)(e_{n-1} - e_{n-2}). \qquad (*)$$

(ii) By setting $v_n = e_n - ne_{n-1}$, or otherwise, use $(*)$ to show that

$$e_n = n!\left(\frac{1}{2!} - \frac{1}{3!} + \ldots + \frac{(-1)^n}{n!}\right) \quad \text{for } n \geq 2.$$

Deduce that, if I assign my letters to the envelopes at random, the probability p_n that no one gets their right letter tends rapidly to e^{-1} as $n \to \infty$. (Thus if two well shuffled packs of cards are dealt, turning over one card of each pack at a time, the chance of no matches is very close to e^{-1}. Do you find this result what you would have expected? We give an alternative treatment in Question 105.3.)

(iii) Find the general solution to the equation

$$u_n = (n-1)(u_{n-1} - u_{n-2})$$

when n ranges over all integers and prove that your solution is indeed the general solution. Is it true that we can choose the value of u_n and u_m at any two distinct integers n and m?

4

THE WEIERSTRASS POLYNOMIAL APPROXIMATION THEOREM

4.1 Suppose $f: \mathbb{R} \to \mathbb{R}$ is n times differentiable and that $f^{(n)}$ is an increasing function. If $f^{(r)}(0) = 0$ for all $0 \leqslant r \leqslant n$ and $f(1) = 1$, show that $f^{(n)}(1) \geqslant n!$ and deduce that $f(3) \geqslant 2^n$. Is it possible to find a non constant, infinitely differentiable function $F: \mathbb{R} \to \mathbb{R}$ such that $F(x) = 0$ for all $x \leqslant 0$ and $F^{(n)}(x) \geqslant 0$ for all $n \geqslant 0$ and all $x \geqslant 0$? Give reasons.

4.2 (Taylor's theorem)

(i) Suppose $g: \mathbb{R} \to \mathbb{R}$ is n times differentiable and $g(0) = g'(0) = \ldots = g^{(n-1)}(0) = 0$. Suppose further that $|g^{(n)}(t)| \leqslant M$ for all $t \in [0, a]$ for some a, $M \geqslant 0$. Use the mean value theorem to show that

$$|g^{(n-1)}(t)| \leqslant Mt \quad \text{for all } t \in [0, a]$$

and then use repeated integration to obtain

$$|g^{(n-r)}(t)| \leqslant Mt^r/r! \quad \text{for all } t \in [0, a]$$

and so, in particular,

$$|g(t)| \leqslant Mt^n/n! \quad \text{for all } t \in [0, a].$$

(ii) Next suppose $f: \mathbb{R} \to \mathbb{R}$ is n times differentiable with $|f^{(n)}(t)| \leqslant M$ for all $t \in [0, a]$. By considering $g(t) = f(t) - \sum_{r=0}^{n-1} f^{(r)}(0)t^r/r!$, or otherwise, show that

$$\left| f(t) - \sum_{r=0}^{n-1} f^{(r)}(0)t^r/r! \right| \leqslant Mt^n/n! \quad \text{for all } t \in [0, a].$$

(iii) Extend the result of (ii) to show that, if $c < a < d$, $M \geqslant 0$ and $f: (c, d) \to \mathbb{R}$ is n times differentiable with $|f^{(n)}(t)| \leqslant M$ for all $t \in (c, d)$, then

$$\left| f(t) - \sum_{r=0}^{n-1} f^{(r)}(a)(t-a)^r/r! \right| \leqslant M(t-a)^n/n! \quad \text{for all } t \in [c, d].$$

This is a very crude form of Taylor's theorem, but it turns out to be sufficient for many practical purposes.

(iv) Use (iii) and standard properties of the exponential function (but not, of course, its Taylor expansion) to show that

$$\left|\exp t - \sum_{r=0}^{n-1} t^r/r!\right| \leq \exp(|t|)|t|^n/n!.$$

By allowing $n \to \infty$, deduce that

$$\exp t = \sum_{r=0}^{\infty} t^r/r!.$$

(v) As stated in (iii), Taylor's theorem is a global theorem, but it also has a local version. (Note however that, when mathematicians refer to 'Taylor's theorem' without qualification, they mean some global version). Suppose $\eta > 0$ and $g: (-\eta, \eta) \to \mathbb{R}$ is $n-1$ times differentiable with $g(0) = g'(0) = \ldots = g^{(n-1)}(0) = 0$. Suppose further that $g^{(n-1)}$ is differentiable at 0 with $g^{(n)}(0) = 0$. Show from the definition of differentiability that, given any $\varepsilon > 0$, we can find a $\delta(\varepsilon)$ with $\eta > \delta(\varepsilon) > 0$ such that

$$|g^{(n-1)}(t)| < \varepsilon|t| \quad \text{for all } |t| < \delta(\varepsilon).$$

Deduce that

$$|g(t)| < \varepsilon|t|^n/n! \quad \text{for all } |t| < \delta(\varepsilon).$$

(vi) Suppose $\eta > 0$ and $f: (-\eta, \eta) \to \mathbb{R}$ is $n-1$ times differentiable. Suppose further that $f^{(n-1)}$ is differentiable at 0. Show that

$$|t|^{-n}\left|f(t) - \sum_{r=0}^{n} f^{(r)}(0)t^r/r!\right| \to 0 \quad \text{as } t \to 0.$$

(We shall use this result in the alternative proof of Lemma 70.4.)

(vii) Use (vi) to prove the following version of L'Hôpital's rule. Suppose f, g are $n-1$ times differentiable near 0 and n times differentiable at 0. Show that, if $f^{(r)}(0) = g^{(r)}(0) = 0$ for $0 \leq r \leq n-1$ but $g^{(n)}(0) \neq 0$, then

$$\frac{f(x)}{g(x)} \to \frac{f^{(n)}(0)}{g^{(n)}(0)} \quad \text{as } x \to 0.$$

(viii) Use (vi) to show that, if f is once differentiable near 0 and twice differentiable at 0, then

$$\frac{f(t) - 2f(0) + f(-t)}{t^2} \to f''(0).$$

Generalise this result to higher derivatives.

(ix) What does (vi) yield when applied to function h of Example 4.2?

4.3 (Taylor's theorem with remainder)

(i) Let $F(x) = (1-x)^{-1}$. Show that

$$F(x) - \sum_{r=0}^{n} x^r = x^{n+1}(1-x)^{-1}$$

and deduce that $F(x) = \sum_{r=0}^{\infty} x^r$ for $|x| < 1$, i.e. that $F(x) = \sum_{r=0}^{\infty} F^{(r)}(0)x^r/r!$ for $|x| < 1$.

(ii) If $0 < a < 1$, compute $M_n(a) = \sup_{|x|<a} |F^{(n)}(x)|$ and show that, if $\frac{1}{2} < t < a < 1$, then $M_n(a)t^n/n! \to \infty$ as $n \to \infty$.

Thus, although (i) shows that F equals its Taylor series on $(-1, 1)$, (ii) shows that the crude form of Taylor's theorem given in Question 4.1(iii) is not powerful enough to prove this. Instead, we must have recourse to one of several theorems which give exact expressions for the remainder term $R_{n-1}(a, t)$ in the equation

$$f(a+t) = f(a) + f'(a)t + f^{(2)}(a)\frac{t^2}{2!} + \ldots + f^{(n-1)}(a)\frac{t^{n-1}}{(n-1)} + R_{n-1}(a, t),$$

from which better estimates can be made. One of these will be obtained in Question 45.2. However, there is one version with a particularly simple proof which is often adequate and which we give in the next part.

(iii) Suppose $c < a < d$ and f is n times continuously differentiable on (c, d). Show by integration by parts that, if $t \in (c, d)$, then

$$\frac{1}{(n-1)!}\int_a^t f^{(n)}(x)(t-x)^{n-1}\,dx = -f^{(n-1)}(a)\frac{(t-a)^{n-1}}{(n-1)!} + \frac{1}{(n-2)!}\int_a^t f^{(n-1)}(x)(t-x)^{n-2}\,dx$$

and hence, by induction, that

$$\frac{1}{(n-1)!}\int_a^t f^{(n)}(x)(t-x)^{n-1}\,dx = f(t) - \sum_{r=0}^{n-1} f^{(r)}(a)\frac{(t-a)^r}{r!},$$

i.e. $f(t) = \sum_{r=0}^{n-1} f^{(r)}(a)(t-a)^r/r! + R_{n-1}(a, t)$ with

$$R_{n-1}(a, t) = \frac{1}{(n-1)!}\int_a^t f^{(n)}(x)(t-x)^{n-1}\,dx.$$

(iv) Show that $0 \leq x \leq t \leq 1$ then $0 \leq (t-x)/(1-x) \leq t$. Hence, or otherwise, show that, if F is as in part (i), then

$$\frac{1}{(n-1)!}\int_0^t F^{(n)}(x)(t-x)^{n-1}\,dx \to 0 \quad n \to \infty$$

for all $0 \leq t < 1$. Also show (using easier estimates) that

$$\frac{1}{(n-1)!}\int_0^t F^{(n)}(x)(t-x)^{n-1}\,dx \to 0 \quad \text{as } n\to\infty$$

for all $-1 < t \leqslant 0$. Hence use (iii) to obtain

$$F(t) = \sum_{r=0}^{\infty} F^{(r)}(0)t^r/r! \quad \text{for all } |t| < 1.$$

(v) More generally show that, if $\alpha \in \mathbb{R}$ and we write

$$\binom{\alpha}{r} = \frac{\alpha(\alpha-1)\ldots(\alpha-r+1)}{r!},$$

we have Newton's binomial series

$$(1+t)^\alpha = \sum_{r=0}^{\infty}\binom{\alpha}{r}t^r \quad \text{for all } |t|<1.$$

(vi) Show by integration by parts, or otherwise, that if $f, \phi\colon \mathbb{R}\to\mathbb{R}$ are n times continuously differentiable, then

$$\int_a^b f^{(n)}(x)\phi(-x) - f(x)\phi^{(n)}(-x)\,dx = \sum_{r=1}^{n}[f^{(r-1)}(x)\phi^{(n-r)}(-x)]_a^b.$$

By choosing ϕ appropriately, obtain a Taylor expansion of the form given in (iii).

4.4 Consider the map $h\colon \mathbb{C}\to\mathbb{C}$ given by $h(z) = z^*$. If C is the circular contour traced out by $e^{i\theta}$ as θ runs from 0 to 2π, show that

$$\frac{1}{2\pi i}\int_C h(z)\,dz = 1.$$

Show however that, if $P(z) = \sum_{r=0}^{n} a_r z^r$ is any polynomial, then

$$\frac{1}{2\pi i}\int_C P(z)\,dz = 0.$$

Deduce that $\sup_{|z|=1}|h(z) - P(z)| \geqslant 1$ and so the obvious extension of the Weierstrass theorem fails when applied to the circle $\{z\in\mathbb{C}\colon |z| = 1\}$ or to the disc $\{z\in\mathbb{C}\colon |z|\leqslant 1\}$.

4.5 (The Peano kernel theorem)

(i) In what follows we assume that $f\colon [a,b]\to\mathbb{R}$ is n times continuously differentiable, so, by Question 4.3,

$$f(t) = f(a) + f'(a)(t-a) + \ldots + \frac{f^{(n-1)}(a)}{(n-1)!}(t-a)^{n-1} + R_n(f,t)$$

with

$$R_n(f, t) = \frac{1}{(n-1)!} \int_a^t f^{(n)}(x)(t-x)^{n-1}\,dx.$$

Show that, if $T\colon C([a, b]) \to \mathbb{R}$ is a linear map such that

(a) $TP = 0$ for all polynomials of degree at most $n - 1$, then

$$Tf = TR_n(f, \cdot).$$

(ii) Let us write $t_+^n = t^n$ if $t \geqslant 0$, $t_+^n = 0$, if $t \leqslant 0$. Show that

$$R_n(f, t) = \frac{1}{(n-1)!} \int_a^b f^{(n)}(x)(t-x)_+^{n-1}\,dx.$$

Suppose now that T commutes with integration. More precisely, suppose that:

(b) if $g\colon [a, b]^2 \to \mathbb{R}$ is continuous, then the function $h\colon [a, b] \to \mathbb{R}$ defined by $h(x) = Tg(x, \cdot)$ is continuous and $T\int_a^b g(x, \cdot)\,dx = \int_a^b h(x)\,dx$.

Show that, if we write $u_n(x, t) = (t - x)_+^{n-1}$ and $K(x) = Tu_n(x, \cdot)/(n-1)!$, then

$$Tf = \int_a^b f^{(n)}(x)K(x)\,dx.$$

(iii) Conclude in particular that

(1) $|T(f)| \leqslant \sup_{x\in[a,b]} |f^{(n)}(x)| \int_a^b |K(x)|\,dx.$

(2) If K does not change sign on $[a, b]$, then

$$|T(f| \leqslant \sup_{x\in[a,b]} |f^{(n)}(x)| \left|\int_a^b K(x)\,dx\right|$$

$$= \sup_{x\in[a,b]} |f^{(n)}(x)|\,|T(Q_n)|$$

where $Q_n(x) = x^n/n!$.

4.6 The content and purpose of the result of the last question only become clear when we use it. In this question we investigate the closeness of the approximation in Simpson's rule

$$\int_{-1}^1 f(x)\,dx \cong \tfrac{1}{3}f(1) + \tfrac{4}{3}f(0) + \tfrac{1}{3}f(-1).$$

(i) Write

$$Tf = \int_{-1}^{1} f(x)\,dx - \tfrac{1}{3}f(1) - \tfrac{4}{3}f(0) - \tfrac{1}{3}f(-1).$$

Show that T obeys the conditions of the previous question with $n = 4$.

(ii) Show that, with the notation of the previous question,

$$K(t) = -72^{-1}(1-t)^3(3t+1) \text{ for } 0 \leqslant t \leqslant 1,$$

$$K(t) = K(-t) \qquad \text{for } -1 \leqslant t \leqslant 0.$$

(iii) Use (iii) (2) of the previous question to deduce that, if $f\colon [-1,1] \to \mathbb{R}$ is four times continuously differentiable, then

$$\left|\int_{-1}^{1} f(x)\,dx - \tfrac{1}{3}f(1) - \tfrac{4}{3}f(0) - \tfrac{1}{3}f(-1)\right| \leqslant \tfrac{1}{90} \sup_{t\in[-1,1]} |f^{(4)}(t)|.$$

Explain why the constant 1/90 is the smallest possible.

(iv) Deduce, by rescaling, that, if $f\colon \mathbb{R} \to \mathbb{R}$ is four times continuously differentiable, then

$$\left|\int_{a-h}^{a+h} f(x)\,dx - \tfrac{1}{3}f(a+h) - \tfrac{4}{3}f(a) - \tfrac{1}{3}f(a-h)\right| \leqslant \frac{h^4}{90} \sup_{|t-a|\leqslant h} |f^{(4)}(t)|.$$

4.7 (i) Find $-1 < u < v < 1$ and c, d such that

$$\int_{-1}^{1} f(x)\,dx = cf(u) + df(v).$$

whenever f is a cubic polynomial. [A more general approach is given in Chapter 41.]

(ii) Use Question 4.5 in the manner of Question 4.6 to show that

$$\left|\int_{a-h}^{a+h} f(x)dx - cf(a+uh) - df(a+vh)\right| \leqslant \frac{h^4}{135} \sup_{|t-a|\leqslant h} |f^{(4)}(t)|$$

for all four times continuously differentiable functions. Explain why the constant 1/135 is best possible.

(Your working will probably involve investigating the sign of a quartic on $[u, v]$. It is not obvious that the sign of the quartic is constant on the interval but with a little work you should be able to prove it.)

4.8 Let $F(x, y) = (y - h(x))(y - 2h(x))$ where h is the function of Example 4.2.

(i) Show that $F\colon \mathbb{R}^2 \to \mathbb{R}$ is infinitely differentiable.

(ii) Show that F does not have a minimum at $(0, 0)$.

(iii) Suppose $f: \mathbb{R} \to \mathbb{R}$ is n times differentiable with $f(0) = f'(0) = \ldots = f^{(n-1)}(0) = 0$ but $f^{(n)}(0) \neq 0$ for some $n \geqslant 1$. By using Question 4.1(vii) show that $x^{-n} f(x) \to f^{(n)}(0) h!$ as $x \to 0$. Conclude that $F(x, f(x))$ has a minimum at 0.

(Thus a function of two variables can have a minimum at $(0, 0)$ along every curve $y = f(x)$ where f satisfied the hypotheses of (iii), and yet not have a minimum at $(0, 0)$.)

5

A SECOND PROOF OF WEIERSTRASS'S THEOREM

5.1 (Alternative treatment of the Tchebychev polynomial)

(i) By using de Moivre's formula $e^{in\theta} = (\cos\theta + i\sin\theta)^n$, show that $\cos n\theta = T_n(\cos\theta)$, $\sin n\theta = \sin\theta\, U_{n-1}(\cos\theta)$ where T_n is a real polynomial of degree at most n and U_{n-1} is a real polynomial of degree at most $n-1$.

(ii) Show that $T'_n(x) = nU_{n-1}(x)$ for $-1 \leqslant x \leqslant 1$ and explain why this means that $T'_n(x) = nU_{n-1}(x)$ for all $x \in \mathbb{R}$.

(iii) By considering the expansion of $(e^{i\theta} + e^{-i\theta})^n$ show that T_n has degree exactly n and find its leading coefficient. Show that $2T_n(t/2)$ is a polynomial with integer coefficients. Show also that there exist polynomials f_n and g_n with integer coefficients such that $2(1 + \cos 2nt) = f_n(\cos t)^2$ and

$$(1 + \cos(2n+1)t)/(1 + \cos t) = g_n(\cos t)^2.$$

(iv) By looking at the real part of $\sum_{n=0}^{\infty}(te^{i\theta})^n$, or otherwise, show that

$$\frac{1 - tx}{1 - 2tx + t^2} = \sum_{n=0}^{\infty} T_n(x)t^n \quad \text{for all } -1 < t < 1,\ -1 \leqslant x \leqslant 1.$$

(v) Show that $T_n(T_m(x)) = T_{nm}(x)$ for all x.

5.2 (Bernstein's proof of Weierstrass's theorem) The following beautiful proof is due to S. Bernstein.

(i) If $1 \geqslant p \geqslant 0$ let $X_n(p)$ be a random variable with binomial distribution

$$Pr(X_n(p) = r/n) = \binom{n}{r} p^r (1-p)^{n-r}.$$

Compute the mean and variance of $X_n(p)$.

(ii) Use Tchebychev's inequality (Theorem 50.4) to show that, if $\varepsilon > 0$ is fixed,

$$Pr(|X_n(t) - t| > \varepsilon) \to 0$$

uniformly for $t \in [0, 1]$.

(iii) Suppose now that $f\colon [0, 1] \to \mathbb{R}$ is continuous. Show that

$$|\mathbb{E}(f(X_n(t)) - f(t))| \leqslant \sup_{|s-t|>\varepsilon} |f(s) - f(t)|\, Pr(|X_n(t) - t| > \varepsilon)$$
$$+ \sup_{|s-t|\leqslant\varepsilon} |f(s) - f(t)|\, Pr(|X_n(t) - t| \leqslant \varepsilon).$$

Conclude, arguing carefully, that $\mathbb{E}(f(X_n(t))) \to f(t)$ uniformly on $[0, 1]$. (You will need, for example, the fact that a continuous function on a closed bounded interval is uniformly continuous.)

(iv) Show that

$$\mathbb{E}(f(X_n(t))) = \sum_{r=0}^{n} f(r/n) \binom{n}{r} t^r(1 - t)^{n-r}$$

and deduce that we can find a sequence of real polynomials P_n with $P_n \to f$ uniformly on $[0, 1]$. (Notice that Bernstein's proof gives us an explicit sequence of approximating polynomials.)

(v) Adapt the proof above to the two-dimensional case, i.e. show that, given any continuous function $f\colon [0, 1]^2 \to \mathbb{R}$ and any $\varepsilon > 0$, we can find $N \geqslant 1$ and $a_{nm} \in \mathbb{R}$ such that $|f(x, y) - \sum_{n=0}^{N} \sum_{m=0}^{N} a_{nm} x^n y^m| < \varepsilon$ for all $0 \leqslant x, y \leqslant 1$.

(vi) Hence show that, if $f\colon [0, 1]^2 \to \mathbb{R}$ is continuous,

$$\int_0^1 \left(\int_0^1 f(x, y)\, dx \right) dy = \int_0^1 \left(\int_0^1 f(x, y)\, dy \right) dx.$$

(This gives a neater proof of Lemma 47.2.)

(vii) Show that there is a continuously differentiable function $f\colon [0, 1] \to \mathbb{R}$ such that $f(r/n) = 0$ for $0 \leqslant r \leqslant n$, $|f'(t)| \leqslant An$ for all $t \in [0, 1]$ (the exact value of A is unimportant, it is easy to get $A = \pi$), yet there exist $x \in [0, 1]$ with $f(x) = 1$. Conclude that for every n there exists a continuously differentiable f such that

$$\sup_{t \in [0,1]} \left| \sum_{r=0}^{n} f(r/n) \binom{n}{r} t^r(1 - t)^{n-r} - f(t) \right| \geqslant A^{-1} \sup_{t \in [0,1]} |f'(t)|\, n^{-1}.$$

(The speed of convergence of Bernstein's polynomials is discussed again in Question 71.1.)

5.3 (A Theorem of Pál) Bernstein's result has an amusing consequence.

(i) Continuing with the notation of Question 5.1 write

$$P_n(t) = \sum_{r=0}^{n} f(r/n) \binom{n}{r} t^r(1 - t)^{n-r},\quad Q_n(t) = \sum_{r=0}^{n} \left[f(r/n) \binom{n}{r} \right] t^r(1 - t)^{n-r},$$

where, as usual, $[x]$ represents the integral part of x. Show that, if $0 < a < b < 1$, then $P_n - Q_n \to 0$ uniformly on $[a, b]$ and so $Q_n \to f$ uniformly on $[a, b]$.

(ii) Deduce that, if $0 < a < b < 1$ and $g: [a, b] \to \mathbb{R}$ is continuous, then we can find polynomials Q_n with integer coefficients such that $Q_n \to g$ uniformly on $[a, b]$.

(iii) Let $c < d$. Find necessary and sufficient conditions to ensure that, whenever $g: [c, d] \to \mathbb{R}$ is continuous, we can find polynomials Q_n with integer coefficients such that $Q_n \to g$ uniformly on $[c, d]$.

(iv) Let $f: [0, 1] \to \mathbb{R}$ be continuous. Show that we can find polynomials Q_n with integer coefficients such that $Q_n \to f$ uniformly on $[0, 1]$ if and only if $f(0)$ and $f(1)$ are integers. (Hint: By considering $g(x) = x - 1$, or otherwise, show how to reduce the problem to one with $f(0) = f(1)$.)

(v) Find necessary and sufficient conditions on a continuous function $f: [-1, 1] \to \mathbb{R}$ to allow us to find a sequence Q_n of polynomials with integer coefficients such that $Q_n \to f$ uniformly on $[-1, 1]$. (Note that you will need a condition on $f(-1) + f(1)$.)

5.4 (An idea of Lebesgue) Lebesgue gave a proof of the theorem of Weierstrass based on a very simple idea which has other applications.

(i) Let $f: [0, 1] \to \mathbb{C}$ be continuous. Define $f_n: [0, 1] \to \mathbb{C}$ to be that function which is linear on each interval $[r/n, (r + 1)/n]$ for $0 \leqslant r \leqslant n - 1$ and such that $f_n(r/n) = f(r/n)$ for $0 \leqslant r \leqslant n$. Sketch f_n in some simple case (e.g. $f(x) = x^2$). By using (with justification) the fact that f is uniformly continuous, or otherwise, show that $f_n \to f$ uniformly on $[0, 1]$.

(ii) Let $\Delta: \mathbb{R} \to \mathbb{R}$, the triangle function, be defined by $\Delta(x) = 1 - |x|$ for $|x| \leqslant 1$, $\Delta(x) = 0$ otherwise, and let us write $\Delta_n(x) = \Delta(nx)$ when n is a strictly positive integer. Sketch $\Delta_n(x - a)$ as a function of x for some typical n and a. Show that

$$f_n(x) = \sum_{r=0}^{n} f(r/n)\Delta_n(x - r/n) \quad \text{for } 0 \leqslant x \leqslant 1.$$

(iii) Show that, if $g: \mathbb{R} \to \mathbb{C}$ is continuous and there exists an $R > 0$ such that $g(x) = 0$ for $|x| \geqslant R$, then

$$\sum_{r=-\infty}^{\infty} g(r/n)\Delta_n(x - r/n) \to g(x)$$

uniformly on $\mathbb{R}$ as $n \to \infty$.

(iv) State with reasons which, if any, of the following are true and which, if any, are false.

(a) If $g: \mathbb{R} \to \mathbb{C}$ is continuous, then

$$\sum_{r=-\infty}^{\infty} g(r/n)\Delta_n(x - r/n) \to g(x)$$

as $n \to \infty$ for each $x \in \mathbb{R}$.

(b) If $g: \mathbb{R} \to \mathbb{C}$ is continuous and bounded, then

$$\sum_{r=-\infty}^{\infty} g(r/n)\Delta_n(x - r/n) \to g(x)$$

uniformly on $\mathbb{R}$ as $n \to \infty$.

(c) If $g: \mathbb{R} \to \mathbb{C}$ is continuous and differentiable with bounded derivative [i.e. $|g'(x)| \leq M$ for all $x \in \mathbb{R}$], then

$$\sum_{r=-\infty}^{\infty} g(r/n)\Delta_n(x - r/n) \to g(x).$$

(d) If $g: \mathbb{R} \to \mathbb{C}$ is continuous and $g(x) \to 0$ as $|x| \to \infty$, then

$$\sum_{r=-\infty}^{\infty} g(r/n)\Delta_n(x - r/n) \to g(x)$$

uniformly on $\mathbb{R}$ as $n \to \infty$.

5.5 (This continues the ideas and notation of Question 5.4.)

(i) Let us write $\mathcal{P}([a, b])$ for the collection of continuous functions $f: [a,b] \to \mathbb{C}$ such that there exists a sequence of polynomials P_n with $P_n(t) \to f(t)$ uniformly on $[a, b]$ and $C([a, b])$ for the set of all continuous functions $f: [a, b] \to \mathbb{C}$. Using Question 5.4, show that, if $\Delta_n \in \mathcal{P}([-1, 1])$ for all $n \geq 1$, then $\mathcal{P}([0, 1]) = C([0, 1])$.

(ii) Let $h: [-2, 2] \to \mathbb{R}$ be given by $h(x) = |x|$ for $x \in [-2, 2]$. Sketch the function given by $h(x + \delta) - h(x - \delta)$ with $|\delta| \leq 1$. Show that, if $h \in \mathcal{P}([-2, 2])$, then $\Delta_n \in \mathcal{P}([-1, 1])$ for all $n \geq 0$ and so $\mathcal{P}([0, 1]) = C([0, 1])$.

(iii) Let $\eta > 0$ and let $h_\eta: [-\eta, \eta] \to \mathbb{R}$ be given by $h_\eta(x) = |x|$ for $|x| \leq \eta$. By using scale changes, or otherwise, show that, if $h_\eta \in \mathcal{P}([-\eta, \eta])$ for some $\eta > 0$, then $h \in \mathcal{P}([-2, 2])$. Conclude that, if $h_\eta \in \mathcal{P}([-\eta, \eta])$ for some $\eta > 0$, then $\mathcal{P}([a, b]) = C([a, b])$ for all $a < b$. In Question 5.6 we give various methods of proving that $h_\eta \in \mathcal{P}([-\eta, \eta])$ for some $\eta > 0$ and thus conclude Lebesgue's proof of the Weierstrass theorem. (Any reader who has done a standard first course in functional analysis will recognise that the usual proof of the Stone Weierstrass theorem is descended from Lebesgue's proof of the Weierstrass theorem.)

5.6 (This continues and concludes Question 5.5.) First we give a proof that $h_1 \in \mathcal{P}([-1, 1])$ which does not use Taylor's theorem.

(i) Let $A_n = \int_{-1}^{1}(1 - x^2)^n\, dx$ and define $f_n: [-1, 1] \to \mathbb{R}$ by $f_n(x) = A_n^{-1}(1 - x^2)^n$. Show that f_n is a polynomial, that $f_n(x) \geq 0$ for all $x \in [-1, 1]$, that $\int_{-1}^{1} f_n(x)\, dx = 1$ and that, if we fix a δ with $1 > \delta > 0$, then $f_n(x) \to 0$ uniformly on $[-1, -\delta] \cup [\delta, 1]$ as $n \to \infty$.

(ii) Let $F_n(x) = \int_{-1}^{x} f_n(t)\, dt$. Show that F_n is a polynomial, that

$0 \leqslant F_n(x) \leqslant 1$ for all $x \in [-1, 1]$ and that, if we fix a δ with $1 > \delta > 0$, then $F_n(x) \to 0$ uniformly on $[-1, -\delta]$ and $F_n(x) \to 1$ uniformly on $[\delta, 1]$ as $n \to \infty$.

(iii) Let $g_n(x) = \int_{-1}^{x} F_n(t)\, dx$ and let $k(x) = \max(0, x)$ for all $x \in [-1, 1]$. Show that g_n is a polynomial and that $g_n(x) \to k(x)$ uniformly on $[-1, 1]$ as $n \to \infty$. By considering $k(x) + k(-x)$, show that $h_1 \in \mathcal{P}([-1, 1])$.

(Next we give a proof using Taylor's theorem.)

(iv) Show by using some form of Taylor's theorem with remainder, or otherwise, that, if $q(x) = \sqrt{(1-x)}$ for $|x| < 1$ (the positive root being chosen) and if $0 < \kappa < 1$, then

$$\sum_{r=0}^{n} (q^{(r)}(0)/r!)x^r \to q(x)$$

uniformly on $[-\kappa, \kappa]$ as $n \to \infty$. Deduce that writing $q_\varepsilon(x) = \sqrt{(1 - (1-\varepsilon)(1-x^2))}$ for $|x| \leqslant 1$, we have $q_\varepsilon \in \mathcal{P}([-1, 1])$ for all $0 < \varepsilon < 1$.

(v) Show that $q_\varepsilon \to h_1$ uniformly on $[-1, 1]$ as $\varepsilon \to 0+$ and deduce that $h_1 \in \mathcal{P}([-1, 1])$.

(Finally we give a proof which depends on solving the equation $f(t)^2 = t^2$ by an iterative procedure.)

(vi) Set $p_0(t) = 0$ and define $p_n(t)$ by

$$p_n(t) = p_{n-1}(t) + \tfrac{1}{2}[t^2 - \{p_{n-1}(t)\}^2] \quad \text{for } n \geqslant 1.$$

Show that p_n is a polynomial. Show by induction, or otherwise, that $|t| \geqslant p_n(t) \geqslant 0$ for all $t \in [-1, 1]$. Show also that, if $t \in [-1, 1]$, $p_n(t) \geqslant p_{n-1}(t)$ for all $n \geqslant 0$.

(vii) By showing that $0 \leqslant t^2 - p_n(t)^2 \leqslant 2/n$ for all $t \in [-1, 1]$ and $n \geqslant 1$, or otherwise, prove that $p_n(t) \to |t|$ uniformly on $[-1, 1]$ as $n \to \infty$ and conclude that $h_1 \in \mathcal{P}([-1, 1])$. Is it true that $p_n(t) \to |t|$ as $n \to \infty$ for all $t \in \mathbb{R}$?

6

HAUSDORFF'S MOMENT PROBLEM

6.1 (Alternative proof of Hausdorff's moment theorem) Give a proof of Theorem 6.1 along the lines of Question 2.3. (If $[a, b] = [0, 1]$ you could consider $h(t) = 1 + \varepsilon - (c - t)^2$.)

6.2 (Algebra required for the next question)

(i) Let $\Delta(x_1, x_2, \ldots, x_n)$ be the determinant of the matrix whose (i, j)th entry is x_i^{j-1}. Show that Δ is a multinomial of degree $n(n-1)/2$. Show further, by considering the leading diagonal, that the coefficient of $x_2 x_3^2 \ldots x_n^{n-1}$ in Δ is 1.

(ii) Explain why $\Delta(x_1, x_2, \ldots, x_n) = 0$, if $x_r = x_s$ for some r and s with $r \neq s$, and conclude that $x_r - x_s$ is a factor of $\Delta(x_1, x_2, \ldots, x_n)$ whenever $r \neq s$. Deduce that $\prod_{1 \leqslant s < r \leqslant n}(x_r - x_s)$ divides $\Delta(x_1, x_2, \ldots, x_n)$ and use the information obtained in (i) to show that

$$\Delta(x_1, x_2, \ldots, x_n) = \prod_{1 \leqslant s < r \leqslant n} (x_r - x_s).$$

(iii) Show that we can find $\lambda_1, \lambda_2, \ldots, \lambda_n \in \mathbb{R}$ such that

$$\sum_{j=1}^{n} \lambda_j\, j^k = 0 \quad \text{for } 1 \leqslant k \leqslant n-1,$$

$$\sum_{j=1}^{n} \lambda_j\, j^n = 1.$$

6.3 (Another function with all its moments zero)

(i) By induction, or otherwise, show that for each integer $n \geqslant 0$ we can find a real polynomial P_n of degree $n + 2$ such that $P_n(0) = P_n(1) = 0$, $\int_0^1 t^r P_n(t)\,dt = 0$ for all $0 \leqslant r \leqslant n - 1$ and $\int_0^1 t^n P_n(t)\,dt = 1$. Deduce, or prove otherwise, that we can find a real polynomial Q_n of degree $n + 2$ such that $Q_n(0) = Q_n(1) = 0$, $\int_0^1 Q_n(t)\,dt = 1$ and $\int_0^1 t^r Q_n(t)\,dt = 0$ for all $1 \leqslant r \leqslant n$.

(ii) Now set $f_n(t) = Q_n(t)$ for $0 \leqslant t \leqslant 1$, $f_n(t) = 0$ otherwise. Observe that $f_n: \mathbb{R} \to \mathbb{R}$ is a continuous function and show by making a change of variable, or otherwise, that $\int_{-\infty}^{\infty} t^r f_n(t-a)\,dt = a^r$ for all $0 \leqslant r \leqslant n$. Show that, if $n \geqslant 1$ and $\lambda_1, \lambda_2, \ldots, \lambda_n$ are as in part (iii) of Question 6.2, then

(a) $\displaystyle\int_{-\infty}^{\infty} t^r \sum_{j=1}^{n} \lambda_j f_n(t-ja)\,dt = 0 \quad \text{for } 0 \leqslant r \leqslant n-1,$

(b) $\displaystyle\int_{-\infty}^{\infty} t^n \sum_{j=1}^{n} \lambda_j f_n(t-ja)\,dt = a^n,$

(c) $\displaystyle\int_{-\infty}^{\infty} \left| t^r \sum_{j=1}^{n} \lambda_j f_n(t-ja)\right| dt \leqslant A|a|^r \quad \text{for all } r,$

where A depends on f_n and on $\lambda_1, \lambda_2, \ldots, \lambda_n$ but not on a.

(iii) Conclude that, given an integer $n \geqslant 1$, a real number $S \geqslant 0$ and some $\varepsilon > 0$, we can find a continuous function F and a real number $R > S$ such that

(a) $\displaystyle\int_{-\infty}^{\infty} t^r F(t)\,dt = 0 \quad \text{for } 0 \leqslant r \leqslant n-1,$

(b) $\displaystyle\int_{-\infty}^{\infty} t^n F(t)\,dt = 1,$

(c) $\displaystyle\int_{-\infty}^{\infty} |t^r F(t)|\,dt < \varepsilon \quad \text{for } 0 \leqslant r \leqslant n-1,$

(d) $|F(t)| < \varepsilon$ for all t,

(e) $F(t) = 0$ for $t \neq [1, R]$.

(iv) Hence show that, choosing g_0 to be a non zero continuous function with $g_0(t) = 0$ for $t \notin [0, 1]$, we can find a sequence of continuous functions $g_n: \mathbb{R} \to \mathbb{R}$ and a sequence $1 < R_1 < R_2 < \ldots$ with $R_n \to \infty$ such that

(a) $\displaystyle\int_{-\infty}^{\infty} t^r \sum_{j=0}^{n} g_j(t)\,dt = 0 \quad \text{for } 0 \leqslant r \leqslant n-1,$

(b) $g_n(t) = 0$ for $t \notin [1, R_n]$,

(c) $\displaystyle\int_{-\infty}^{\infty} |t^r g_n(t)|\,dt \leqslant 2^{-n} \quad \text{for } 0 \leqslant r \leqslant n-1,$

(d) $|g_n(t)| \leqslant 2^{-n}$ for all t.

Deduce that $\sum_{m=0}^{\infty} g_m(t)$ converges uniformly to a continuous function G which is not identically zero. Remark that $G(t) = 0$ for $t \leqslant 0$ and $G(t) \to 0$ as $t \to \infty$. By examining the behaviour of $\int_{-R}^{R} |t^r \sum_{m=0}^{n} g_m(t)|\,dt$ and $\int_{-R}^{R} t^r \sum_{m=0}^{n} g_m(t)\,dt$ for fixed $R \geqslant R_r$ as $n \to \infty$ show that $\int_{-\infty}^{\infty} |t^r G(t)|\,dt$ converges and $\int_{-\infty}^{\infty} t^r G(t)\,dt = 0$ for all $r \geqslant 0$.

7

THE IMPORTANCE OF LINEARITY

7.1$^{(-)}$ A worm cannot survive if the temperature of the soil falls to 0 °C. Suppose the diffusivity (the K of Lemma 7.1) is 5×10^{-7} metres^2 per second and that the temperature of a winter day varies between −8 °C and 12 °C reaching its minimum at 4 a.m. Estimate the depth to which worms must burrow in order to survive, and the time at which the worms should be at their greatest depth.

7.2 Consider a homogeneous sphere centre $\mathbf{0}$ radius a whose surface temperature $\theta(\boldsymbol{r}, t)$ $[\|\boldsymbol{r}\| = a]$ depends only on t and not on $\boldsymbol{r}$. It is natural to look for a radially symmetric solution to the three dimensional heat equation $\partial\theta/\partial t = K\nabla^2\theta$, i.e. a solution $\theta(\boldsymbol{r}, t) = \phi(r, t)$ where $r = \|\boldsymbol{r}\|$. Verify that

$$\nabla^2\theta(\boldsymbol{r}, t) = \frac{1}{r^2}\frac{\partial}{\partial r^2}\left(r^2\frac{\partial\phi}{\partial r}\right)$$

and that, if we set $\eta(r, t) = r\phi(r, t)$, the heat equation reduces to $K\,\partial^2\eta/\partial r^2 = \partial\eta/\partial t$. Explain why we must demand that $\eta(0, t) = 0$.

Suppose the surface temperature is given by $\phi(a, t) = \cos\omega t$. By adapting the arguments of Lemma 7.1, or otherwise, show that, if $0 < r \leqslant a$,

$$\theta(\boldsymbol{r}, t) = \phi(r, t) = \frac{aA(r)}{r}\cos(\omega t + \tau(r))$$

where

$$A(r) = \left|\frac{\sin((\omega/2K)^{1/2}r(1+i))}{\sin((\omega/2K)^{1/2}a(1+i))}\right|,$$

$$\tau(r) = \arg\left\{\frac{\sin((\omega/2K)^{1/2}r(1+i))}{\sin((\omega/2K)^{1/2}a(1+i))}\right\}.$$

Show also that

$$\theta(\mathbf{0}, t) = \frac{(\omega/K)^{1/2}}{|\sin((\omega/2K)^{1/2}a(1+i))|}\cos\omega t.$$

By considering what happens if $\omega a^2/K$ is very large and $(a - r)/a$ is small, recover the result of Lemma 7.1. If on the other hand $\omega a^2/K$ is small compared with 1, show that to a first approximation $\theta(r, t) = a\cos(\omega t + \tau(r))$ for $0 \leqslant r \leqslant a$.

(Although the physical mechanism is different, the diffusion equation also describes the strength of radio waves penetrating a suitable body in microwave cookery. Taking heat conduction in to account, we see that microwaves cook from the inside out where-as old-fashioned heating cooks from the outside in. Those who like a joint well done on the outside and less well done inside should not use a microwave.

Incidently the results of Lemma 7.1 are also, and for a similar reason, relevant to the strength of radiowaves penetrating the sea. For wireless communication with submarines we must use low frequencies whose effect dies off less rapidly with depth than high frequencies. In order to communicate with submarines at depths of 100 metres the major navies of the world use very low frequency (300 Hertz to 3 kiloHertz) radio waves requiring transmission aerials many tens of kilometres long. (For one problem this raises see Question 60.16.)

Finally we note that in Sweden experimental systems have been set up in which solar energy is used during the summer to raise the temperature of heat reservoirs deep under ground. Except that the heat flows in the opposite direction, this is exactly the same kind of situation as that considered above and gives rise to the same equations. Unfortunately such systems seem to be uneconomic at present, though only marginally so.)

8

COMPASS AND TIDES

8.1 It is tempting to extend the ideas of Lemma 8.1 as follows. Let $h(t) = A_0 + \sum_{r=1}^{N} A_r \cos \omega_r t + \sum_{r=1}^{N} B \sin \omega_r t$ with $0 = \omega_0 < \omega_1 < \dots < \omega_N$ and $\omega_r - \omega_{r-1} > \delta$ for $1 \leqslant r \leqslant N$ and some $\delta > 0$. Suppose now that $T\delta$ is very large compared with 1 and set $\alpha(k) = (2/T)\int_0^T h(t) \cos kt \, dt$ for $k \geqslant 0$. Show that

(a) $\alpha(k) \approx A_r \dfrac{\sin(k - \omega_r)T}{(k - \omega_r)T}$ for $|k - \omega_r| \leqslant \delta/2$

(what value do you assign to $\alpha(k)$ when $k = \omega_r$?),

(b) $\alpha(k)$ is small otherwise.

Sketch the behaviour of $\alpha(k)$ in the region $|k - \omega_r| \leqslant \delta/2$.

Suppose now that you know that h has the form given, but you do not know N, A_0, A_1, ..., A_N, B_1, ..., B_N and ω_1, ω_2, ..., ω_N. Describe how you would estimate the quantities from a graph of $\alpha(k)$.

The way now seems clear to apply this method to any h whatsoever to uncover its 'hidden periodicities' ω_1, ω_2, ... In the early part of the twentieth century it was applied to economic and meteorological statistics with what seemed to be great success in uncovering all sorts of unexpected periodicities. (There are some periods one would expect to find in, say, American economic statistics: a yearly one corresponding to the seasons and a four yearly one corresponding to presidential elections.) Unfortunately these periodicities did not continue beyond the period analysed. Patterns without theoretical explanation are apt to prove mirages.

8.2 (Dirichlet's lemma. Both the result and the method of proof are more useful than may at first appear.)

(i) Suppose $\lambda_1, \lambda_2, \dots, \lambda_n \in \mathbb{T}$ and $N \geqslant 1$ is an integer. Observe that $\mathbb{T}^n$ is the disjoint union of precisely N^n sets ('pigeon holes') of the form $\prod_{j=1}^{n}[2\pi k_j/N, 2\pi(k_j + 1)/N)$ where the k_j are integers with $0 \leqslant k_j \leqslant N - 1$.

Thus at least those two of the N^n+1 points $(r\lambda_1, r\lambda_2, \ldots, r\lambda_n)$ with $0 \leq r \leq N^n$ must lie in the same 'pigeon hole'. (This is an example of the justly famous pigeon hole principle introduced by Dirichlet.) In other words, we can find integers r, s with $0 \leq r < s \leq N^n$ and integers $k_1, k_2, \ldots, k_n$ such that

$$(r\lambda_1, r\lambda_2, \ldots, r\lambda_n) \in \prod_{j=1}^{n} [2\pi k_j/N, 2\pi(k_j+1)/N),$$

$$(s\lambda_1, s\lambda_2, \ldots, s\lambda_n) \in \prod_{j=1}^{n} [2\pi k_j/N, 2\pi(k_j+1)/N),$$

i.e. such that $r\,\lambda_j$, $s\,\lambda_j \in [2\pi k_j/N, 2\pi(k_j+1)/N)$ for $1 \leq j \leq n$.

If we now set $m = s - r$, show that $0 < m \leq N^n$ and $m\lambda_j \in (-2\pi/N, 2\pi/N)$, i.e. $|m\lambda_j| < 2\pi/N$ for all $1 \leq j \leq n$.

(ii) Conclude that, given $\lambda_1, \lambda_2, \ldots, \lambda_n \in \mathbb{T}$, $\varepsilon > 0$ and N an integer with $N > 2\pi\varepsilon^{-1}$, we can find an integer m with $0 < m \leq N^n$ and $|m\lambda_j| < \varepsilon$ for $1 \leq j \leq n$.

(iii) Show that, given $z_1, z_2, \ldots, z_n \in \mathbb{C}$ with $|z_j| = 1$ $[1 \leq j \leq n]$, $\varepsilon > 0$ and N an integer with $N > 2\pi\varepsilon^{-1}$, we can find an integer m with $0 < m \leq N^n$ and $|z_j^m - 1| < \varepsilon$ for $1 \leq j \leq n$.

8.3 (i) Let us continue with the hypotheses and notation of Question 8.2(i). Let us call a 'pigeon hole' $I = \prod_{j=1}^{n}[2\pi k_j/N, 2\pi(k_j+1)/N)$ an 'occupied pigeon hole', if there exists some integer r such that $(r\lambda_1, r\lambda_2, \ldots, r\lambda_n) \in I$. Show that there exists an integer $M \geq 1$ with the property that, if I is an occupied pigeon hole, then there exists an r with $M \geq |r|$ and $(r\lambda_1, r\lambda_2, \ldots, r\lambda_n) \in I$. Deduce that, given any integer m_0, we can find an m with $|m - m_0| \leq M$ and $m\lambda_j \in (-2\pi/N, 2\pi/N)$ $[1 \leq j \leq n]$.

(ii) Conclude that, given $\lambda_1, \lambda_2, \ldots, \lambda_n \in \mathbb{T}$ and $\varepsilon > 0$, we can find an integer M such that, given any integer m_0, we can find an integer m with $|m - m_0| \leq M$ and $|m\lambda_j| < \varepsilon$ for $1 \leq j \leq n$.

(iii) By considering the special case when $n = 1$ and $|\lambda_1|$ is small but non zero, show that M cannot be taken to be a function of n and ε alone but depends on the λ_j. (Contrast Question 8.2(ii).)

(iv) Let $P: \mathbb{R} \to \mathbb{C}$ be given by $P(t) = \sum_{j=1}^{n} a_j \exp(i\,\lambda_j t)$. Show that, with the notation of (ii), $|P(t) - P(t+m)| \leq \sum_{j=1}^{n} |a_j|\varepsilon$ for all $t \in \mathbb{T}$. Conclude specifically that, given any $\eta > 0$, we can find a real number L (depending on η and P) such that every interval $[a, a+L]$ of length L contains a point τ with $|P(t) - P(t+\tau)| < \eta$ for all $t \in \mathbb{R}$.

(v) Modify part (i) as follows. Let $L > 0$ and suppose that we are given x_j, $y_j \in \mathbb{T}$ such that $jL \leq x_j \leq y_j < (j+1)L$ for each $j \in \mathbb{Z}$. Show that, given any $\varepsilon > 0$, we can find an integer $M \geq 1$ such that, given any integer m_0, we can find $j, k \in \mathbb{Z}$ with $|(j-k) - m_0| \leq M$ and $|(x_j - x_k) - (y_j - y_k)| < \varepsilon$.

8.4 The preceding questions suggest the following definition. Let us say that a function $f: \mathbb{R} \to \mathbb{C}$ is *almost periodic* if (a) it is continuous and (b) given any $\varepsilon > 0$, we can find $L = L(\varepsilon, f)$ such that, given any interval $[a, a + L)$ of length L, we can find $\tau \in [a, a + L)$ with $|f(t + \tau) - f(t)| < \varepsilon$ for all $t \in \mathbb{R}$. (We call τ an *ε-almost period* of f.) The following exercises are intended to illustrate this definition.

(i) Show that, if $P: \mathbb{R} \to \mathbb{C}$ is given by $P(t) = \sum_{j=1}^{n} a_j \exp(i\lambda_j t)$, then P is almost periodic.

(ii) Show that, if $f_n: \mathbb{R} \to \mathbb{C}$ is almost periodic and $f_n \to f$ uniformly, then f is almost periodic.

(It is natural to conjecture that all almost periodic functions arise as the uniform limit of trigonometric polynomials P of the kind given in (i). This conjecture is correct, but hard to prove with the tools at our disposal.)

(iii) By considering $P(t) = \exp(2\pi i t) + \exp((2\pi)^{1/2} it)$ and using Weyl's equidistribution theorem, or otherwise (less powerful techniques will suffice), show that there exist almost periodic functions which are not periodic.

(iv) Suppose $f: \mathbb{R} \to \mathbb{R}$ is continuous with $f(t) = 0$ for $|t| > T$ and $|f(t)| \leqslant 1$ for $|t| \leqslant T$. Suppose $S > 2T$ and $\varepsilon > N^{-1} > 0$. Set $g(t) = N^{-1}\sum_{r=-N}^{N}(N - |r|)f(t + rS)$. Show that $g(t) = f(t)$ for $|t| \leqslant T$, $g(t) = 0$ for $|t| \geqslant (N + 1)S$, that g is continuous and that g has an ε-almost period S.

(v) By iterating the construction of (iv), or otherwise, show that we can find a continuous function f such that

(a) $f(0) = 1$,
(b) for each integer $n \geqslant 1$, f has an almost period τ_n with $\tau_n \geqslant n$,
(c) given any integer $n \geqslant 1$, we can find an $a \geqslant 1$ such that $f(t) = 0$ for all $t \in [a, a + n]$.

(vi) Let $f_1(t) = 1$ if $t \in \{2\pi n: n \in \mathbb{Z}\}$, $f_1(t) = 0$ otherwise, and let $f_2(t) = 1$ if $t \in \{(2\pi)^{1/2} n: n \in \mathbb{Z}\}$, $f_2(t) = 0$ otherwise. Show that f_1, f_2 satisfy condition (b) of the definition of an almost periodic function, but $f_1 + f_2$ does not.

8.5 (i) Show that a continuous periodic function $f: \mathbb{R} \to \mathbb{C}$ is bounded and uniformly continuous.

(ii) By modifying your proof of (i), or otherwise, show that any almost periodic function is bounded and uniformly continuous.

(iii) Show that, if f is almost periodic, so are the complex conjugate $\bar{f}$, $|f|$ and af where $a \in \mathbb{C}$.

(iv) Show that, if f is almost periodic and $g(x) = f(ax + b)$ for all $x \in \mathbb{R}$ $[a, b \in \mathbb{R}, a \neq 0]$, then g is almost periodic.

(v) Show that, if f is almost periodic, so is f^2.

(vi) By using Question 8.3(v), or otherwise, show that if f and g are almost periodic, then so is $f + g$. (This is our first non trivial result.)

(vii) By using the relation $fg = 4^{-1}((f+g)^2 - (f-g)^2)$, or otherwise, show that if f and g are almost periodic then so is fg.

(It is clear that many astronomical events follow an almost periodic pattern. To take an example, the earth rotates with a fixed period, the moon rotates round the earth with a fixed period and the earth rotates round the sun with a fixed period, so we have eclipses in an almost periodic fashion.

Many other phenomena exhibit short-term predictability and long-term patterns without, apparently, long-term predictability (this is sometimes referred to as *chaotic* behaviour). The English weather has clear seasons, but we cannot tell a year in advance if the next winter will be harsh or mild. For some time almost periodic functions provided the only recognised mathematic objects with similar properties and were much used as physical models. Nowadays we have a wider variety of mathematical objects to use and prefer those which assign a greater rôle to chance (as seems reasonable in our imperfect sublunary sphere). However we are still in the middle of prolonged controversy between crystallographers as to whether there exist materials in which the atoms lie in almost periodic arrays.)

8.6 The object of this question is to show that every almost periodic function f has a long-term average, i.e. that $\lim_{S,T\to\infty}(1/(T+S))\int_{-S}^{T} f(x)\,dx$ exists.

(i) Suppose that f is a bounded continuous function and that $T > 0$. Show that

$$\sup_{(n+1)T \geqslant X \geqslant nT}\left|\frac{1}{nT}\int_0^{nT} f(x)\,dx - \frac{1}{X}\int_0^X f(x)\,dx\right| \to 0 \quad \text{as } n \to \infty.$$

(ii) Suppose that f is continuous function such that $|f(t)| \leqslant A$ for all $t \in \mathbb{R}$ and that $T > \tau > 0$, $\varepsilon > 0$ and $r \in \mathbb{Z}$ are such that $(r+1)T > \tau \geqslant rT$ and $|f(x+rT) - f(x+\tau)| < \varepsilon$ for all $x \in \mathbb{R}$. Show that

$$\left|\int_{rT}^{(r+1)T} f(x)\,dx - \int_0^T f(x)\,dx\right| \leqslant T\varepsilon + \int_0^T |f(x+\tau) - f(x)|\,dx$$
$$\leqslant T\varepsilon + 2AT.$$

Conclude, by choosing T and τ carefully (observe that you will need $\tau \in [rT, (r+\delta)T]$ for some small $\delta > 0$), that, given any almost periodic function f and any $\eta > 0$, we can find a $T > 0$ such that

$$\left|\frac{1}{nT}\int_0^{nT} f(x)\,dx - \frac{1}{T}\int_0^T f(x)\,dx\right| < \frac{\eta}{3}.$$

(iii) Now use (i) and (ii) to show that, if f is an almost periodic function, then given any $\eta > 0$ we can find $X_0(\eta) > 0$ such that

$$\left|\frac{1}{X}\int_0^X f(x)\,dx - \frac{1}{Y}\int_0^Y f(x)\,dx\right| < \eta$$

for all $X, Y > X_0$.

Show that this means that $\frac{1}{n}\int_0^n f(x)\,dx$ converges to some limit $M(f)$ say as $n \to \infty$ and deduce that $\frac{1}{T}\int_0^T f(x)\,dx \to M(f)$ as $T \to \infty$.

(iv) Show that, with the notation and hypotheses on (iii)

$$(1/(T+S))\textstyle\int_{-S}^{T} f(x)\,dx \to M(f) \text{ as } T, S \to \infty.$$

8.7 In this question we derive some simple properties of the long-term average introduced in Question 8.6.

(i) Show that, if f and g are almost periodic, $M(f+g) = Mf + Mg$.

(ii) Show that, if f is almost periodic and $a \in \mathbb{C}$, then $M(af) = aMf$.

(iii) Suppose f is almost periodic and $g(x) = f(ax+b)[a,\ b \in \mathbb{R},\ a \neq 0]$. Find Mg in terms of Mf.

(iv) If f is a real valued almost periodic function and $f(x) \geq 0$ for all $x \in \mathbb{R}$, show that $Mf \geq 0$.

(v) (a) If $h: [a, b] \to \mathbb{R}$ is continuous, $h(x) \geq 0$ for all $x \in \mathbb{R}$, and $\int_a^b h(t)\,dt = 0$ show that $h = 0$.

(b) If $f: \mathbb{R} \to \mathbb{R}$ is continuous and periodic, $f(x) \geq 0$ for all $x \in \mathbb{R}$ and $Mf = 0$, show that $f = 0$.

(c) If f is a real valued almost periodic function, $f(x) \geq 0$ for all $x \in \mathbb{R}$ and $Mf = 0$, show that $f = 0$.

(vi) If f is almost periodic, explain why for each $\lambda \in \mathbb{R}$ we can define the Fourier coefficient $(\mathcal{F}_M f)(\lambda)$ by $(\mathcal{F}_M f)(\lambda) = M(fe_{-\lambda})$ where $e_{-\lambda}(t) = \exp(i\lambda t)$ $[t \in \mathbb{R}]$. Show further that, if $P(t) = \sum_{j=1}^{n} a_j \exp(i\lambda_j t)$ $[t \in \mathbb{R}]$ where $\lambda_1, \lambda_2, \ldots, \lambda_n \in \mathbb{R}$ are distinct, then $(\mathcal{F}_M P)(\lambda_j) = a_j$ $[1 \leq j \leq n]$, $\mathcal{F}_M P(\lambda) = 0$ otherwise. [It is usual to write $\hat{f}(\lambda) = (\mathcal{F}_M f)(\lambda)$ but this can give rise to inconsistent notations unless care is exercised.]

(vii) Suppose that we have a sequence f_n of almost periodic functions such that $f_n \to f$ uniformly as $n \to \infty$. Show that $Mf_n \to Mf$. Show also that $(\mathcal{F}_M f_n)(\lambda) \to (\mathcal{F}_M f)(\lambda)$ for each $\lambda \in \mathbb{R}$.

(viii) Suppose $f: \mathbb{T} \to \mathbb{C}$ is continuous. Define a 2π periodic function $\tilde{f}: \mathbb{R} \to \mathbb{C}$ by $\tilde{f}(t) = f(t)$ $[t \in \mathbb{R}]$. (There is a slight abuse of notation here; if the reader is disturbed by it she may correct it along the lines of Appendix A.) Show by using Fejér's theorem (Theorem 2.3(ii)), or otherwise, that $(\mathcal{F}_M \tilde{f})(n) = \hat{f}(n)$ if $n \in \mathbb{Z}$, $(\mathcal{F}_M \tilde{f})(\lambda) = 0$ if $\lambda \notin \mathbb{Z}$. (We shall continue our study of almost periodic functions in Question 33.1.)

9

THE SIMPLEST CONVERGENCE THEOREM

9.1 (i) Show that $(3x^2 - \pi^2)/12 = \sum_{n=1}^{\infty}(-1)^n n^{-2} \cos nx$ for $-\pi \leqslant x \leqslant \pi$.

(ii) By integrating term by term (you must justify this), or otherwise, evaluate $\sum_{n=1}^{\infty}(-1)^n\ n^{-3} \sin nx$ for $-\pi \leqslant x \leqslant \pi$.

(iii) By using the formula $\sin a \sin b = 2^{-1}(\cos(a-b) - \cos(a+b))$, or otherwise, show that $f(x, y) = \sum_{n=1}^{\infty}(-1)^n n^{-2}\ \sin nx\ \sin ny$ is a well-defined continuous function $f: \mathbb{R}^2 \to \mathbb{R}$ such that $\{((x, y): f(x, y) = 0\}$ consists of two systems of lines at right angles dividing the coordinate plane into squares of area π^2.

(iv) Show that, if the point $(x, y, z) \in \mathbb{R}^3$ lies inside the octohedron bounded by the planes $\pm x \pm y \pm z = \pi$, then

$$-4\sum_{n=1}^{\infty}(-1)^n\ n^{-3} \sin nx \sin ny \sin nz = xyz.$$

What happens elsewhere?

(Parts (iii) and (iv) are taken from the exercise sets in Whittaker and Watson *Modern Analysis*, where it is indicated that this sort of question (without our extensive hints and with further elaboration) was popular in Cambridge exams around 1900.)

9.2 (i) Prove the following version of Weierstrass M-test. Let X be a set and f_1, $f_2, \ldots$ a sequence of complex valued functions on X. Suppose that there exists a sequence $M_1, M_2, \ldots$ of non negative real numbers such that $\sum_{n=1}^{\infty} M_n$ converges and $|f_n(x)| \leqslant M_n$ for all $x \in X$ and $n \geqslant 1$. Then $\sum_{n=1}^{\infty} f_n$ converges uniformly on X.

(ii) According to Halmos the following question should be asked more frequently. Does there exist a sequence of continuous functions $f_1, f_2, \ldots$ with $f_n: [0, 1] \to \mathbb{R}$ such that

(a) $f_n(x) \geqslant 0$ for all $x \in [0, 1]$,

(b) $\sum_{n=1}^{N} f_n(x)$ converges uniformly on $[0, 1]$ as $N \to \infty$,

but the Weierstrass M-test fails (i.e. $\sum_{n=1}^{\infty} \sup_{x\in[0,1]}|f_n(x)|$ diverges)? Give a proof or a counter example.

10
THE RATE OF CONVERGENCE

10.1 Jackson begins his book *The Theory of Approximation* with a reminiscence of the time when he did his PhD thesis with Landau.

> One day about twenty years ago I was admitted to the study of Professor Landau, seeking advice as to a subject for a thesis. After some preliminary inquiries as to my experience and preferences, he handed me a long sheet of paper, and directed me to take notes as he enumerated some dozen or fifteen topics in various fields of analysis and number theory, with a few words of explanation of each. He told me to think about them for a few days, and to select one of them, or any other problem of my own choosing, with the single reservation that I should *not* prove Fermat's theorem, an injunction which I have observed faithfully. Guided partly by natural inclination, perhaps, and partly by recollection of a course on methods of approximation which I had taken with Professor Bôcher a few years earlier, I committed myself to one of the topics which Landau had proposed, an investigation of the degree of approximation with which a given continuous function can be represented by a polynomial of given degree. When I reported my choice, he said meditatively, in words which I remember vividly in substance, if not perfectly as to idiom: 'Das ist ein schönes Thema, ich beneide Sie um das Thema. . . Nein, ich beneide Sie nicht, aber es ist ein wunderschönes Thema!' [That is a beautiful topic, I envy you that topic. . . No, I do not envy you, but it is a wonderful topic.]

In the next few questions, we develop some complementary results of Jackson and Bernstein on how well certain functions can be approximated by trigonometric polynomials of low degree. If $f: \mathbb{T} \to \mathbb{C}$ is continuous, we write

$$E_n(f) = \inf\{\sup_{t\in\mathbb{T}}|f(t) - P(t)|: P \text{ a trigonometric polynomial of degree at most } n\}.$$

Bernstein's work depends on his inequality which states that there is a constant A such that whenever P is a trigonometric polynomial of degree n, we have $|P'(x)| \leqslant An \sup_{t\in\mathbb{T}}|P(t)|$ (see Question 2.5 or Question 43.8).

(i) Let $f: \mathbb{T} \to \mathbb{C}$ be continuous. Suppose $m \geqslant 1$ is an integer, $m < \alpha \leqslant m+1$ and $E_n(f) \leqslant Bn^{-\alpha}$ for all $n \geqslant 1$. Explain why we can find polynomials P_n of degree 2^n such that $\sup_{t\in\mathbb{T}}|P_n(t) - f(t)| \leqslant 2B2^{-n\alpha}$. Show that, if we write $Q_n = P_1$ and $Q_n = P_{n+1} - P_n$ for $n \geqslant 1$, then

$$\sup_{t\in\mathbb{T}}|Q_n(t)| \leqslant 4B2^{-n\alpha}$$

and $\sum_{r=0}^{N} Q_r(t) \to f(t)$ uniformly on $\mathbb{T}$ as $N \to \infty$.

(ii) Using Bernstein's inequality to estimate $\sup_{t\in\mathbb{T}}|Q_n^{(k)}(t)|$ for $1 \leqslant k \leqslant m$, show that $\sum_{r=0}^{\infty} Q_r^{(k)}(t)$ converges uniformly for each $0 \leqslant k \leqslant m$ and deduce (I assume the reader will not need to be directed to Lemma 53.2) that f is m times continuously differentiable with $f^{(m)}(t) = \sum_{r=0}^{\infty} Q_r^{(m)}(t)$. (Thus if a function is well approximated by trigonometric polynomials, it must be smooth.)

(iii) By estimating $\sup_{t\in\mathbb{T}}|f^{(m)}(t) - \sum_{r=0}^{N-1} Q_r^{(m)}(t)|$, show that

$$E_n(f^{(m)}) \leqslant C_m B n^{m-\alpha} \quad \text{for all } n \geqslant 1$$

where C_m is a constant depending only on m.

10.2 Bernstein showed that much finer information about the smoothness of f can be obtained using these ideas.

(i) Let $f\colon \mathbb{T} \to \mathbb{C}$ be continuous. Suppose $0 < \alpha < 1$ and $E_n(f) \leqslant Bn^{-\alpha}$ for all $n \geqslant 1$. Show that we can find polynomials Q_n of degree 2^{n+1} such that $\sup_{t\in\mathbb{T}}|Q_n(t)| \geqslant 4B2^{-n\alpha}$ and $\sum_{r=0}^{N} Q_n(t) \to f(t)$ uniformly on $\mathbb{T}$ as $N \to \infty$.

(ii) Let $|h| \leqslant \pi$, $t \in \mathbb{T}$ and $N \geqslant 0$. Show that

$$|f(t+h) - f(t)| \leqslant \sum_{n=0}^{N} |Q_n(t+h) - Q_n(t)| + C_1(\alpha)\, B2^{-\alpha N}$$

where $C_1(\alpha)$ is a constant depending only on α. By using the mean value theorem together with Bernstein's inequality, show that

$$|Q_n(t+h) - Q_n(t)| \leqslant 8BA2^{n(1-\alpha)}\,|h|$$

and deduce that

$$|f(t+h) - f(t)| \leqslant B(C_2(\alpha)2^{N(1-\alpha)}\,|h| + C_1(\alpha)2^{-\alpha N})$$

where $C_2(\alpha)$ is a constant depending only on α.

(iii) By choosing N in an appropriate manner (depending on h), show that

$$|f(t+h) - f(t)| \leqslant C_3(\alpha)B|h|^{\alpha}$$

where $C_3(\alpha)$ is a constant depending only on α. This result is usually expressed using the notion of the modulus of continuity $\omega(f, h)$ defined by

$$\omega(f, h) = \sup_{t\in\mathbb{T}} |f(t+h) - f(t)|$$

so that our result becomes

$$\omega(f, h) \leqslant C_3(\alpha)B|h|^{\alpha}.$$

(iv) Graph $f(t) = |t|^\alpha\ [|t| \leqslant \pi]$ for various values of α. Do you agree that we have a quantitative version of the following statement from Acton's *Numerical Methods That Work*: '... while polynomials can wriggle in a number of satisfactory ways there are geometries they are notoriously poor at representing. No polynomial, for example, ever had a [vertical] asymptote.'?

(v) Examine the reasoning of (ii). Where does it break down if $\alpha = 1$? Show, however, that, if $\alpha = 1$, we do have

$$\omega(f, h) \leqslant C_4 B|h| \log |h|^{-1} \quad \text{for } |h| < 1.$$

(vi) Let $f: \mathbb{T} \to \mathbb{C}$ be continuous. Suppose $m \geqslant 0$ is an integer, $m < \alpha \leqslant m + 1$ and $E_n(f) \leqslant Bn^{-\alpha}$ for all $n \geqslant 1$. What can we say about $\omega(f^{(m)}, h)$?

10.3 We now turn to results in the opposite direction. All the theorems we shall obtain are due to Jackson.

(i) Let σ_n and K_n be as in Chapter 2. Show that, if f is continuous, then, using the notation of the previous question,

$$|f(t) - \sigma_n(f, t)| \leqslant \frac{1}{2\pi} \int_{\mathbb{T}} K_n(x)\omega(f, |x|)\, dx.$$

(ii) Show by simple calculus that $s \geqslant \sin s \geqslant 2s/\pi$ for $0 \leqslant s \leqslant \pi/2$. Hence, or otherwise, show that

$$K_n(s) \leqslant (n+1)^{-1} \pi^2 s^{-2} \quad \text{for } 0 < |s| < \pi.$$

Show also that $0 \leqslant K_n(s) \leqslant n + 1$ for all s. By estimating the two integrals on the right hand side of the equality

$$\frac{1}{2\pi} \int_{\mathbb{T}} |x|^\alpha K_n(x)\, dx = \frac{1}{2\pi} \int_{|x| \leqslant n^{-1}} |x|^\alpha K_n(x)\, dx + \frac{1}{2\pi} \int_{|x| \geqslant n^{-1}} |x|^\alpha K_n(x)\, dx,$$

show that, if $0 < \alpha < 1$ then, if $n \geqslant 1$,

$$\frac{1}{2\pi} \int_{\mathbb{T}} |x|^\alpha K_n(x)\, dx \leqslant A_1(\alpha) n^{-\alpha}$$

where $A_1(\alpha)$ is a constant depending only on α.

(iii) Conclude that, if $0 < \alpha < 1$ and $\omega(f, h) \leqslant B|h|^\alpha$ for $|h| \leqslant \pi$, then

$$E_n(f) \leqslant A_1(\alpha) B n^{-\alpha} \quad \text{for } n \geqslant 1.$$

(iv) Where does the proof break down if $\alpha = 1$?

10.4 (i) To get round the difficulty we observed in Question 10.3 (iv), Jackson introduced a new function now called the Jackson kernel and defined by

$$L_n(t) = \lambda_n^{-1} K_n(t)^2$$

where λ_n is chosen so that $(1/2\pi)\int_{\mathbb{T}} L_n(t)\,dt = 1$. By observing that

$$\lambda_n \geqslant \frac{1}{2\pi}\int_{|t|<4/n\pi} K_n(t)^2\,dt,$$

or otherwise, show that $\lambda_n \geqslant Cn^{-1}$ for some constant C. Hence, arguing much as in the previous question, show that

$$\frac{1}{2\pi}\int_{\mathbb{T}} |x|\,L_n(x)\,dx \leqslant A_2 n^{-1}$$

for some constant A_2.

(ii) If $f\colon \mathbb{T} \to \mathbb{C}$ is continuous, show that, if

$$J_n(f, x) = \frac{1}{2\pi}\int_{\mathbb{T}} L_n(t) f(x-t)\,dt,$$

then $J_n(f) = J_n(f,.)$ is a trigonometric polynomial of degree at most $2n$. (Observe that L_n is a trigonometric polynomial of degree $2n$.) Deduce that, if $\omega(f, h) \leqslant B|h|$ for all $|h| < \pi$, then $E_{2n}(f) \leqslant A_2 B n^{-1}$ and so

$$E_n(f) \leqslant A_3 B n^{-1}$$

for all $n \geqslant 1$ where A_3 is some fixed constant.

10.5 (This question completes Question 10.4 and complements Question 10.2.)

(i) Show that, if $f\colon \mathbb{T} \to \mathbb{C}$ is a continuous function with $\omega(f, h) \leqslant B|h|^\alpha$ for some $\alpha > 1$ and some B, then f is constant.

(ii) If $f\colon \mathbb{T} \to \mathbb{C}$ is differentiable with bounded derivative, explain why $\omega(f, h) \leqslant \sup_{t\in\mathbb{T}}|f'(t)|\,|h|$.

(iii) If $f\colon \mathbb{T} \to \mathbb{C}$ is continuous with $\hat{f}(0) = 0$, show that $J_n(f)\hat{\ }(0) = 0$.

(iv) If $f\colon \mathbb{T} \to \mathbb{C}$ is continuous with $\hat{f}(0) = 0$, let us write

$$e_n(f) = \inf\{\sup_{t\in\mathbb{T}}|f(t) - P(t)| : P \text{ is a trigonometric polynomial of degree at most } n \text{ with } \hat{P}(0) = 0\}.$$

Use the methods and results of Question 10.4 together with parts (ii) and (iii) of this one to show that, if $f\colon \mathbb{T} \to \mathbb{C}$ is a continuously differentiable function with $\hat{f}(0) = 0$, then

$$e_n(f) \leqslant A_3 \sup_{t\in\mathbb{T}} |f'(t)|\, n^{-1}$$

for some constant A_3.

(v) Now suppose that f is as in (iv) and, in addition, there is a trigonometric polynomial Q of degree at most n with $\hat{Q}(0) = 0$ such that $|Q(t) - f'(t)| \leqslant \varepsilon$ for all $t \in \mathbb{T}$. Show that there is a trigonometric polynomial R of degree at most n with $\hat{R}(0) = 0$ such that $R'(t) = Q(t)$ and that

$$e_n(f) = e_n(f - R) \leqslant A_3 \varepsilon n^{-1}.$$

(vi) Deduce that, if f is as in (iv), then

$$e_n(f) \leqslant A_3 e_n(f')\, n^{-1} \quad \text{for } n \geqslant 1.$$

If f is m times continuously differentiable show that $e_n(f) \leqslant A_3^m\, e_n(f^{(m)}) n^{-m}$.

(vii) Now suppose that f is m times continuously differentiable, but we no longer assume that $\hat{f}(0) = 0$. Show that it remains true that

$$e_n(f) \leqslant A_3^m\, e_n(f^{(m)})\, n^{-m}.$$

(viii) At what point in the earlier argument would we run into problems if we did not take $\hat{f}(0) = 0$?

(ix) Show that, if f is m times continuously differentiable with $\omega(f^{(m)}, h) \leqslant B|h|^\alpha$ for $|h| \leqslant \pi$ [$m \geqslant 0$ an integer, $1 \geqslant \alpha > 0$], then $E_n(f) \leqslant A_4(m, \alpha)\, n^{-m-\alpha}$ for $n \geqslant 1$, where $A_4(m, \alpha)$ is a constant depending on m and α alone.

(The reader may feel there is nothing more to be said. However looking at Question 10.3 (iv) we see that our Jackson and Bernstein theorems do not match exactly in the important case $\alpha = 1$. Also we have made no attempt to find best possible constants. If the reader wishes to know how these questions are resolved, she can start by consulting Cheney's *Introduction to Approximation Theory*. (Questions 11.2 and 15.1 are also related to the results above.)

The first generation of American mathematicians learned their trade in the German Universities. Wiener in his autobiography 'Ex Prodigy' recalls that two generations of Harvard mathematicians had a collection of shared mannerisms, including the way they speared their cigars with a pen knife, all derived from Felix Klein, the uncrowned king of German mathematics.)

11

A NOWHERE DIFFERENTIABLE FUNCTION

11.1 In this question and the ones that follow we give a more sophisticated approach to the function discussed in Chapter 11. (There is some overlap in this question with Question 10.4 but it is not necessary to have done that question.)

(i) Let K_n be the Fejér kernel defined in Chapter 2. Show that $K_n(t) \leq \pi^2/((n+1)t^2)$ for all $t \neq 0$. Show also that $(1/2\pi)\int_{\mathbb{T}} K_n(t)^2\, dt = \sum_{r=-n}^{n} \hat{K}_n(r)^2 > n/2$.

(ii) Deduce that, if we define the Jackson kernel by $L_n(t) = \lambda_n^{-1} K_n(t)^2$ with λ_n chosen so that $(1/2\pi)\int_{\mathbb{T}} L_n(t)\, dt = 1$, then $0 \leq L_n(t) \leq 2\pi^4 n^{-3} t^{-4}$ for all $t \neq 0$.

(iii) Suppose that $f\colon \mathbb{T} \to \mathbb{C}$ is a continuous function such that $\hat{f}(r) = 0$ for $0 \neq |r| \leq 2n$. Show, by considering the degree of L_n as a trigonometric polynomial, that $\hat{f}(0) = (1/2\pi)\int_{\mathbb{T}} f(t) L_n(t)\, dt$. Deduce that $|\hat{f}(0)| \leq I_1 + I_2 + I_3$ where

$$I_1 = \frac{1}{2\pi}\int_{|t| \leq n^{-1}} |f(t)|\ L_n(t)\, dt,$$

$$I_2 = \frac{1}{2\pi}\int_{n^{-1} < t \leq n^{-1/4}} |f(t)|\ L_n(t)\, dt,$$

$$I_3 = \frac{1}{2\pi}\int_{n^{-1/4} < t} |f(t)|\ L_n(t)\, dt.$$

(iv) Show that

$$I_1 \leq \sup_{|t| < n^{-1}} |f(t)| \int_{\mathbb{T}} L_n(t)\, dt \leq n^{-1} \sup_{|t| < n^{-1}} |t^{-1} f(t)|,$$

$$I_2 \leq \pi^3 n^{-3} \sup_{|t| < n^{-1/4}} |t^{-1} f(t)| \int_{n^{-1}}^{n^{-1/4}} t^{-3}\, dt \leq \pi^3 n^{-1} \sup_{|t| < n^{-1/4}} |t^{-1} f(t)|,$$

$$I_3 \leq 2\pi^4 n^{-2} \frac{1}{2\pi}\int_{\mathbb{T}} |f(t)|\, dt.$$

(v) Conclude that, if $\sup_{t\in\mathbb{T}}|t^{-1}f(t)|$ exists, then

$$|\hat{f}(0)| \leqslant 2\pi^4\left(n^{-1}\sup_{|t|<n^{-1/4}}|t^{-1}f(t)| + \frac{n^{-2}}{2\pi}\int_{\mathbb{T}}|f(t)|\,dt\right).$$

(vi) Suppose that $g:\mathbb{T}\to\mathbb{C}$ is a continuous function such that $\sup_{t\in\mathbb{T}}|t^{-1}\,g(t)|$ exists. Show that, if $\hat{g}(r) = 0$ for $0 \neq |r-m| < 2n$, then

$$|\hat{g}(m)| \leqslant 2\pi^4\left(n^{-1}\sup_{|t|<n^{-1/4}}|t^{-1}\,f(t)| + \frac{n^{-2}}{2\pi}\int_{\mathbb{T}}|g(t)|\,dt\right).$$

11.2 We call a sequence of integers $0 \leqslant \lambda(1) < \lambda(2) < \ldots$ lacunary, if there exists a $q > 1$ and an $N \geqslant 1$ such that $\lambda(n+1) \geqslant q\lambda(n)$ for all $n \geqslant N$.

(i) Which of the following sequences are lacunary? Give reasons.

(a) $\lambda(j) = 2^j$,
(b) $\lambda(j) = (j!)^2$,
(c) $\lambda(j) = j(j+1)(j+2)$.

(ii) Suppose $\lambda(j)$ is a lacunary sequence, $a_{\lambda(j)} \in \mathbb{R}$, $\sum_{j=1}^{\infty}|a_{\lambda(j)}|$ converges and $g(t) = \sum_{j=1}^{\infty}a_{\lambda(j)}\cos\lambda(j)t$.

(a) Show, using Question 11.1 (vi), that if $g(0) = g'(0) = 0$, then

$$\lambda(j)a_{\lambda(j)} \to 0 \quad \text{as } j \to \infty.$$

(b) By considering $g(t) - g(0) - g'(0)\sin t$, show that the conclusion of (a) holds if g is merely differentiable at 0.

(c) Show that the conclusion of (a) holds if g is merely differentiable at one point.

(iii) Conclude that, if $\lambda(j)$ is a lacunary sequence, $a_{\lambda(j)} \in \mathbb{T}$ and $\sum_{j=1}^{\infty}|a_{\lambda(j)}|$ converges but $\lambda(j)a_{\lambda(j)} \not\to 0$ as $j \to \infty$ then $g(t) = \sum_{j=1}^{\infty}a_{\lambda(j)}\cos\lambda(j)t$ defines a continuous nowhere differentiable function.

(iv) Obtain the classical result of Weierstrass and Hardy that $f(x) = \sum_{n=1}^{\infty}b^{-n\alpha}\cos b^n x$ with b an integer $b \geqslant 2$ defines a continuous nowhere differentiable function whenever $0 < \alpha < 1$. What happens if $\alpha > 1$? [When α is small this is the original result of Weierstrass; the extension to all $\alpha < 1$ is due to Hardy.]

11.3 A very important moral to be drawn from the existence of nowhere differentiable continuous functions is that repeated differentiation may result in extremely bad behaviour even for apparently well-behaved functions. (Notice that, in contrast, integration improves the behaviour of functions.)

(i) By choosing A, M and θ appropriately, obtain, for each $\varepsilon > 0$, $K > 0$ and integer n, a periodic infinitely differentiable function f of the form

$f(x) = A\sin(Mx + \theta)$ such that $|f^{(r)}(x)| \leq \varepsilon$ for all $x \in \mathbb{R}$ and all $0 \leq r < n$ but $|f^{(n)}(0)| \geq K$.

(ii) Suppose $\kappa(1), \kappa(2), \ldots$ are positive numbers with $\kappa(r) \geq 1$. Show that we can find periodic infinitely differentiable functions $f_1, f_2, f_3, \ldots$ such that

(a) $|f_n^{(r)}(x)| \leq 2^{-r-n}$ for all $x \in \mathbb{R}$ and all $0 \leq r < n$.

(b) $$|f_n^{(n)}(0)| \geq \kappa(n)\left(\sum_{r=1}^{n-1} \sup_{x\in\mathbb{R}} |f_r^{(n-1)}(x)| + 2\right) + \left(\sum_{r=1}^{n-1} \sup_{x\in\mathbb{R}} |f_r^{(n)}(x)| + 2\right).$$

Prove, stating precisely all the theorems that you use, that $\sum_{n=1}^{\infty} f_n(x)$ converges uniformly to an infinitely differentiable function F. Show that

$$|F^{(n-1)}(x)| \leq \sum_{r=1}^{n-1} |f_r^{(n-1)}(x)| + 2 \quad \text{for all } x$$

and that

$$|F^{(n)}(x)| \geq |f_n^{(n)}(x)| - \left(\sum_{r=1}^{n-1} |f_r^{(n)}(x)| + 2\right) \quad \text{for all } x.$$

Deduce that $|F^{(n)}(0)| \geq \kappa(n) \sup_{x\in\mathbb{R}} |F^{(n-1)}(x)|$ for each $n \geq 0$.

12

REACTIONS

12.1 One of the simplest constructions of a nowhere differentiable continuous function is that of Van der Waerden. It echos the ideas behind the construction in Chapter 11 but avoids much of the technical detail.

(i) Let f be given by $f(x) = x$ if $0 \leqslant x < 1/2$, $f(x) = 1 - x$ if $1/2 \leqslant x < 1$ and $f(x+1) = f(x)$ for all x. Sketch f and observe (no proof is required) that f is a continuous function with $0 \leqslant f(x) \leqslant 1/2$ for all x. Prove that $2^n f(m2^{-n}) \in \mathbb{Z}$ whenever $m, n \in \mathbb{Z}$ and $n \geqslant 1$.

(ii) Show, stating any theorems that you require, that $\sum_{r=1}^{\infty} 2^{-r} f(4^r x)$ converges uniformly to a continuous function $F(x)$. Show that $0 \leqslant F(x) \leqslant 1/2$ for all x and that F is periodic with period 1/4.

(iii) Show that, if $w = m4^{-n}$ with $m, n \in \mathbb{Z}$ and $n \geqslant 1$, then $f(4^r w) = 0$ for $r \geqslant n$ and so

$$F(w) = \sum_{r=1}^{n-1} 2^{-r} f(m4^{r-n}).$$

Show that if $u = (4m)4^{-n}$ and $v = (4m+2)4^{-n}$ with $m, n \in \mathbb{Z}$ and $n \geqslant 1$ then $2^n F(u)$ is an even integer, $2^n F(v)$ is an odd integer and so $|2^n(F(u) - F(v))| \geqslant 1$.

(iv) Deduce that, given $x \in \mathbb{R}$, we can find $w \in \mathbb{R}$ such that $|w - x| \leqslant 4^{-n+1}$ and $|2^n(F(w) - F(x))| \geqslant 1/2$. Conclude that F is nowhere differentiable.

12.2 (A Peano curve) The construction in the last question serves as an introduction to the elegant space filling curve of Liu Wen (see bibliography).

(i) Let $f(x) = 1$ for $0 \leqslant x \leqslant 3$, $f(x) = 4 - x$ for $3 \leqslant x < 4$, $f(x) = 0$ for $4 \leqslant x < 7$, $f(x) = x - 7$ for $7 \leqslant x < 8$, $f(x+8) = f(x)$ for all x. Sketch f and observe (no proof is required) that f is continuous and periodic with period 8. Set $g(x) = f(x+2)$ and observe that

$$0 \leqslant f(x), g(x) \leqslant 1 \quad \text{for all } x,$$

$$f(x) = 1, g(x) = 1 \quad \text{for } 0 \leqslant x \leqslant 1,$$

$$f(x) = 1,\ g(x) = 0 \quad \text{for } 2 \leqslant x \leqslant 3,$$

$$f(x) = 0,\ g(x) = 0 \quad \text{for } 4 \leqslant x \leqslant 5,$$

$$f(x) = 0,\ g(x) = 1 \quad \text{for } 6 \leqslant x \leqslant 7.$$

(ii) Show, stating any theorems that your need, that $\sum_{r=1}^{\infty} 2^{-r} f(8^{r+1}x)$ converges uniformly to a continuous function $F(x)$ with $0 \leqslant F(x) \leqslant 1$ and that $\sum_{r=1}^{\infty} 2^{-r} g(8^{r+1}x)$ converges uniformly to a continuous function $G(x)$ with $0 \leqslant G(x) \leqslant 1$.

(iii) If $0 \leqslant a,\ b \leqslant 1$, write a and b in non terminating binary expansions $a = \sum_{r=1}^{\infty} a_r 2^{-r}$, $b = \sum_{r=1}^{\infty} b_r 2^{-r}$ with $a_r, b_r \in \{0, 1\}$. Define

$$x_r = 0 \quad \text{if } a_r = 1,\ b_r = 1,$$

$$x_r = 2 \quad \text{if } a_r = 1,\ b_r = 0,$$

$$x_r = 4 \quad \text{if } a_r = 0,\ b_r = 0,$$

$$x_r = 6 \quad \text{if } a_r = 0,\ b_r = 1,$$

(there are thus four cases to consider) and set $x = \sum_{r=1}^{\infty} x_r 8^{-r-1}$. Show that, if $a_r = 1$, $b_r = 0$, then $[8^{r+1}x] \equiv 2 \bmod 8$ (i.e. the integer part of $8^{r+1}x$ is congruent to 2 modulo 8) and deduce that $f(8^{r+1}x) = 1$, $g(8^{r+1}x) = 0$. State what happens in the other 3 cases and show that $F(x) = a$, $G(x) = b$. Conclude that we have a continuous function $x \mapsto (F(x),\ G(x))$ which maps $[0, 1]$ onto $[0, 1]^2$.

(Peano's result casts doubt on the dictum that a curve is one dimensional. It shows that only one parameter is needed to find the position of a point in $[0, 1]^2$. New ideas are needed to make the idea of dimension precise. Such ideas, supplied by Poincaré, Brouwer, Lebesgue and others, lead on to algebraic topology. Many constructions of Peano curves give very pretty computer graphics (the graphics for the construction just given are convincing but not pretty). It has been suggested that the problem of finding appropriate standards for television screens of various resolutions could be resolved by making the scanning beam follow the path corresponding to

$$x \mapsto \left(\sum_{r=1}^{N} 2^{-r} f(8^{r+1}x),\ \sum_{r=1}^{N} 2^{-r} g(8^{r+1}x) \right)$$

(or some similar expression). The higher the resolution required the greater the N.)

13

MONTE CARLO METHODS

13.1$^{(-)}$ Let $X_1, X_2, \ldots$ be independent random variables each uniformly distributed over $[0, 1]$. Let f: $[0, 1] \to \mathbb{R}$ be a non constant continuous function. Show that $\operatorname{var} f(X_1) > 0$ and write $\sigma^2 = \operatorname{var} f(X_1)$, $\mu = \int_0^1 f(x)\, dx$, $S_n = \sum_{r=1}^n f(X_r)$. Use the central limit theorem (Chapter 70) to show that

$$Pr(|n^{-1}S_n - \mu| < \lambda\, \sigma n^{-1/2}) \to \frac{1}{\sqrt{(2\pi)}} \int_{-\lambda}^{\lambda} \exp(-t^2/2)\, dt.$$

(This shows how to make precise the statement that the accuracy for Monte Carlo integration improves as the square root of the number of trials.)

13.2$^{(-)}$ Our discussion of the Monte Carlo method is informal. In particular, it might be objected that we have not shown that the particle ever hits the boundary. The following simple argument resolves the problem. We use the notation of Lemma 13.2.

(i) Show that there exists an M such that, whichever grid point the particle starts from in Ω, there is a possible path involving at most M grid points leading from the initial grid point to the boundary.

(ii) Show that the probability that the particle hits the boundary in $(k+1)M$ jumps or less, given that it has not hit the boundary in kM jumps, is greater than or equal to 2^{-M}.

(iii) Show

$$Pr\text{ (particle does not hit boundary in } kM \text{ jumps or less)} \leq (1 - 2^{-M})^k$$

and deduce that the probability of not hitting the boundary is zero.

(iv) Show further that

$$\mathbb{E}\text{ (number of jumps before hitting boundary)} \leq \sum_{k=1}^{\infty} M(k+1)(1 - 2^{-M})^k < \infty,$$

so that the average number of jumps required to hit the boundary is finite.

13.3$^{(-)}$ Let $n \geq 1$ be an integer and let $\lambda \geq 0$. Develop a Monte Carlo method for solving the partial differential equations $\nabla^2 \phi = \lambda \phi$ on the bounded region

$\Omega \subseteq \mathbb{R}^n$ subject to the boundary condition $\phi(\mathbf{x}) = G(\mathbf{x})$ for all $\mathbf{x}$ on the boundary $\partial\Omega$. How does your method fail if $\lambda < 0$? By considering the special case when $n = 1$, $\Omega = (0, \pi)$ and $\lambda = -1$, show that this failure is not the fault of the method, but reflects deeper problems.

15

POINTWISE CONVERGENCE

15.1 Let $f: \mathbb{T} \to \mathbb{C}$ be continuous. In Chapter 10 we raised the question of finding a sequence P_n of trigonometric polynomials of degree n such that $P_n \to f$ uniformly as fast as possible. In this question we show that the sequence $\sigma_{n,2n}f$ of de la Vallée Poussin sums, although not optimal, has good properties along these lines. Let us write $\mathcal{P}_n$ for the space of trigonometric polynomials of degree at most n, $C(\mathbb{T})$ for the space of continuous functions, $\|f\|_\infty = \sup_{t\in\mathbb{T}}|f(t)|$ if $f \in (\mathbb{T})$ and

$$E_n(f) = \inf\{\|f - P\|_\infty : P \in \mathcal{P}_n\}.$$

(i) We start with the key general remark. Suppose $T: C(\mathbb{T}) \to C(\mathbb{T})$ is a linear map with $\|Tf\|_\infty \leq A\|f\|_\infty$ for all $f \in C(\mathbb{T})$ and such that $TP = P$ for all $P \in \mathcal{P}_n$. Show that $\|Tf - P\|_\infty \leq A\|f - P\|_\infty$ for all $P \in \mathcal{P}_n$ and conclude that $\|Tf - f\|_\infty \leq (A + 1)E_n(f)$.

(ii) Show that, if $f \in C(\mathbb{T})$, then $\|\sigma_n f\|_\infty \leq \|f\|_\infty$.

(iii) By considering $Tf = \sigma_{n,2n}f$ show that, if $f \in C(\mathbb{T})$, then $\sigma_{n,2n}f \in \mathcal{P}_{2n}$ and $\|\sigma_{n,2n}f - f\|_\infty \leq 4E_n(f)$.

(iv) Explain why the argument which leads to the final formula of (iii) fails for $\sigma_n f$ and why it fails for $S_n f$.

(v) Show that $\sigma_{n,2n}(f, t) = n^{-1}\sum_{r=n}^{2n-1} S_r(f, t)$ and so $f(t) - \sigma_{n,2n}(f, t) = n^{-1}\sum_{r=n}^{2n-1}(f(t) - S_r(f, t))$.

(vi) By applying (iii) and (v) to the function h discussed in Chapter 10, show that

$$E_n(h) \geq 2^{-1}\pi^{-1}/(n + 2).$$

Thus no trigonometric polynomial of degree n can be a very good uniform approximation to h.

(vii) Combine the results of this question with those of Questions 10.1 to 10.5.

15.2 (The first two parts of this question use the ideas of Chapter 11.) Choose a sequence of integers such that $N(1) \geqslant 2$ and $N(r+1) > (rN(r))^r$. Show that $f(t) = \sum_{r=1}^{\infty} N(3r)^{-1} \sin(N(2r+1)t)$ is a well-defined continuous function.

(i) By looking at $h = N(3r+2)^{-1}\pi$ show that

$$\liminf_{h\to 0} \left(|h|^{-\alpha} \sup_{t\in\mathbb{T}} |f(t+h) - f(t)|\right) = 0 \quad \text{for all } 1 > \alpha > 0.$$

(ii) By looking at h with $3N(3r+1)^{-1}\pi \geqslant h \geqslant N(3r+1)^{-1}\pi$ show that, for each $t \in \mathbb{T}$,

$$\limsup_{h\to 0} \left(|h|^{-\alpha}|f(t+h) - f(t)|\right) = \infty \quad \text{for all } 1 > \alpha > 0.$$

(iii) Let us define $E_n(f)$ as in the previous question. By looking at $P(t) = \sum_{r=1}^{k} N(3r)^{-1} \sin(N(3r+1)t)$ as a trigonometric polynomial of degree at most $N(3k+1)$ show that

$$\liminf_{n\to\infty} n^m E_n(f) = 0 \quad \text{for all integers } m \geqslant 0.$$

(iv) By looking at $E_n(f)$ when $n = N(3r)$ and applying Question 15.1 show that

$$\limsup_{n\to\infty} n^{-m} E_n(f) = \infty \text{ for all integers } m \geqslant 1.$$

(Thus the results of Questions 10.1 to 10.5 do not tell the whole story.)

16

BEHAVIOUR AT POINTS OF DISCONTINUITY I

16.1 (Functions of bounded variation) Theorem 16.4 can be extended to a wider and (in some sense) more natural class of function. Let us say that $f: [a, b] \to \mathbb{C}$ is of bounded variation if there exists a V such that for any $a = x_0 < x_1 < \ldots < x_n = b$ we have

$$\sum_{j=1}^{m} |f(x_j) - f(x_{j-1})| \leqslant V.$$

(i) Show that $f: [a, b] \to \mathbb{C}$ is of bounded variation if and only if $\operatorname{Re} f$ and $\operatorname{Im} f$ are. (Thus for most purposes it suffices to look at real valued functions of bounded variation.)

(ii) Show that, if $f, g: [a, b] \to \mathbb{C}$ are of bounded variation, so is $\lambda f + \mu g$ (where $\lambda, \mu \in \mathbb{C}$). If $f: [a, b] \to \mathbb{R}$ is increasing, show that f is of bounded variation. Deduce that, if $f_1, f_2: [a, b] \to \mathbb{R}$ are increasing, then $f_1 - f_2$ is of bounded variation.

(iii) Conversely suppose $f: [a, b] \to \mathbb{R}$ is of bounded variation. Show that

$$f_1(t) = \sup\left\{\sum_{j=1}^{m} |f(x_j) - f(x_{j-1})|: a = x_0 \leqslant x_1 \leqslant \ldots \leqslant x_m = t,\ m \geqslant 1\right\}$$

gives a well-defined increasing function $f_1: [a, b] \to \mathbb{R}$. Show that $f_2 = f_1 - f$ is also an increasing function and so $f = f_1 - f_2$ is the difference of two increasing functions.

(iv) Show that, if $f: [a, b] \to \mathbb{C}$ is of bounded variation, we can find f_1, f_2, f_3, f_4: $[a, b] \to \mathbb{R}$ increasing such that $f = (f_1 - f_2) + i(f_3 - f_4)$. Is it true that f_1 is unique provided $f_1(a)$ is specified?

(v) Show that the following are of bounded variation.

(a) A function $f: [a, b] \to \mathbb{C}$ with bounded derivative on (a, b).

(b) A function $f: [a, b] \to \mathbb{C}$ with only a finite number of maxima and minima.

16.2 (i) Show directly from Lemma 16.1 by translation that

$$\pi - t = \sum_{n=1}^{\infty} \frac{2}{n} \sin nt \quad \text{for all } 0 < t < 2\pi.$$

(ii) By considering the Fourier sum associated with a function $f: \mathbb{T} \to \mathbb{C}$ given by $f(t) = \exp -i\alpha t$ for $0 < t < 2\pi$ show that

$$e^{i(\pi - t)\alpha} = \frac{\sin \pi \alpha}{\pi} \lim_{N \to \infty} \sum_{n=-N}^{N} \frac{e^{int}}{\alpha + n} \quad \text{for all } 0 < t < 2\pi,$$

for all non integers $\alpha \in \mathbb{C}$. Setting $t = \pi$ deduce that

$$\frac{\pi \alpha}{\sin \pi \alpha} = 1 + 2\,\alpha^2 \sum_{n=1}^{\infty} \frac{(-1)^n}{\alpha^2 - n^2} \quad \text{(a result of Euler).}$$

(iii) If $|\alpha| \leqslant 1/4$, show that

$$\left| \frac{e^{int}}{\alpha + n} + \frac{e^{-int}}{\alpha - n} - \frac{2i}{n} \sin nt \right| \leqslant \frac{8|\alpha|}{n^2}$$

for all $t \in \mathbb{R}$ and all integers $n \neq 0$. Hence show that

$$\left| \lim_{N \to \infty} \sum_{1 \leqslant |n| \leqslant N} \frac{e^{int}}{\alpha + n} - \sum_{n=1}^{\infty} \frac{2i}{n} \sin nt \right| \to 0 \quad \text{as } \alpha \to 0$$

for all $0 < t < 2\pi$. By allowing $\alpha \to 0$ in part (ii), deduce part (i).

(iv) Show that

$$\lim_{N \to \infty} \sum_{n=-N}^{N} \frac{\sin (n + \alpha)t}{n + \alpha} = \pi, \quad \lim_{N \to \infty} \sum_{n=-N}^{N} \frac{\cos (n + \alpha)t}{n + \alpha} = \pi \cot \pi \alpha$$

for all $0 < t < 2\pi$ and all non integers $\alpha \in \mathbb{C}$.

(If you do not find the last formula but one rather odd, you are either very clever or have no sense of wonder.)

16.3 (Pointwise convergence for functions of bounded variation) Suppose $f: [a, b] \to \mathbb{R}$ is increasing.

(i) Show that, if $a = x_0 \leqslant x_1 \leqslant \ldots \leqslant x_n = b$, then

$$\begin{aligned} \sum_{j=1}^{n} \sup_{x \in [x_{j-1}, x_j]} f(x)\, (x_j - x_{j-1}) - \sum_{j=1}^{n} \inf_{x \in [x_{j-1}, x_j]} f(x)\, (x_j - x_{j-1}) \\ = \sum_{j=1}^{n} (f(x_j) - f(x_{j-1}))\, (x_j - x_{j-1}) \\ \leqslant (f(b) - f(a)) \max_{1 \leqslant j \leqslant n} (x_j - x_{j-1}) \end{aligned}$$

and deduce that f is Riemann integrable.

(ii) If $a < c \leqslant b$ show that, if $a < y_1 < y_2 < \ldots y_m \leqslant c$ and $y_m \to c$, then $f(y_m)$ tends to a limit as $m \to \infty$. Deduce that $\lim_{x \to c-} f(x)$ exists. Show also that, if $a \leqslant c < b$, then $\lim_{x \to c+} f(x)$ exists and that, if $a < c < b$, $\lim_{x \to c+} f(x) \geqslant \lim_{x \to c-} f(x)$.

(iii) If $[a, b] = [0, 2\pi]$ and we write

$$\hat{f}(n) = \frac{1}{2\pi} \int_0^{2\pi} f(t) \exp(-int)\, dt$$

show by writing

$$\hat{f}(n) = \frac{1}{2\pi} \sum_{r=0}^{n-1} \int_{2r\pi/n}^{2(r+1)/n} f(t) \exp(-int)\, dt$$

and proving

$$\left| \frac{1}{2\pi} \int_{2r/\pi}^{2(r+1)\pi/n} f(t) \exp(-int)\, dt \right| \leqslant n^{-1}(f(2(r+1)\pi/n) - f(2r\pi/n))$$

that $\hat{f}(n) = O(|n|^{-1})$.

(iv) Let us say that $f\colon \mathbb{T} \to \mathbb{C}$ is of bounded variation if there exists a V such that for any $0 = x_0 \leqslant x_1 \leqslant \ldots \leqslant x_n = 2\pi$ we have $\sum_{j=1}^{n} |f(x_j) - f(x_{j-1})| \leqslant V$. By using Question 16.1 (iv) show that

(a) f is Riemann integrable,
(b) $\lim_{t \to x-} f(t)$ and $\lim_{t \to x+} f(t)$ exist everywhere,
(c) $\hat{f}(n) = O(|n|^{-1})$.

Deduce from Theorem 16.3 that

$$S_m(f, x) \to \left(\lim_{t \to x+} f(t) + \lim_{t \to x-} f(t) \right) \Big/ 2 \quad \text{as } m \to \infty$$

for all x.

(v) Deduce Theorem 16.4 as a particular case of (iv) (See also Question 52.9 and 52.10 below.)

16.4 Question 16.1 (v) shows that most of the functions encountered in 'mathematical methods' courses are of bounded variation. Here is an example of a function f that is not, even though $\hat{f}(n) = 0(|n|^{-1})$.

(i) Let $g\colon \mathbb{T} \to \mathbb{R}$ be given by $g(t) = 1 - \varepsilon^{-1}|t|$ for $|t| \leqslant \varepsilon$, $g(t) = 0$ otherwise.

(a) Explain why $|\hat{g}(n)| \leqslant \varepsilon\pi^{-1}$ for all n,
(b) Compute $\hat{g}(n)$ (make use of symmetry) and show that $|\hat{g}(n)| \leqslant 2(\pi\varepsilon)^{-1} n^{-2}$ for all $n \neq 0$.

(ii) Let $f_j(t) = j^{-1}(1 - 2^{j+2}|t - 2^{-j}|)$ for $|t - 2^{-j}| \leqslant 2^{-j-2}$ and $f_j(t) = 0$ otherwise. Show that $\sum_{j=1}^{n} f_j$ converges uniformly to a continuous function f as $n \to \infty$ and sketch f.

(iii) Show that f is not of bounded variation.
(iv) If $2^k \leqslant |n| < 2^{k+1}$, show, using (i) (a), that

$$\sum_{j=k}^{\infty} |\hat{f}_j(n)| \leqslant k^{-1}2^{-k-1}$$

and, using (i) (b), that

$$\sum_{j=1}^{k-1} |\hat{f}_j(n)| \leqslant 2\pi^{-1} \sum_{j=1}^{k-1} (j2^{j+2})^{-1} n^{-2}.$$

Hence show that

$$n \sum_{j=1}^{\infty} |\hat{f}_j(n)| \to 0 \quad \text{as } |n| \to \infty.$$

(v) Explain why $\hat{f}(n) = \sum_{j=1}^{\infty} \hat{f}_j(n)$ and conclude that $n\hat{f}(n) \to 0$ as $|n| \to \infty$. (Thus not only does $\hat{f}(n) = 0(|n|^{-1})$ but, in fact, $\hat{f}(n) = o(|n|^{-1})$.)

17

BEHAVIOUR AT POINTS OF DISCONTINUITY II

17.1 Let K_n be the Fejér kernel discussed in Chapter 2.

(i) Show that, if $\kappa(n) \to \infty$ as $n \to \infty$, then $\int_{|s| \geq \kappa(n)/n} K_n(s)\, ds \to 0$ as $n \to \infty$. (We shall be interested in slowly diverging $\kappa(n)$ such as $\kappa(n) = \log(n+2)$.)

(ii) Show that if h is the saw tooth function of Chapter 17 then

(a) $\sup_{|t| \geq \kappa(n)/n} |h(t) - \sigma_n(h, t)| \to 0$ as $n \to \infty$,
(b) $\sigma_n(h, t) \geq 0$ for $0 \leq t \leq \pi$, $\sigma_n(h, t) \leq 0$ for $-\pi \leq t \leq 0$,
(c) $|\sigma_n(h, t)| \leq \pi$ for all t.

Conclude that nothing resembling the Gibbs phenomenon can occur for $\sigma_n(h, t)$. Extend this observation to cover well-behaved functions with only a finite number of discontinuities. (If you actually graph $\sigma_n(h, t)$ you may find that, in spite of what we have proved, $\sigma_n(h,)$ is not a very good copy of h for small n, being a bit 'round shouldered' near π. Small degree trigonometric polynomials just do not look very much like discontinuous functions.)

17.2 As C. H. Su has pointed out to me, the result of Theorem 17.1 can be interpreted by using a technique from boundary layer theory (and elsewhere) and 'blowing up' the independent variable x near π. More precisely, instead of considering $S_n(h, x)$ we introduce a new variable $y = n(\pi - x)$.

(i) Show that

$$S_n(h, \pi - y/n) \to 2 \int_0^y \frac{\sin t}{t}\, dt$$

as $h \to \infty$.

(ii) If you have access to a computer, plot $F_n(y) = S_n(h, \pi - y/n)$ for various values of n.

(iii) Write $u_k = \displaystyle\int_0^{k\pi} \frac{\sin x}{x}\, dx$. Show that $u_1, u_3, u_5, \ldots$ form a decreasing sequence and that $u_2, u_4, u_6, \ldots$ form an increasing sequence both tending to the same limit, u say.

(iv) Show that $\int_0^X \frac{\sin x}{x}\,dx \to u$ as $X \to \infty$ (i.e. the integral $\int_0^\infty \frac{\sin x}{x}\,dx$ converges).

(v) For which real α does the integral $\int_0^\infty \frac{\sin x}{x^\alpha}\,dx$ converge? (The problem of evaluation when $\alpha \neq 1$ is dealt with in Question 75.20.)

17.3 There are many ways of evaluating $\int_0^\infty \frac{\sin x}{x}\,dx$. Several are discussed in two papers by Hardy (pages 528 and 615 of Volume 5 of his *Collected Works*).

(i) The most obvious way to any one who has done a complex variable course is contour integration. Do this.

(ii) However, if we want a method which uses as few advanced results as possible, the following (taken from the 1922 Tripos) is of interest. Let

$$a_n = \int_0^{\pi/2} \sin 2nx \cot x\,dx, \quad b_n = \int_0^{\pi/2} \frac{\sin 2nx}{x}\,dx.$$

Show that $a_{n+1} = a_n$ for $n \geqslant 1$ and deduce that $a_n = \pi/2$. Show that $b_n \to \int_0^\infty \frac{\sin x}{x}\,dx$ as $n \to \infty$. Finally use integration by parts to show that $a_n - b_n \to 0$ as $n \to \infty$. Conclude that $\int_0^\infty \frac{\sin x}{x}\,dx = \frac{\pi}{2}$. (You can handle behaviour near $x = 0$ directly or by considering $\lim_{\varepsilon \to 0+} \int_\varepsilon^{\pi/2}$.)

17.4 (i) Using the previous question, show that $\int_0^\infty \frac{\sin \lambda t}{t}\,dt$ converges for all real λ and takes the values $\pi/2$, 0 or $-\pi/2$ according as λ is strictly positive, zero or strictly negative.

(ii) Show that $\int_0^\infty \frac{\sin t \cos t}{t}\,dt$ converges with value $\pi/4$ and use integration by parts to show that $\int_0^\infty (\sin x/x)^2\,dx$ converges with value $\pi/2$. (This integral is computed by a different method in the proof of Lemma 49.2 (iii).)

(iii) Show that the graph of $f(x) = \frac{2}{\pi}\int_0^\infty \frac{\sin^3 t \cos xt}{t}\,dt$ consists of parts of the lines $y = -1/4$, $y = 0$ and $y = 1/2$ together with four isolated points.

(iv) If $a > 0$, $b > 0$ and $\delta > 0$, show that $\int_\delta^\infty \frac{e^{-ax} - e^{-bx}}{x}\,dx$ exists and has value $\int_{a\delta}^{b\delta} \frac{e^{-t}}{t}\,dt$. Deduce that $\int_0^\infty \frac{e^{-ax} - e^{-bx}}{x}\,dx$ exists and find its value.

17.5 (The object of this question is to exercise skill in accurate argument, so make sure your answer is water tight.)

(i) Show that, if $|h| < \pi/2$, then $\lim_{a\to\infty} \int_0^h \frac{\sin ax}{x}\,dx$ exists and find its value.

(ii) Extend the results of (i) to the case $|h| < \pi$.

(iii) Show that, whatever the value of h, $\lim_{n\to\infty} \int_0^h \frac{\sin(2n+1)x}{x}\,dx$ and $\lim_{n\to\infty} \int_0^h \frac{\sin 2nx}{x}\,dx$ exist and find their value.

17.6 (i) Suppose $f: \mathbb{R} \to \mathbb{C}$ has a continuous derivative. By applying Theorem 16.4 show that

$$\sum_{n=-N}^{N} \frac{1}{2\pi} \int_0^{2\pi} f(t) \exp(int)\,dt \to \frac{1}{2}(f(0) + f(2\pi))$$

as $N \to \infty$.

(ii) Use the ideas of (i) to show that, if p and q are integers with $p > q$, then

$$\sum_{n=-N}^{N} \frac{1}{2\pi} \int_{2q\pi}^{2p\pi} f(t) \exp(int)\,dt \to \frac{1}{2} f(2q\pi) + f(2(q+1)\pi) + \dots + f(2(p-1)\pi) + \frac{1}{2} f(2p\pi).$$

(iii) Show that, if $f: \mathbb{R} \to \mathbb{C}$ has a continuous derivative and there exists an R with $f(t) = 0$ for $|t| \geq R$, then

$$\sum_{n=-N}^{N} \frac{1}{2\pi} \int_{-\infty}^{\infty} f(t) \exp(int)\,dt \to \sum_{m=-\infty}^{\infty} f(2m\pi) \quad \text{as } N \to \infty$$

or, more briefly,

$$\sum_{n=-\infty}^{N} \frac{1}{2\pi} \int_{-\infty}^{\infty} f(t) \exp(int)\,dt = \sum_{m=-\infty}^{\infty} f(2m\pi).$$

(This is the simplest case of the Poisson summation formula, see Question 60.4.)

(iv) By applying a scale change to (ii), show that, if $g: \mathbb{R} \to \mathbb{C}$ has a continuous derivative, $a \in \mathbb{R}$, $\omega > 0$, $k \in \mathbb{Z}$, $k \geq 1$ and $b = a + k\omega$,

$$\frac{1}{2} g(a) + g(a+\omega) + \dots + g(a + (k-1)\omega) + \frac{1}{2} g(b) = \frac{1}{\omega} \int_a^b g(x)dx + R,$$

where the error term R is given by

$$R = \sum_{n=1}^{\infty} \frac{2}{\omega} \int_a^b g(x) \cos\left(\frac{2\pi n(x-a)}{\omega}\right) dx.$$

(R is thus the error term for approximate integration (numerical quadrature) using the trapezoidal rule.)

(v) Show from (ii) that if $g: \mathbb{R} \to \mathbb{R}$ is continuously differentiable

$$\frac{1}{2}g(0) + g(2\pi) + \ldots + g(2(n-1)\pi) + \frac{1}{2}g(2n\pi)$$

$$= \frac{1}{2\pi}\int_0^{2\pi n} g(t)dt + 2\sum_{r=1}^{\infty} \frac{1}{2\pi}\int_0^{2\pi n} g(t)\cos rt\,dt.$$

Now suppose that g is $2m+1$ times continuous differentiable. By integrating $\int_0^{2\pi n} g(t)\cos rt\,dt$ by parts $2m$ times, show that

$$\frac{1}{2}g(0) + g(2\pi) + \ldots + g(2(n-1)\pi) + \frac{1}{2}g(2n\pi)$$

$$= \frac{1}{2\pi}\int_0^{2\pi n} g(t)dt + \frac{2}{2\pi}\sum_{k=1}^{m}(-1)^m \zeta(2k)(g^{(2h-1)}(2n\pi) - g^{(2k-1}(0)) + S_{m,n}$$

where $S_{m,n} = \frac{2}{2\pi}(-1)^m \sum_{r=1}^{\infty} \int_0^{2\pi n} g^{(2m)}(t)\cos rt\,dt$.

(By definition $\zeta(2k) = \sum_{r=1}^{\infty} r^{-2k}$. We return to these ideas in Questions 18.6 onwards.)

18

A FOURIER SERIES DIVERGENT AT A POINT

18.1 Let us return to the saw tooth function h of Chapter 17. Life would be easy if we could leave this as a simple counter example. Unfortunately, the Fourier expansion has a useful part to play in several parts of number theory. We must thus show that the behaviour of $S_n(h, t)$ is no worse than it seems.

(i) By integration show that if $|t| < \pi$

$$h(t) - S_n(h, t) = \int_0^t \sum_{r=-n}^{n} \exp(ir(x + \pi))\, dx = \int_0^t D_n(x + \pi)\, dx.$$

(ii) Show that $D_n(x)$ and $D_n(x + 2\pi/(n + 1))$ are of opposite sign for $0 \leqslant x \leqslant \pi - 2\pi/(2n + 1)$ and that $|D_n(x) \geqslant |D_n(x + 2\pi/(2n + 1))|$ under the same conditions. Show also that $D_n(x)$ is of constant sign on $[2r\pi/(2n + 1), 2(r + 1)\pi/(2n + 1)]$. Hence show that if $u_r = (1/2\pi) \int_{2r\pi/(2n+1)}^{2(r+1)\pi/(2n+1)} D_n(x + \pi)\, dx$, then

$$|h(t) - S_n(h, t)| \leqslant |u_r|$$

for all $t \in [2r\pi/(2n + 1), 2(r + 1)\pi/(2n + 1)]$ for all $0 \leqslant r \leqslant n$. (Remember the idea of the alternating series test.)

(iii) Deduce that, if $\delta > 0$, then $S_n(h, t) \to h(t)$ uniformly for all $|t| \leqslant \pi - \delta$ and that in any case

$$|h(t) - S_n(h, t)| \leqslant |u_n| \leqslant 2\pi.$$

(iv) Conclude (and this is the result we shall need) that, if g is Riemann integrable,

$$\frac{1}{2\pi} \int_{\mathbb{T}} g(t) S_n(h, t)\, dt \to \frac{1}{2\pi} \int_{\mathbb{T}} g(t) h(t)\, dt.$$

(v) More generally show that, if f satisfies the hypotheses of Theorem 16.4 with points of discontinuity $x_1, x_2, \ldots, x_n$, then

(a) if $\delta > 0$, $S_n(f, t) \to f(t)$ uniformly for $|t - x_j| \geqslant \delta\, [1 \leqslant j \leqslant n]$,

(b) there exists a K such that $|S_n(f,t)| \leqslant K$ for all $t \in \mathbb{T}$ and all $n \geqslant 0$,

(c) if g is Riemann integrable,

$$\frac{1}{2\pi}\int_{\mathbb{T}} g(t)\, S_n(f,t)\, dt \to \frac{1}{2\pi}\int_{\mathbb{T}} g(t) f(t)\, dt.$$

18.2 Here is an immediate application of the previous result.

(i) Suppose $g\colon \mathbb{T} \to \mathbb{C}$ is Riemann integrable. Show that

$$\sum_{n \geqslant |k| \geqslant 1} \frac{\hat{g}(k)}{ik} \to -\frac{1}{2\pi}\int_0^{2\pi} g(x)\,(\pi - x)\, dx.$$

(ii) Now define $g_t(x) = g(x+t)$. Show that

$$\frac{1}{2\pi}\int_0^{2\pi} g_t(x)\,(\pi - x)\, dx = \frac{t}{2\pi}\int_0^{2\pi} g(x)\, dx + \frac{1}{2\pi}\int_0^{2\pi} g(x)\,(\pi - x)\, dx$$
$$- \int_0^t g(x)\, dx$$

and hence deduce, using (i), that

$$\sum_{N \geqslant |k| \geqslant 1} \hat{g}(k)\, \frac{\exp(ikt) - 1}{ik} \to \int_0^t g(t)\, dt - t\hat{g}(0) \quad \text{as } N \to \infty$$

i.e. that $\int_0^t g(t)\, dt = t\hat{g}(0) + \sum_{k \neq 0} \hat{g}(k) \exp(ikt)/ik$ for $-\pi \leqslant t < \pi$. Thus a Fourier series can be integrated term by term.

(iii) Show that the formula holds for general t if $\int_0^t$ is suitably interpreted. What is special about the case $\hat{g}(0) = 0$?

(iv) Use the result of (ii) to given another proof that, if $f\colon \mathbb{T} \to \mathbb{C}$ is once continuously differentiable, then $\sum_{k=-N}^{N} f(k) \exp ikt \to f(t)$ as $N \to \infty$.

(v) By carefully examining the arguments of Questions 18.1 and 18.2 show that the convergence in part (ii) is uniform.

18.3 (Rate of divergence of Fourier sums.) We can actually extract some positive information from the ideas of Chapter 18.

(i) Show that

$$|D_n(s)| \leqslant \pi/s \quad \text{for } s \neq 0,$$
$$|D_n(s)| \leqslant 2n + 1 \quad \text{for all } s.$$

(ii) By writing

$$S_n(f,0) = \frac{1}{2\pi}\int_{|x| \leqslant 1/n} f(x) D_n(-x)\, dx + \frac{1}{2\pi}\int_{|x| > 1/n} f(x) D_n(-x)\, dx$$

and estimating the two integrals separately, show that, if f is Riemann integrable (and so, in particular, bounded),

$$|S_n(f, 0)| \leqslant A \sup_{x \in \mathbb{T}} |f(x)| \log n$$

where A is some constant independent of n. ($A = 10$ will certainly do the trick, but the precise value of A that you obtain is irrelevant.)

(iii) Deduce that

$$\sup_{x \in \mathbb{T}} |S_n(f, x)| \leqslant A \sup_{x \in \mathbb{T}} |f(x)| \log n.$$

18.4 We now show that part (iii) of the previous question cannot be improved. The arguments are of the type used in Lemmas 18.3 to 18.8.

(i) Show that if $\varepsilon > 0$ is fixed we can find an $n_0(\varepsilon) \geqslant 2$ depending on ε such that

$$\frac{1}{2\pi} \int_{-\varepsilon}^{\varepsilon} |D_n(s)| \, ds \geqslant \frac{3}{\pi^2} \log n$$

for all $n \geqslant n_0(\varepsilon)$.

(ii) Deduce that for ε and $n_0(\varepsilon)$ as in (i) we can find a continuous function $g_n: \mathbb{T} \to \mathbb{R}$ such that

(a) $|g_n(s)| \leqslant 1$ for all s,
(b) $|S_n(g_n, 0)| \geqslant (2/\pi^2) \log n$,
(c) $g_n(s) = 0$ for $|s| \geqslant \varepsilon$.

(iii) Deduce that for ε and $n_0(\varepsilon)$ as in (i) we can find a trigonometric polynomial $G_n: \mathbb{T} \to \mathbb{R}$ such that

(a) $|G_n(s)| \leqslant 2$ for all $s \in \mathbb{T}$,
(b) $|S_n(G_n, 0)| \geqslant (2/\pi^2) \log n$,
(c) $|G_n(s)| \leqslant \varepsilon$ for $|s| \geqslant \varepsilon$.

(iv) Show that, for fixed m,

$$S_m(G_n, s) \to 0 \quad \text{uniformly as } n \to \infty.$$

Deduce that, given $\eta_1, \eta_2 > 0$ and $N \geqslant 1$, we can find a real trigonometric polynomial G and an integer M such that

(a) $|G(s)| \leqslant 2$ for all $s \in \mathbb{T}$,
(b) $|S_M(G, 0)| \geqslant (2/\pi^2) \log M$,
(c) $|G(s)| \leqslant \eta_1$ for $|s| \geqslant \eta_1$,
(d) $|S_m(G, s)| \leqslant \eta_2$ for all $0 \leqslant m \leqslant N$ and all $s \in \mathbb{T}$.

(v) Show, using (iv) and induction, that we can find real trigonometric polynomials $P_1, P_2, \ldots$ of degree $N(1), N(2), \ldots$ and integers $M(1), M(2), \ldots$ such that

$$0 = N(0) < M(1) < N(1) < M(2) < N(2) < M(3) < \ldots.$$

and

(a)$_k$ $|P_k(s)| \leq 2$ for all $s \in \mathbb{T}$,
(b)$_k$ $|S_{M(k)}(P_k, 2^{-k})| \geq (2/\pi^2) \log M(k)$,
(c)$_k$ $|P_k(s)| \leq 2^{-k}$ for $|s - 2^{-k}| \geq 2^{-k-3}$
(d)$_k$ $|S_m(P_k, s)| \leq 2^{-k}$ for all $0 \leq m \leq N(k-1)$ and all $s \in \mathbb{T}$.

(vi) Show that $\sum_{k=1}^{\infty} P_k(s)$ converges uniformly on each set of the form $\mathbb{T}\backslash\{s: 0 < s < 2^{-j}\}$ to a function F_j with $|F_j(s)| \leq 3$. Conclude that $\sum_{k=1}^{\infty} P_k(s)$ converges everywhere to a function F which is continuous except at the origin and has $|F(s)| \leq 3$ for all $s \in \mathbb{T}$. Show that this implies that F is Riemann integrable.

(vii) Show further that, if N is large enough (depending on j),

$$\left| F(s) - \sum_{k=1}^{N} P_k(s) \right| \leq 2^{-j} \quad \text{for } s \notin (0, 2^{-j}),$$

$$\left| F(s) - \sum_{k=1}^{N} P_k(s) \right| \leq 6 \quad \text{for } s \in (0, 2^{-j}),$$

and that it follows that

$$\left| \hat{F}(r) - \sum_{k=1}^{N} \hat{P}_k(r) \right| \leq 7.2^{-j}$$

for all sufficiently large N. Conclude that

$$\hat{F}(r) = \sum_{k=1}^{\infty} \hat{P}_k(r).$$

(viii) Show that for $j \geq 100$

$$|S_{M(j)}(F, 2^{-j})| \geq \pi^{-2} \log M(j).$$

(For further developments see Question 52.6.)

18.5 (i) Suppose $a_1, a_2, \ldots, a_n$, $b_1, b_2, \ldots, b_n \in \mathbb{C}$. Let $s_k = \sum_{j=1}^{k} a_j$ for $1 \leq k \leq n$ and let $b_{n+1} = 0$. Prove the Abel summation formula which states that $\sum_{j=1}^{k} a_j b_j = \sum_{j=1}^{k} (b_j - b_{j+1}) s_j$.

(ii) Suppose now that $|s_j| \leq M$ for all $1 \leq j \leq n$ and $b_1, b_2, \ldots, b_n$ are real with $b_1 \geq b_2 \geq \ldots \geq b_n \geq 0$. Show that $|\sum_{j=1}^{n} a_j b_j| \leq M b_1$.

(iii) More generally show that, if we have sequences $a_1, a_2, \ldots \in \mathbb{C}$, $b_1, b_2, \ldots \in \mathbb{R}$ with $b_1 \geq b_2 \geq \ldots \geq b_n \geq 0$ for all n, then, if $|\sum_{j=p}^{q} a_j| \leq M$ for all $q \geq p$, it follows that

$$\left|\sum_{j=m}^{n} a_j b_j\right| \leqslant M b_m \quad \text{for all } n \geqslant m \geqslant p.$$

(iv) Suppose E is some set and we have a sequence of functions a_j: $E \to \mathbb{C}$ together with a decreasing sequence of real positive numbers b_j. Show

(a) if there exists an M such that $|\sum_{j=1}^{n} a_j(x)| \leqslant M$ for all $n \geqslant 1$ and all $x \in E$ and if $b_j \to 0$, then $\sum_{j=1}^{\infty} b_j a_j(x)$ converges uniformly on E;

(b) if $\sum_{j=1}^{\infty} a_j(x)$ converges uniformly on E, then so does $\sum_{j=1}^{\infty} b_j a_j(x)$.

(Result (a) is often called Abel's test and (b) Dirichlet's test. Since few people can remember which is which and they are similarly obtained consequences of (i), there is a good case for following Stromberg's suggestion in his *An Introduction to Classical Real Analysis* and referring to them all as Abel's tests.)

(v) Abel used his result to obtain results on the behaviour of power series on their circle of convergence. Suppose $a_n \in \mathbb{C}$ and $\sum_{n=0}^{\infty} a_n$ converges. Show that $\sum_{n=0}^{\infty} a_n r^n$ converges uniformly on $[0, 1]$, deduce that $\sum_{n=0}^{\infty} a_n r^n$ is continuous on $[0, 1]$ and conclude that $\sum_{n=0}^{\infty} a_n r^n \to \sum_{n=0}^{\infty} a_n$ as $r \to 1-$. This gives us the original proof of the first sentence of Question 27.3.

(vi) Suppose f: $[a, b] \to \mathbb{R}$ and g: $[a, b] \to \mathbb{R}$ are continuous. By using results of the form

$$n^{-1} \sum_{r=1}^{n} f(a + r(b-a)/n) \to \int_a^b f(x)\,dx$$

and part (iii) of this question, show that, if g is increasing, then

$$\left|\int_a^b f(x)g(x)\,dx\right| \leqslant 4 \max(|g(a)|, |g(b)|) \sup \left|\int_a^t f(x)\,dx\right|.$$

(This is a weaker result than the second mean value theorem proved in Question 52.7 but is strong enough for most purposes.)

(vii) Show that $\sum_{k=1}^{N} \sin kx/\log(k+1)$ converges uniformly as $N \to \infty$ on $\{x \in \mathbb{T}: |x| \geqslant \delta\}$ for each $\delta > 0$. Deduce that $\sum_{k=1}^{\infty} \sin kx/\log(k+1)$ converges everywhere to some function g_0: $\mathbb{T} \to \mathbb{R}$. Show that g_0 is continuous everywhere except possibly at 0 and that $g_0(0) = 0$.

(viii) Define a_k by the formulae $a_0 = 0$ and

$$a_k e^{ikx} + a_{-k} e^{-ikx} = \sin kx/\log(k+1)$$

for $k \geqslant 1$. By using Question 18.2(ii), show that there is no Riemann integrable function g such that $\hat{g}(k) = a_k$ for all $k \in \mathbb{Z}$. (That is, $\sum_{k=-N}^{N} a_k \exp(ikx)$ converges everywhere as $N \to \infty$, but the a_k are not the Fourier coefficients of a Riemann integrable function. The reader may ask what in this case are the Fourier coefficients of g_0? General theorems show that $(1/2\pi)\int_{|t|\geq\delta}|g_0(t)|\,dt \to \infty$ as $\delta \to 0+$, so there are problems in defining $\hat{g}_0(r)$ as an integral. More general tools are needed such as the Cantor uniqueness theorem proved in Question 30.2.)

18.6 (The Bernoulli polynomials) (i) By considering the saw tooth function of Chapter 16, or otherwise, show that

$$-\frac{1}{\pi}\sum_{r=1}^{\infty}\frac{\sin 2\pi rt}{r} = \phi_1(t) \quad [0 < t < 1]$$

where $\phi_1(t) = t - 1/2$.

(ii) By repeated integration show that if $m \geq 1$ and $0 \leq t \leq 1$

$$(-1)^{m-1}\frac{2}{(2\pi)^{2m}}\sum_{r=1}^{\infty}\frac{\cos 2\pi rt}{r^{2m}} = \phi_{2m}(t),$$

$$(-1)^{m-1}\frac{2}{(2\pi)^{2m+1}}\sum_{r=1}^{\infty}\frac{\sin 2\pi rt}{r^{2m+1}} = \phi_{2m+1}(t),$$

where $\phi_0, \phi_1, \ldots$ is a sequence of polynomials such that

(a) $\phi_0 = 1$,
(b) $\phi_n' = \phi_{n-1}$ for all $n \geq 1$,
(c) $\int_0^1 \phi_n(t)\,dt = 0$ for all $n \geq 1$.

(iii) Show that (a), (b), (c) define the sequence ϕ_n uniquely. The ϕ_n are called the Bernoulli polynomials.

(iv) Show that the coefficients of ϕ_n are rational and deduce, by setting $t = 1$ in (ii), that, if $m \geq 1$, then $\sum_{r=1}^{\infty} r^{-2m}$ is a rational multiple of π^{2m}. Find the sum explicitly for $m = 1, 2, 3$.
(The zeta function $\zeta(m) = \sum_{r=1}^{\infty} r^{-m}$ was first considered by Euler for m an integer with $m \geq 2$. He was the first to show that $\zeta(2m)$ was a rational multiple of π^{2m}, but left open the question whether $\zeta(2m+1)$ was transcendental, algebraic or even rational. The only progress on this question in 300 years is reported in Question 43.5.)

18.7 (The Bernoulli numbers) We continue with the ideas and notation of Question 18.6.

(i) By induction, or otherwise, establish that:

(a) ϕ_n has degree exactly n,
(b) $\phi_n(t) = (-1)^n\phi_n(1-t)$ for all $t \in \mathbb{R}$ and all $n \geq 0$,
(c) $\phi_{2n+1}(0) = \phi_{2n+1}(1) = 0$ for $n \geq 1$,
(d) $\phi_{2n}(0) = \phi_{2n}(1)$ for $n \geq 0$.

(ii) By induction, or otherwise, show that

$$\phi_m(t+1) - \phi_m(t) = t^{m-1}/(m-1)! \quad \text{for all } t \in \mathbb{R} \text{ and all } m \geq 1,$$

and deduce that

$$\sum_{r=1}^{n} r^m = m!(\phi_{m+1}(n+1) - \phi_{m+1}(1)).$$

(iii) We define the Bernoulli numbers B_n by setting $B_n = n!\phi_n(0)$. (Older texts give a different definition, so you must be careful when consulting them.) Show that

$$\phi_n(t) = \frac{1}{n!}\sum_{k=0}^{n}\binom{n}{k}B_k t^{n-k}.$$

Using (i)(c) and (d) show that $B_{2n+1} = 0$ for $n \geq 1$ and

$$B_n = -\frac{1}{n+1}\sum_{k=0}^{n-1}\binom{n+1}{k}B_k \quad \text{for all } n \geq 1. \tag{*}$$

What is the value of B_1?

(iv) Using 18.6 (ii) show that

$$\sum_{r=1}^{\infty}\frac{1}{r^{2m}} = \frac{(-1)^{m-1}(2\pi)^{2m}B_{2m}}{2(2m)!}.$$

Deduce that $(-1)^{m-1}B_{2m} > 0$ for $m \geq 0$.

(v) The Bernoulli numbers are easy to compute from (*). Compute B_m for $1 \leq m \leq 14$. [$B_1 = -1/2$, $B_2 = 1/6$, $B_4 = -1/30$, $B_6 = 1/42$, $B_8 = -1/30$, $B_{10} = 5/66$, $B_{12} = -691/2730$, $B_{14} = 7/6$].

(vi) Using (ii), (iii) and (v), show that

$$\sum_{r=1}^{n} r^{10} = (1/11)n^{11} + (1/2)n^{10} + (5/6)n^9 - n^7 + n^5 - (1/2)n^3 + (5/66)n.$$

In the work which introduced the Bernoulli numbers, Jacob Bernoulli boasts that in seven or eight minutes ('intra semi-quadrantem horae') he found $\sum_{r=1}^{1000} r^{10}$. Verify his answer of

91 409 924 241 424 243 424 241 924 242 500

(a) by hand, using the formula above,
(b) directly, using a computer.

Time the computer. If the sum were taken to 10^{12} terms rather than 10^3 who would win, Bernoulli or the computer?

(vii) Show that

$$\sum_{r=1}^{\infty} r^{-10} = \pi^{10}/(93555).$$

18.8 This question is not required later but should appeal to anyone who has tried to find the Taylor series for $\tan x$.

(i) What is the Taylor series of $\sin x$?

(ii) Compute the first few terms of the Taylor series for $\tan x$. Can you see any pattern?

(iii) Explain why $f(z) = z/(e^z - 1)$ for $0 < |z| < 2\pi$, $f(0) = 1$ defines an analytic function $f: \{z \in \mathbb{C}: |z| < 2\pi\} \to \mathbb{C}$. Explain why $f(z) = \sum_{n=0}^{\infty} a_n z^n$ for $|z| < 2\pi$, for some $a_n \in \mathbb{C}$. By multiplying both sides by $e^2 - 1$ and equating coefficients, show that

$$\frac{z}{e^z - 1} = \sum_{n=0}^{\infty} \frac{B_n}{n!} z^n \quad \text{for } |z| < 2\pi.$$

(iv) Show that $\sum_{n=0}^{\infty} (B_n/n!) z^n$ has radius of convergence exactly 2π and deduce that, if $1 > \varepsilon > 0$, we have

$$\limsup \frac{|B_n|}{n!}(2\pi + \varepsilon)^n = \infty; \quad \frac{B_n}{n!}(2\pi - \varepsilon)^n \to 0 \text{ as } n \to \infty.$$

(We shall obtain much more exact results later in Question 18.16 but this shows already that some of the B_n must be very large.)

(v) Show that

$$\frac{ze^{tz}}{e^z - 1} = \sum_{n=0}^{\infty} \phi_n(t) z^n \quad \text{for all } t \in \mathbb{C}, |z| < 2\pi.$$

(vi) Starting from (iii), show that

$$\frac{z}{e^z - 1} = 1 - \frac{z}{2} + \sum_{n=1}^{\infty} \frac{B_{2n}}{(2n)!} z^{2n}$$

and deduce that

$$\frac{z}{2} \frac{e^{z/2} + e^{-z/2}}{e^{z/2} - e^{-z/2}} = \sum_{n=0}^{\infty} \frac{B_{2n}}{(2n)!} z^{2n} \quad \text{for } |z| < 2\pi.$$

Conclude that

$$z \cot z = \sum_{n=0}^{\infty} (-1)^n \frac{2^{2n} B_{2n}}{(2n)!} z^{2n}.$$

(vii) Show that $\tan z = \cot z - 2\cot(2z)$ and deduce that

$$\tan z = \sum_{n=1}^{\infty} (-1)^{n-1} \frac{2^{2n}(2^{2n}-1)}{(2n)!} B_{2n} z^{2n-1} \quad \text{for } |z| < \pi/2.$$

(viii) Show that $\operatorname{cosec} z = \cot z + \tan z/2$ and deduce the Taylor series for $z \operatorname{cosec} z$. State the range of validity.
(These formulae are all due to Euler. The method of proof is due to Cauchy.)

18.9 (The Euler–Maclaurin summation formula) In this question we continue themes from Questions 17.5, 18.6 and 18.7. Let us write b_n for the function $b_n: \mathbb{R} \to \mathbb{R}$ which is periodic with period 1 and has $b_n(t) = \phi_n(t)$ for $0 \leqslant t < 1$.

(i) Show that if $n \geqslant 2$ then b_n is $(n-1)$ times differentiable but that $b_n^{(n-1)}$ is discontinuous at the integers.

(ii) Using Question 18.6 find $a_{nr} \in \mathbb{C}$ such that, if $n \geqslant 2$,

$$b_n(t) = \sum_{r=-\infty}^{\infty} a_{nr} \exp(2\pi i r t) \quad \text{for all } t.$$

What can you say along these lines if $n = 1$?

(iii) Let $f: \mathbb{R} \to \mathbb{R}$ be once continuously differentiable. Show by integration by parts that

$$\frac{1}{2}(f(0) + f(1)) = \int_0^1 f(t)\,dt + \int_0^1 \phi_1(t) f'(t)\,dt.$$

By repeated integration by parts, show that, if f is M times continuously differentiable,

$$\frac{1}{2}(f(0) + f(1)) = \int_0^1 f(t)\,dt + \sum_{k=2}^{M} \frac{B_k}{k!}(f^{(k-1)}(1) - f^{(k-1)}(0)) + S_M$$

where

$$S_M = -\int_0^1 \phi_M(t) f^{(M)}(t)\,dt.$$

(iv) Hence deduce the Euler–Maclaurin formula

$$\sum_{r=0}^{n} f(r) = \int_0^n f(t)dt + \frac{1}{2}(f(0) + f(n)) + \sum_{k=2}^{M} \frac{B_k}{k!}(f^{(k-1)}(n) - f^{(k-1)}(0)) + R_{M,n}$$

with $R_{M,n} = -\int_0^n b_M(t) f^{(M)}(t)\,dt$.

(v) Show that the Euler–Maclaurin formula may be deduced from Question 17.6(v) by setting $g(t) = f(t/2\pi)$. To what extent are our derivations of the formula actually different?

(Note that, from the point of view of the numerical analyst, the Euler–Maclaurin formula has an air of lunacy about it. We attempt to link the behaviour of f in the interior of $[0, n]$ with derivatives $f^{(k)}(0)$, $f^{(k)}(n)$ at its end points. Such a procedure can only be justified if we know far more about f than numerical analysts can normally know about the functions they deal with.)

(vi) If P is a polynomial of degree M, then

$$\sum_{r=0}^{n} P(r) = \int_0^n P(t)\,dt + \frac{1}{2}(P(n) + P(0)) + \sum_{k=2}^{M} \frac{B_k}{k!}(P^{(k)}(n) - P^{(k)}(0)).$$

Prove this

(a) using the Euler–Maclaurin formula of (iv),
(b) using Question 18.7 (ii) and (iii).

18.10 In this question we examine a typical use of the Euler–Maclaurin formula.

(i) By using the Euler–Maclaurin formula of Question 18.9 (iv) show that, if $M \geqslant 2$, $m \geqslant n \geqslant 1$,

$$\sum_{r=n}^{m} \frac{1}{r^2} = \frac{1}{n} - \frac{1}{m} + \frac{1}{2n^2} + \frac{1}{2m^2} + \sum_{k=2}^{M} (-1)^{k-1} B_k \left(\frac{1}{n^{k+1}} - \frac{1}{m^{k+1}}\right) + R_M(n, m)$$

where $R_M(n, m) = (-1)^M (M+1)! \int_n^m \frac{b_M(t)}{t^{M+2}}\,dt$.

(ii) Observing that b_M is a continuous periodic function and therefore bounded, show that $R_M(n, m)$ converges to a finite limit $R_M(n)$, say. Show that

$$\sum_{r=1}^{\infty} \frac{1}{r^2} = \sum_{r=1}^{n-1} \frac{1}{r^2} + \frac{1}{n} + \frac{1}{2n^2} + \sum_{k=2}^{M} (-1)^{k-1} \frac{B_k}{n^{k+1}} + R_M(n).$$

(iii) We know (see e.g. Question 18.6 (iv)) that $\sum_{r=1}^{\infty} r^{-2} = \pi^2/6$. Using your calculator, examine the approximation obtained by setting $n = 1$ for $M = 2, 4, 6, 8, 10, 12, 14$ to give

$$\sum_{r=1}^{\infty} r^{-2} \approx 1 + 2^{-1} + \sum_{k=2}^{M} (-1)^{k-1} B_k.$$

Examine the approximation

$$\sum_{r=1}^{\infty} r^{-2} \approx \sum_{r=1}^{n-1} r^{-2} + n^{-1} + (2n^2)^{-1} + \sum_{k=2}^{M} (-1)^{k-1} B_k n^{-(k+1)}$$

when $n = 2$, $M = 2, 4, 6, 8, 10, 12, 14$ and when $n = 10$, $M = 2, 4, 6, 8, 10, 12, 14$.

(iv) Estimate how large N must be for $\sum_{r=1}^{N} r^{-2}$ to approximate $\sum_{r=1}^{\infty} r^{-2}$ to 10 places of decimals.

(For $n = 1, 2$ the calculations show a remainder term $R_M(n)$ which decreases in magnitude as M increases up to a certain point and then starts to increase again. If we think a little, this is not unexpected. In the case $n = 1$ we attempt to sum $\sum_1^\infty r^{-2}$ using only correction terms and in the case $n = 2$ we attempt to sum $\sum_1^\infty r^{-2}$ using only the first term 1^{-2} and correction terms. As the northern proverb goes 'You get now't for now't and damn little for sixpence'. In view of this, it becomes important to examine the behaviour of the remainder term and this we shall do in the next question.

For those who like to live dangerously this question leads on to a method known as Richardson's deferred approach to the limit. Like so many of L. F. Richardson's ideas it is simple but thought provoking. Consider the problem of summing $S = \sum_{r=1}^{\infty} f(r)$ numerically. To do this we form $S(1) = f(1)$, $S(2) = S(1) + f(1), \ldots, S(k+1) = S(k) + f(k)$ until we reach $S(n)$. We then hope that $S(n)$ is a good approximation to S. Richardson observed that, if f is well behaved, $S(k) = A + Bk^{-1} + Ck^{-2} + \theta(k)$ where $k^{-2}\theta(k) \to 0$ and $k^{-2}\theta(k)$ becomes small quite fast. But A, B, C can be estimated as α, β, γ, say, by looking at the values of $S(k)$ and treating $\theta(k)$ as an unknown but small error term. Since allowing $k \to \infty$ shows that $A = S$ we may hope that α will be a much better approximation to S than $S(n)$ was. This simple idea has many generalisations and, if used wisely, furnishes a very effective tool for numerical analysis. We give another example in Question 18.12.

Incidentally, if the reader's university library does not contain Richardson's *Arms and Insecurity* and *Statistics of Deadly Quarrels*, she should give the librarian no peace until it does.)

18.11 We now investigate the nature of the remainder $R_M = R_{M,n} = \int_0^n b_M(t) f^{(M)}(t)\,dt$ in the Euler–Maclaurin formula. Such an investigation was first carried out by Jacobi who introduced the polynomials ϕ_n in this context.

(i) Explain why $R_{2M} = -R_{2M+1}$ for $M \geqslant 1$.

(ii) By integration by parts show that (provided f is continuously $M + 1$ times differentiable)

$$R_M = -\int_0^n (b_{M+1}(t) - b_{M+1}(0)) f^{(M+1)}(t)\,dt.$$

(iii) Part (ii) suggests we should look at $\psi_m = \phi_m - \phi_m(0)$. Compute and sketch $\psi_1, \psi_2, \psi_3, \psi_4, \psi_5, \psi_6$. Show in particular that $\psi_2(x) = x(x-1)/2$.

(iv) Show (using the results of Questions 18.6 and 18.7 if you require) that $\psi_m(0) = \psi_m(1) = 0$ for $m \geqslant 2$ and that $\psi_{2m+1}(1/2) = 0$ for $m \geqslant 1$.

(v) Show, using Question 18.7 (iii) and (iv), that, if $m \geqslant 1$, then $(-1)^m \psi_{2m}(t) > 0$ and $(-1)^{m+1} \psi_{2m+1}(t) > 0$ for t small and strictly positive.

(vi) Observe that $\psi'_{2m}(t) = \psi_{2m-1}(t)$, $\psi'_{2m+1}(t) = \psi_{2m}(t) + B_{2m}(2m)!$.

(vii) Consider the statements

$$(a)_{2m} \quad (-1)^m \psi_{2m}(t) > 0 \quad \text{for all } 0 < t < 1,$$

$$(a)_{2m+1} \quad (-1)^{m+1} \psi_{2m+1}(t) > 0 \quad \text{for all } 0 < t < 1/2,\ (-1)^m \psi_{2m+1}(t) > 0$$

for all $1/2 < t < 1$.

Show by induction that $(a)_m$ is true for all $m \geqslant 2$. (The results of (iii), (iv), (v) and (vi) are intended to be useful.)

(viii) Suppose f is infinitely differentiable. Experience shows that the Euler–Maclaurin formula is at its most effective when $f^{(2k)}(t) \geqslant 0$ for all $t \in [0, n]$ and all $k \geqslant 1$. (A similar discussion covers the case when $f^{(2k)}(t) \leqslant 0$ for all $t \in [0, n]$.) Show, using (i), (ii) and (vii) $(a)_{2m}$, that $(-1)^{M-1} R_{2M} \geqslant 0$, i.e that the remainders $R_2, R_4, \ldots$ alternate in sign. Setting

$$T_{2M} = \int_0^n f(t)dt + \frac{1}{2}(f(0) + f(n)) + \sum_{k=1}^{M} \frac{B_{2k}}{(2k)!}(f^{(2k-1)}(n) - f^{(2k-1)}(0))$$

deduce that T_{2M-2} lies between $\sum_{r=0}^{n} f(r)$ and

$$\sum_{r=0}^{n} f(r) - \frac{B_{2M}}{(2M)!}(f^{(2M-1)}(n) - f^{(2M-1)}(0))$$

and so

$$\left| T_{2M-2} - \sum_{r=0}^{n} f(r) \right| \leqslant \left| \frac{B_{2M}}{(2M)!}(f^{(2M-1)}(n) - f^{(2M-1)}(0)) \right|.$$

This is usually expressed by saying that (under the hypotheses stated) the error involved in using the Euler–Maclaurin sum as an approximation is less than the first term omitted. Note, however, that nothing is said about whether that term is large or small.

(ix) Use a similar analysis to show that in Question 18.10

$$\left| \sum_{r=1}^{\infty} r^{-2} - \sum_{r=1}^{n-1} r^{-2} + n^{-1} + (2n^2) - \sum_{k=1}^{M-1} B_{2k} n^{-(2k+1)} \right| \leqslant |B_{2M}| n^{-2M-1}.$$

Discuss parts (iii) and (iv) of Question 18.10 in this context. Show in particular that, provided your calculator is sufficiently accurate, you can get $\sum_{r=1}^{\infty} r^{-2}$ to a guaranteed decimal accuracy of 10^{-10} with less than 100 button presses.

(x) State and prove an appropriate general theorem which has the formula of (ix) as a special case.

(xi) Compute $\sum_{r=1}^{\infty} r^{-4}$ to a guaranteed accuracy of 10^{-6}.

(For related questions see Questions 107.2 and 108.5.)

18.12 (i) Let $g: [a, b] \to \mathbb{R}$ be continuously M times differentiable. Show by using a simple change of variable that, if $N \geqslant 1$ is an integer with $a + Nh = b$, the Euler–Maclaurin formula yields

$$\int_a^b g(t)dt = h\left(\frac{1}{2}g(a) + \sum_{r=1}^{N-1} g(a+rh) + \frac{1}{2}g(b)\right)$$
$$- \sum_{k=1}^{M} \frac{B_{2k}}{(2k)!} h^{2k}(g^{(2k-1)}(b) - g^{(2k-1)}(a)) + R(h),$$

where $|R(h)| \leqslant A|h|^{2M+1}$ for some A depending on g, a, b and M.

(ii) Explain why the approximation

$$\int_a^b g(t)dt \doteqdot h\left(\frac{1}{2}g(a) + \sum_{r=1}^{N-1} g(a+rh) + \frac{1}{2}g(b)\right)$$

('the trapezoidal rule') is extremely accurate (provided h is not too large):

(a) when g is periodic with period $b - a$,
(b) when g is symmetric (i.e. $g(x) = g(-x)$) and $a = -b$.

Why are the approximations not perfect?

(iii) Illustrate (ii)(a) by calculating both sides of the approximation when $g(t) = \cos^2 t$ for various values of N in the case when $[a, b] = [-\pi, \pi]$ and in the case when $[a, b] = [0, 1]$.

(iv) Observe that $\exp(-x^2/2) \leqslant R^{-1} x \exp(-x^2/2)$ for $x \geqslant R$. Hence estimate $\int_R^\infty \exp(-x^2/2)\,dx$. Using the ideas of (ii)(b) write a short program that will obtain $\int_u^v \exp(-x^2/2)\,dx$ correct to a guaranteed error of less than 10^{-10} for any given $\infty \geqslant u \geqslant v \geqslant -\infty$.

(Like the ideas of Question 18.10, the ideas of this question also lend themselves to Richardson's deferred approach to the limit. Observe that, if g is well behaved we expect

$$S(h) = h\left(\frac{1}{2}g(a) + \sum_{r=1}^{N-1} g(a+rh) + \frac{1}{2}g(b)\right)$$

to satisfy

$$S(h) = a_0 + a_2h^2 + a_4h^4 + \ldots, \text{ with } a_0 = \int_a^b g(t)dt.$$

It is customary in numerical integration to halve step lengths, finding sucessively $S(h)$, $S(h/2)$, $S(h/4)$, say. We can now estimate a_4, a_2 and a_0. Since the problem is one of interpolation rather than extrapolation we are on safer ground than in Question 18.10 and this technique is much used in numerical analysis under the name of Romberg integration.)

18.13 In this question we obtain another estimate for the remainder $R_{2M,n} = \int_0^n b_{2M}(t)f^{(2M)}(t)dt$. Apart from supposing that f is $2M+2$ times continuously differentiable, we make no further assumptions on f.

(i) Show, by using Question 18.11(vii), or otherwise, that

$$0 \leqslant (-1)^m \psi_{2m}(t) \leqslant (-1)^m \psi_{2m}(1/2) \text{ for all } 0 \leqslant t \leqslant 1,\ m \geqslant 1.$$

(ii) By setting $t = 0$ and $t = 1/2$ in Question 18.8(v), show that

$$\sum_{n=0}^{\infty} (\phi_n(1/2) + \phi_n(0))z^n = \frac{z(e^{z/2}+1)}{e^z - 1} = \frac{z}{e^{z/2}-1} = 2\sum_{n=0}^{\infty} \phi_n(0)\left(\frac{z}{2}\right)^n.$$

(iii) Using Question 18.11(ii), show that

$$|R_{2M,n}| \leqslant \int_0^n |f^{(2M+2)}(t)\psi_{2M+2}(t)|\, dt \leqslant 2B_{2M+2}\int_0^n |f^{(2M+2)}(t)|\, dt.$$

(The reader may object that it does not look as though $R_{2M,n}$ is small. She is quite right. Only in special circumstances, in effect when f is 'slowly varying', will $R_{2M,n}$ be small. But when f is 'slowly varying' sums like $\sum_0^\infty f(r)$ are hard to calculate directly and the Euler–Maclaurin method comes into its own. Note also that, although under these circumstances $R_{2M,n}$ may be small for suitably chosen M, it will still usually be the case that $|R_{2M,n}|$ becomes unbounded as $M \to \infty$.)

18.14 (Stirling's formula)

(i) Let $f(t) = \log t$. Use the Euler–Maclaurin formula to obtain (for $M \geqslant 2$)

$$\sum_{r=1}^{n} \log r = \int_1^n \log t dt + \frac{1}{2}\log n + \sum_{k=2}^{M} \frac{(-1)^k B_k}{k(k-1)}\left(\frac{1}{n^k} - 1\right) + \int_1^n b_M(t)f^{(M)}(t)\, dt.$$

(ii) Using the fact that b_M is periodic and so bounded, show that $\int_1^\infty b_M(t)f^{(M)}(t)\, dt$ converges to $C_M(1)$ say and $|\int_n^\infty b_M(t)f^{(M)}(t)\, dt| \leqslant K'_M n^{-M+1}$ for some constant K'_M.

(iii) By integration by parts, or otherwise, show that $\int_1^n \log t\, dt = n\log n - n + 1$.

(iv) Conclude that, if we write

$$C_M = C_M(1) - \sum_{k=2}^{M} \frac{(-1)^k B_k}{k(k-1)} + 1,$$

we can find a constant K_M such that

$$\left|\sum_{r=1}^{n} \log r - n\log n + n - \frac{1}{2}\log n - \sum_{k=2}^{M} \frac{(-1)^k\, B_k n^{1-k}}{k(k-1)} - C_M\right| \leqslant K_M n^{-M+1}$$

for all $n \geqslant 1$.

(v) By considering what happens as $n \to \infty$, show that C_M does not depend on M, i.e. there is a constant C' with $C_M = C'$ for all M.

(vi) Deduce that there is a constant C such that

$$\left|\log\left(\frac{n!}{Cn^{n+1/2}e^{-n}}\right) - \sum_{k=2}^{M} \frac{(-1)^k B_k n^{1-k}}{k(k-1)}\right| \leqslant K_M n^{-M+1}$$

and so, if n is large enough (depending on M), then

$$Cn^{n+1/2}e^{-n} \exp\left(\sum_{k=2}^{M} \frac{(-1)^k B_k n^{1-k}}{k(k-1)}\right)$$

will be a very good approximation for $n!$.

(vii) Prove, in particular, that

$$\frac{n!e^n}{n^{n+1/2}} \to C \quad \text{as } n \to \infty.$$

(Part (vi) puts us in the embarassing position of knowing the correction terms for our formula for $n!$ without knowing the main term C. The next questions remedy the defect.)

18.15 (The Wallis formula)

(i) Let $I_n = \int_0^{\pi/2} \sin^n x dx$ for $n \geqslant 0$. Show that $I_{n+1} = (n/(n+1))I_{n-1}$ for $n \geqslant 1$ and hence compute I_{2n+1} and I_{2n} for all $n \geqslant 0$. Deduce that

$$\frac{I_{2n+1}}{I_{2n}} = \frac{2}{1}\cdot\frac{2}{3}\cdot\frac{4}{3}\cdot\frac{4}{5}\cdot\frac{6}{5}\cdot\frac{6}{7}\cdots\frac{2n}{2n-1}\cdot\frac{2}{\pi}.$$

(ii) Show that $0 < I_{n+1} < I_n < I_{n-1}$ for $n \geqslant 1$ and deduce that $1 < I_n I_{n+1}^{-1} < 1 + n^{-1}$.

(iii) Combining (i) and (ii), show that

$$\frac{2}{1}\cdot\frac{2}{3}\cdot\frac{4}{3}\cdot\frac{4}{5}\cdot\frac{6}{5}\cdot\frac{6}{7}\cdots\frac{2n}{2n-1} \to \frac{\pi}{2} \quad \text{as } n \to \infty.$$

(This beautiful formula is due to Wallis.)

(iv) Use (iii) to obtain the result $\prod_{r=1}^{n} (1 + 1/(4r^2 - 1)) \to \pi/2$ as $n \to \infty$. Show, by taking reciprocals, that $\prod_{r=1}^{n} (1 - 1/4r^2) \to 2/\pi$ as $n \to \infty$.

18.16 (Stirling's formula concluded)

(i) By writing

$$\frac{2}{1}\cdot\frac{2}{3}\cdot\frac{4}{3}\cdot\frac{4}{5}\cdot\frac{6}{5}\cdot\frac{6}{7}\cdots\frac{2n}{2n-1} = \frac{1}{2n}\,\frac{(2^n n!)^4}{((2n)!)^2}$$

and using Question 18.13 (iii), show that the constant C in Question 18.14 is $\sqrt{2\pi}$. We thus have Stirling's formula $n!e^n n^{-(n+1/2)} \to \sqrt{2\pi}$ as $n \to \infty$. Write down a version of Stirling's formula as in Question 18.14 (vi).

(ii) Using 18.7 (iv), show that

$$\frac{(-1)^{m-1}(2\pi)^{2m} B_{2m}}{2(2m)!} \to 1 \quad \text{as } m \to \infty$$

and using Stirling's formula deduce that

$$\frac{(-1)^{m-1} e^{2m} (2\pi)^{2m-1/2}}{2(2m)^{2m+1/2}} B_{2m} \to 1 \quad \text{as } m \to \infty.$$

Thus (although the first few terms give no hint of this) the even Bernoulli numbers grow explosively fast. Abramowitz and Stegun *Handbook of Mathematical Functions* gives $B_{60} = -2.134 \times 10^{34}$. Verify, using a calculator, that the limiting result just given yields a good approximation for B_{60}.

18.17 (De Moivre's theorem) Suppose $X_1, X_2, \ldots, X_n$ are independent random variables with $Pr(X_j = 1) = Pr(X_j = -1) = 1/2$. The object of this question is to show that

$$Pr\left(\frac{X_1 + X_2 + \ldots + X_n}{n^{1/2}} \in [a, b]\right) \to \frac{1}{(2\pi)^{1/2}} \int_a^b \exp(-x^2/2)\, dx$$

as $n \to \infty$ for all $a < b$.

(i) Show that this will follow if we show that

$$Pr\left(\frac{X_1 + X_2 + \ldots + X_{2n}}{(2n)^{1/2}} \in [a, b]\right) \to \frac{1}{(2\pi)^{1/2}} \int_a^b \exp(-x^2/2)\, dx$$

as $n \to \infty$ for all $a < b$.

(ii) Show that (i) will follow if we show that

$$Pr\left(\frac{X_1 + X_2 + \ldots + X_{2n}}{(2n)^{1/2}} \in [-a, a]\right) \to \frac{1}{(2\pi)^{1/2}} \int_{-a}^a \exp(-x^2/2)\, dx$$

as $n \to \infty$ for all $a > 0$.

(iii) Show that (ii) will follow if we show that

$$\sum_{|k| \leqslant a n^{1/2}/2^{1/2}} \binom{2n}{n+k} 2^{-2n} \to \frac{1}{(2\pi)^{1/2}} \int_{-a}^a \exp(-x^2/2)\, dx.$$

(iv) Show that (iii) will follow if we show that

$$\sum_{x \in \Lambda(n) \cap [-a,a]} (2^{-1}n)^{-1/2} f_n(x) \to \frac{1}{(2\pi)^{1/2}} \int_{-a}^a \exp(-x^2/2)\, dx$$

as $n \to \infty$ where $\Lambda(n) = \{2^{1/2} k n^{-1/2}: k \in \mathbb{Z}\}$ and f_n is the simplest piecewise linear function with $f_n(2^{1/2} k n^{-1/2}) = \binom{2n}{n+k} 2^{-2n} (2^{-1}n)^{1/2}$ for $|k| \leqslant n$, $f_n(2^{1/2} k n^{-1/2}) = 0$ for $|k| \geqslant n + 1$.

(v) Suppose $g_n: [-a, a] \to \mathbb{R}$ and $g: [-a, a] \to \mathbb{R}$ are continuous functions such that $g_n \to g$ uniformly on $[-a, a]$. Show that

$$\left(\frac{2}{n}\right)^{1/2} \sum_{x \in \Lambda(n) \cap [-a,a]} g_n(x) \to \int_{-a}^{a} g(x)\, dx,$$

(vi) If $x \in [-a, a]$, write $k_n(x)$ for one of the (at most two) integers which minimises $|2^{1/2} kn^{-1/2} - x|$. Using (iv) and (v), show that the theorem stated at the beginning of the question will follow if we can show that, when $f(x) = (2\pi)^{-1/2} e^{-x^2/2}$, $f_n(2^{1/2} k_n(x) n^{-1/2}) \to f(x)$ uniformly on $[-a, a]$.

(vii) Conclude that the theorem will follow if we show that

$$\left| \binom{2n}{n+k} 2^{-2n} (2^{-1} n)^{1/2} - \frac{1}{\sqrt{2\pi}} \exp\left(\frac{-k^2}{n}\right) \right| \leqslant \varepsilon(n)$$

for all $|k| \leqslant 2^{1/2} n^{1/2} a$ and some $\varepsilon(n) \to 0$ as $n \to \infty$.

(viii) Show that (vii) and thus the theorem will follow if we show that for any fixed $R > 0$

$$\sup_{|k| \leqslant R\sqrt{n}} \left| \log\left(\binom{2n}{n+k} 2^{-2n} (2^{-1} n)^{1/2} \right) - \log\left(\frac{1}{\sqrt{(2\pi)}} \exp\left(\frac{-k^2}{n}\right) \right) \right| \to 0$$

as $n \to \infty$.

(ix) Prove that there exists a constant A such that

$$|\log(1 + \delta) - \delta + \delta^2/2| \leqslant A\delta^3 \quad \text{for } |\delta| < 1/2.$$

By using Stirling's formula and results like $\log(n + k) = \log n + \log(1 + k/n)$, prove the result of (viii) and thus De Moivre's theorem in the form stated in the question.

(x) Using Tchebychev's inequality (Theorem 50.4), show that given any $\varepsilon > 0$ there exists an $R_0(\varepsilon)$ such that

$$1 \geqslant Pr\left(\frac{X_1 + X_2 + \ldots + X_n}{n^{1/2}} \in [-R, R] \right) \geqslant 1 - \varepsilon$$

for all $R \geqslant R_0(\varepsilon)$. Deduce from De Moivre's theorem that $\int_{-\infty}^{\infty} \exp(-x^2/2) dx$ converges with value $(2\pi)^{1/2}$.

(Conversely, if we already know that $\int_{-\infty}^{\infty} \exp(-x^2/2) dx = (2\pi)^{1/2}$, the arguments above enable us to identify the constant C in Question 18.14 (vii) without using Question 18.16.)

18.18 (A more general De Moivre's theorem) Suppose $X_1, X_2, \ldots, X_n$ are independent random variables with $Pr(X_j = 1) = p$, $Pr(X_j = 0) = 1 - p$ where $0 < p < 1$. If we write $\sigma^2 = p(1 - p)$, $\sigma > 0$, then the general De Moivre's theorem states that

$$Pr\left(\frac{X_1+\ldots+X_n-np}{n^{1/2}\sigma}\in[a,b]\right)\to\frac{1}{(2\pi)^{1/2}}\int_a^b \exp(-x^2/2)\,dx$$

as $n\to\infty$ for all $a<b$.

(i) If $p=1/2$ explain why this is, essentially, the result of Question 18.17.

(ii) State, giving reasons for your answer, whether it is true that if a and b are fixed

$$Pr\left(\frac{X_1+\ldots+X_n-np}{n^{1/2}\sigma}\in[a,b]\right)\to\frac{1}{(2\pi)^{1/2}}\int_a^b \exp(-x^2/2)\,dx$$

uniformly for $0<p<1$ as $n\to\infty$.

(iii) Show that the result of this question will follow if we can show

$$\sup_{r\in\Lambda(n)}\left|A(p)\binom{n}{r}p^r(1-p)^{n-r}n^{1/2}-\frac{1}{\sqrt{(2\pi)}}\exp\left(\frac{-(r-np)^2}{2n\sigma^2}\right)\right|\to 0$$

as $n\to\infty$ where

$$\Lambda(n)=\{r\colon (r-np)/(n^{1/2}\sigma)\in[a,b]\}$$

and $A(p)$, which is to be calculated, depends on p alone.

(iv) Prove the result of (iii) and hence the general De Moivre theorem. (The computations in (iii) and (iv) are just a little tedious and complicated, but tedious and complicated calculations are part of mathematical life.)

18.19 Fejér gave a very neat proof of the existence of a continuous function whose Fourier series diverges at a point.

(i) Use Abel summation (see Question 18.5 (iii)) to show that, if $0<x<\pi$ and m is the integer with $m\leqslant \pi^{1/2}x^{-1}<m+1$, then

$$\left|\sum_{k=m+1}^{n}k^{-1}\sin kx\right|\leqslant\pi^{1/2}\quad\text{for all } n\geqslant m.$$

Show, by using almost trivial inequalities, that $|\sum_{k=m+1}^{n}k^{-1}\sin kx|\leqslant 2\pi^{1/2}$ for all $x\in\mathbb{T}$ and all $n\geqslant m\geqslant 1$.

(ii) Let $A_n(x)=\sum_{r=1}^{n}r^{-1}\cos(n-r+1)x$ and $B_n(x)=\sum_{r=1}^{n}r^{-1}\cos(n+r+1)x$.
Show that, if we set $\phi_n(x)=A_n(x)-B_n(x)$, then

$$\phi_n(x)=2\sin(n+1)x\sum_{k=1}^{n}k^{-1}\sin kx$$

and deduce that $|\phi_n(x)|\leqslant 4\pi^{1/2}$ for all $x\in\mathbb{T}$.

(iii) By writing down $S_p(\phi_n,x)$ explicitly, or otherwise, show that $S_p(\phi_n,0)\geqslant 0$ for all n, $p\geqslant 0$ and that $S_n(\phi_n,0)\geqslant\log n$.

(iv) Let us write $N(q) = 2^{q^3}$. Show that $\sum_{q=1}^{\infty} q^{-2}\phi_{N(q)}$ converges uniformly on $\mathbb{T}$ to a continuous function f, say. Explain why

$$\hat{f}(r) = \sum_{q=1}^{\infty} q^{-2}\hat{\phi}_{N(q)}(r) \quad \text{for all } r \in \mathbb{Z}$$

and deduce that $S_n(f,x) = \sum_{q=1}^{\infty} q^{-2} S_n(\phi_{N(q)}, x)$. Conclude, in particular, that $S_{N(q)}(f,0) \to \infty$.

18.20$^{(-)}$ We have seen that, even if $f\colon \mathbb{T} \to \mathbb{C}$ is continuous, $S_n(f,x)$ need not converge. Show, however, that, if $S_n(f,x)$ does converge, it must converge to the right answer. More precisely, use Lemma 1.4 (i) and Theorem 2.3 to show that, if $S_n(f,x) \to l$ as $n \to \infty$, then $l = f(x)$.

19

POINTWISE CONVERGENCE, THE ANSWER

19.1 The next three questions outline the proof of the theorem of Kahane and Katznelson stated as Theorem 19.5. In this question we dispose of some simple combinatorial preliminaries. If I is an interval, we write $|I|$ for its length.

(i) Let $I_1, I_2, \ldots, I_m$ be intervals. Show that we can find an integer $N \geqslant 1$, a real number $\delta > 0$ and intervals $J_1, J_2, \ldots, J_N$ each of length δ such that

$$\text{(a)} \quad \bigcup_{k=1}^{N} J_k \supseteq \bigcup_{r=1}^{m} I_r,$$

but $\quad$ (b) $\quad N\delta \leqslant 2 \sum_{r=1}^{m} |I_r|.$

(ii) Let E be a set of measure zero and let $\varepsilon \geqslant 0$. Observe that (by Definition 19.3) we can find intervals $I(r, s)$ such that $\bigcup_{s=1}^{\infty} I(r, s) \supseteq E$ yet $\sum_{s=1}^{\infty} |I(r, s)| \leqslant 2^{-r}\varepsilon$ for each $r \geqslant 1$. Hence conclude that we can find intervals $J(k)$ such that

(a) each point $e \in E$ belongs to infinitely many $J(k)$,

(b) $\sum_{k=1}^{\infty} |J(k)| \leqslant \varepsilon.$

(iii) Suppose $b_k \geqslant 0$ for all $k \geqslant 1$ and $\sum_{k=1}^{\infty} b_k \leqslant \varepsilon$. Explain briefly why, if $\varepsilon = \varepsilon(1) > \varepsilon(2) > \ldots$ and $\varepsilon(m) > 0$ for all m, we can find integers $1 = n(1) < n(2) < \ldots$ such that $\sum_{k=n(j)}^{\infty} b_k \leqslant \varepsilon(j)$ and so $\sum_{k=n(j)}^{n(j+1)-1} b_k \leqslant \varepsilon(j)$.

19.2 This question gives the central idea of the proof.

(i) Let $1 > \delta > 0$. Show that, if $|z| \leqslant 1 + \delta/2$, then $\operatorname{Re}(1 + \delta - z) > 0$ and so $\operatorname{Re}(1 + \delta - z)^{-1} > 0$. Show also that $\operatorname{Re}(1 + \delta - z)^{-1} \geqslant (3\delta)^{-1}$ whenever $z = e^{i\theta}$ with $|\theta| \leqslant \delta/2$, $\theta \in \mathbb{R}$.

(ii) Suppose $\theta_1, \theta_2, \ldots, \theta_N \in \mathbb{R}$. Write

$$\phi(z) = (1+\delta)N^{-1}\sum_{k=1}^{N}(1+\delta - ze^{-i\theta_k})^{-1}.$$

Show that ϕ is a well-defined analytic function on $\{z: |z|<1+\delta/2\}$, that $\operatorname{Re}\phi(z)>0$ for $|z|\leqslant 1+\delta/2$, $\phi(0)=1$ and $\operatorname{Re}\phi(e^{i\theta})\geqslant(3N\delta)^{-1}$ for $|\theta-\theta_j|\leqslant\delta/2$ $[1\leqslant j\leqslant n]$.

(iii) Define $\log: \mathbb{C}\backslash\{x: x\leqslant 0\}\to\mathbb{C}$ by $\log(re^{i\theta}\} = \log r + i\theta$ $[r>0, \pi>0>-\pi]$. Explain why $\psi=\log\phi$ is a well-defined analytic function on $\{z: |z|<1+\delta/2\}$ with $|\operatorname{Im}\psi(z)|<\pi$ for $|z|<1+\delta/2$, $\psi(0)=0$ and $|\psi(e^{i\theta})|\geqslant\log(3N\delta)^{-1}$ for $|\theta-\theta_j|\leqslant\delta/2$ $[1\leqslant j\leqslant N]$.

(iv) Explain why we can write $\psi(z)=\sum_{n=1}^{\infty}a_nz^n$ for $|z|<1+\delta/2$ and why we can find an M such that, if $\psi_M(z)=\sum_{n=1}^{M}a_nz^n$, then $|\psi_M(z)-\psi(z)|\leqslant\log 4/3$ for $|z|\leqslant 1$. Show that $|\operatorname{Im}\psi_M(z)|<2\pi$ for $|z|\leqslant 1$ and $|\psi_M(e^{i\theta})|\geqslant\log(4N\delta)^{-1}$ for $|\theta-\theta_j|\leqslant\delta/2$ $[1\leqslant j\leqslant N]$.

(v) By considering $\psi_M-\psi_M^*$, or otherwise, show that, if T is an integer with $T\geqslant 0$ and we write

$$P(\theta)=e^{i(T+M)\theta}\left(\sum_{n=1}^{M}a_ne^{in\theta}-\sum_{n=1}^{M}a_n^*e^{-in\theta}\right),$$

then P is a trigonometric polynomial with

(a) $\hat{P}(r)=0$ for $r\leqslant T-1$,
(b) $|P(\theta)|\leqslant 4\pi$ for all $\theta\in\mathbb{R}$,
(c) $|S_{T+M}(P,\theta)|\geqslant\log(4N\delta)^{-1}$ for $|\theta-\theta_j|\leqslant\delta/2$, $1\leqslant j\leqslant N$.

(vi) Conclude, using part (i) of Question 19.1, that, if $I_1, I_2, \ldots, I_m$ are intervals with $\sum_{r=1}^{m}|I_r|\leqslant\varepsilon$ and T is any positive integer, we can find a trigonometric polynomial P with

(a) $\hat{P}(r)=0$ for $r\leqslant T-1$,
(b) $|P(\theta)|\leqslant 4\pi$ for all $\theta\in\mathbb{R}$,
(c) $|S_{T+M}(P,\theta)|\geqslant\log(8\varepsilon)^{-1}$ for $\theta\in\bigcup_{r=1}^{m}I_r$.

19.3 (Theorem of Kahane and Katznelson) The remainder of the proof is entirely routine. Let E have measure zero.

(i) Pick $\delta(j)>0$ such that $\log(8\delta(j))^{-1}\geqslant 4^j$. Use Question 19.1 to show that we can find intervals $I(1), I(2), \ldots$ and integers $1=n(1)<n(2)<\ldots$ such that

(a) each point $e\in E$ belongs to infinitely many $I(k)$,
(b) $\sum_{k=n(j)}^{n(j+1)-1}|I(k)|<\delta(j)$.

(ii) Use Question 19.2 to show that we can find integers $0<T(1)<T(2)<\ldots$ and trigonometric polynomials P_j such that

(a) $\hat{P}_j(r)=0$ unless $T(j)\leqslant r<T(j+1)$,

(b) $|P_j(\theta)| \leq 4\pi$ for all $\theta \in \mathbb{R}$,
(c) $\max_{T(j+1)>n\geq T(j)} |S_n(P_j, \theta)| \geq 4^j$ for all $\theta \in \bigcup_{k=n(j)}^{n(j+1)-1} I_k$.

(iii) Show that $\sum_{j=1}^{\infty} 2^{-j} P_j$ converges uniformly to some continuous function f. Show carefully that $\hat{f}(r) = 2^{-j}\hat{P}_j(r)$ for $T(j) \leq r < T(j+1)$ and $\hat{f}(r) = 0$ for $r < T(1)$ and that $S_n(f, \theta) = \sum_{k=1}^{j-1} 2^{-k} P_k(\theta) + 2^{-j} S_n(P_j, \theta)$ for $T(j) \leq n < T(j+1)$.

(iv) Conclude that f is a continuous function with $\limsup_{n\to\infty} |S_n(f, \theta)| = \infty$ for all $\theta \in E$.

20

THE UNDISTURBED DAMPED OSCILLATOR DOES NOT EXPLODE

20.1 In this question and the succeeding ones we discuss the existence and uniqueness of solutions of differential equations.

(i) Let $F: \mathbb{R}^2 \to \mathbb{R}$ be a continuous function. Show that any solution of the problem:

(A) Find a continuously differentiable function $f: \mathbb{R} \to \mathbb{R}$ such that $f(0) = y_0$ and

$$f'(t) = F(t, f(t)) \quad \text{for all } t \in \mathbb{R}$$

is also a solution of the problem:

(B) Find a continuous function $f: \mathbb{R} \to \mathbb{R}$ such that

$$f(t) = y_0 + \int_0^t F(x, f(x))\,dx \quad \text{for all } t \in \mathbb{R},$$

and vice versa.

(ii) Write the following problems in form (B).

(a) Solve $(x^2 + y^2 + 1)\dfrac{dy}{dx} = y, \quad y(0) = 1.$

(b) Solve $\dfrac{dy}{dx} = \sin xy, \quad y(0) = -1.$

(iii) One way of trying to solve a problem of form (B) goes under the generic name of successive approximation. Let us start with some guess for f, say f_0. Now let us try to refine our guess inductively, feeding in our old guess to obtain our new one as

$$f_{n+1}(t) = y_0 + \int_0^t F(x, f_n(x))\,dx.$$

In the case when $F(t, s) = st$, $y_0 = 1$ and $f_0(t) = 1$, show that

$$f_1(t) = 1 + t^2/2,$$

$$f_2(t) = 1 + t^2/2 + t^4/(2^2.2!)$$

and obtain the general form of $f_n(t)$. Show that $f_n(t) \to \exp(t^2/2)$ as $n \to \infty$ and verify that the associated differential equation

$$f'(t) = tf(t), \quad f(0) = 1,$$

has $f(t) = \exp(t^2/2)$ as a solution.

If you have access to a computer and know how to integrate numerically, try a few randomly chosen examples. They should convince you that this method can work like a charm.

(iv) In the case when F satisfies the condition

$$|F(t, s_1) - F(t, s_2)| \leqslant K|s_1 - s_2| \quad \text{for all } t, s_1, s_2 \in \mathbb{R}, \tag{$*$}$$

for some K, it is easy to show that the method will always work as follows:

(a) Observe that, if $t > 0$,

$$f_{n+1}(t) - f_n(t) = \int_0^t F(x, f_n(x)) - F(x, f_{n-1}(x))\, dx$$

and hence, using $(*)$, that

$$|f_{n+1}(t) - f_n(t)| \leqslant K \int_0^1 |f_n(x) - f_{n-1}(x)|\, dx.$$

Thus prove inductively that

$$|f_{n-1}(t) - f_n(t)| \leqslant \frac{(Kt)^n}{n!} \sup_{0 \leqslant x \leqslant t} |f_1(x) - f_0(x)|.$$

(b) Conclude that, if $b > 0$,

$$\sup_{0 \leqslant x \leqslant b} |f_{n+1}(x) - f_n(x)| \leqslant \frac{(Kb)^n}{n!} \sup_{0 \leqslant x \leqslant b} |f_1(x) - f_0(x)|$$

and deduce that, if $m > n$,

$$\sup_{0 \leqslant x \leqslant b} |f_m(x) - f_n(x)| \leqslant \left(\sum_{r=n}^{m-1} \frac{(Kb)^r}{r!} \right) \sup_{0 \leqslant x \leqslant b} |f_1(x) - f_0(x)|.$$

Now use the general principle of uniform convergence to show that f_n converges uniformly to a function f on $[0, b]$.

(c) Explain why f is continuous and show that

$$\left| \int_0^t F(x, f_n(x)) - F(x, f(x))\, dx \right| \leqslant K \sup_{0 \leqslant x \leqslant b} |f_n(x) - f(x)|$$

for all $0 \leqslant t \leqslant b$. By letting $n \to \infty$ in the formula

$$f_{n+1}(t) = y_0 + \int_0^t F(x, f_n(x))\,dx,$$

show that

$$f(t) = y_0 + \int_0^t F(x, f(x))\,dx$$

for all $0 \leqslant t \leqslant b$.

(d) Extend the above result to show that, if $0 < b$, then f_n tends uniformly to a continuous function f with

$$f(t) = y_0 + \int_0^t F(x, f(x))\,dx \quad \text{for all } |t| < b.$$

(v) Suppose now that $g, h\colon \mathbb{R} \to \mathbb{R}$ are continuous functions with

$$g(t) = y_0 + \int_0^t F(x, g(x))\,dx,$$

$$h(t) = y_0 + \int_0^t F(x, h(x))\,dx.$$

By imiting the argument of (iv) (a), show that, if $t \geqslant 0$,

$$|g(t) - h(t)| \leqslant \frac{(Kt)^n}{n!} \sup_{0 \leqslant x \leqslant t} |g(x) - h(x)|$$

and hence by letting $n \to \infty$ that $g(t) = h(t)$ for all $t \geqslant 0$. Extend this result to show that $g(t) = h(t)$ for all $t \in \mathbb{R}$.

20.2 The condition (*) in Question 20.1 (iv) can be easily and usefully relaxed.

(i) Let $F\colon \mathbb{R}^2 \to \mathbb{R}$ and $K\colon \mathbb{R} \to \mathbb{R}$ be continuous functions such that

$$|F(t, s_1) - F(t, s_2)| \leqslant K(t)|s_1 - s_2| \quad \text{for all } t, s_1, s_2 \in \mathbb{R}. \qquad (*)$$

Recalling that a continuous function is bounded on any closed bounded interval, show that the equation

$$f(t) = y_0 + \int_0^t F(x, f(x))\,dx \quad [t \in \mathbb{R}]$$

with f continuous has exactly one solution and that, if f_0 is any continuous function and

$$f_{n+1}(t) = y_0 + \int_0^t F(x, f_n(x))\,dx,$$

then $f_n \to f$ uniformly on each interval $[-a, a]$.

(ii) Conditions of the type (*) are called Lipschitz conditions. They are closely related to, but slightly weaker than, differentiability conditions

(a) Use the mean value theorem to show that, if $F: \mathbb{R}^2 \to \mathbb{R}$ has partial derivative $F_2 = \partial F/\partial s$ satisfying $|F_2(t, s)| \leqslant K(t)$, then F satisfies $(*)$.

(b) If $F(t, s) = t|s|$, show that F satisfies $(*)$ with $K(t) = |t|$, but that F_2 does not exist everywhere.

(iii) Show that, if $\alpha, \beta: \mathbb{R} \to \mathbb{R}$ are continuous, the equations

$$y'(x) = \alpha(x)y(x) + \beta(x), \quad y(0) = y_0$$

have a unique solution. One use of this result is to allow us to define functions as solutions of differential equations as in the next question.

20.3 Let $e: \mathbb{R} \to \mathbb{R}$ be the unique solution of

$$e'(t) = e(t), \quad e(0) = 1.$$

(i) Show that, if a is fixed, $d(e(a - x)e(x))/dx = 0$ and deduce that $e(a - x)e(x) = e(a)$ for all $x \in \mathbb{R}$. By choosing a suitable value for a, deduce that $e(x)e(y) = e(x + y)$ for all $x, y \in \mathbb{R}$.

(ii) By taking $y = -x$ in (i), or otherwise, show that $e(x) \neq 0$ for all $x \in \mathbb{R}$ and hence using the intermediate value theorem, deduce that $e(x) > 0$ for all x.

(iii) By considering $d(e(x) - x)/dx$, or otherwise, show that $e(x) \geqslant x$ for all $x \geqslant 0$ and so that $e(x) \to \infty$ as $x \to \infty$. Show also that $e(x) \to 0$ as $x \to -\infty$ and deduce that

$$\{e(x): x \in \mathbb{R}\} = \{t \in \mathbb{R}: t > 0\}.$$

(iv) Show that e is strictly increasing on $\mathbb{R}$ and deduce that there is a unique well-defined function $l: (0, \infty) \to \mathbb{R}$ such that $l(e(x)) = x$ for all $x \in \mathbb{R}$ and $e(l(x)) = x$ for all $x \in (0, \infty)$.

(v) Show that l is differentiable with $l'(x) = 1/x$ and that l is strictly increasing.

(vi) Show also that $l(xy) = l(x) + l(y)$ for all $x, y \in (0, \infty)$.

20.4 (i) If $\boldsymbol{u} = (u_1, u_2, \ldots, u_m) \in \mathbb{R}^m$, let us write $\|\boldsymbol{u}\| = \sqrt{(\sum_{j=1}^m u_j^2)}$. Generalise the argument of Questions 20.1 and 20.2 to obtain the following result. Let $\mathbf{F}: \mathbb{R} \times \mathbb{R}^m \to \mathbb{R}^m$ and $K: \mathbb{R} \to \mathbb{R}$ be continuous functions such that

$$\|\mathbf{F}(t, \boldsymbol{u}) - \mathbf{F}(t, \boldsymbol{v})\| \leqslant K(t)\|\boldsymbol{u} - \boldsymbol{v}\| \text{ for all } t \in \mathbb{R},\ \boldsymbol{u}, \boldsymbol{v} \in \mathbb{R}^m. \qquad (*)$$

Then there is one and only one continuous solution $\mathbf{f}: \mathbb{R} \to \mathbb{R}^m$ to the equations

$$f_j(t) = y_{0j} + \int_0^t F_j(x, \mathbf{f}(x))\, dx \quad [t \in \mathbb{R}, 1 \leqslant j \leqslant m].$$

(This system can, of course, be written more suggestively as

$$\mathbf{f}(t) = \mathbf{y}_0 + \int_0^t \mathbf{F}(x, \mathbf{f}(x))\, dx \quad [t \in \mathbb{R}].)$$

Express the result in terms of differential equations.

(ii) We are interested in (i), not for the sake of generalisation, but for a particular special case. Show that, if $b, a_0, a_1, \ldots, a_{m-2}: \mathbb{R} \to \mathbb{R}$ are continuous functions, then, if the system of equations

$$\left.\begin{aligned} f'_1(t) &= f_2(t)\\ f'_2(t) &= f_3(t)\\ &\vdots\\ f'_{m-1}(t) &= f_m(t)\\ f'_m(t) &= -\sum_{j=1}^{m-1} a_{j-1}(t)f_j(t) + b(t) \end{aligned}\right\}$$

with $\qquad f_j(\theta) = y_{j-1} \quad \text{for } 1 \leqslant j \leqslant m$

has a continuously differentiable solution $\mathbf{f}$, it follows that the equation

$$\left.\begin{aligned} f^{(m-1)}(t) + \sum_{j=0}^{m-2} a_j(t) f^{(j)}(t) &= b(t)\\ \text{with} \qquad f^{(j-1)}(0) &= y_{j-1} \quad [1 \leqslant j \leqslant m] \end{aligned}\right\}$$

has an $(m-1)$ times continuously differentiable solution f with $f^{(j)}(t) = f_{j+1}(t)$ $[0 \leqslant j \leqslant m-1]$ and vice versa.

(iii) Use (i) and (ii) to show that, if $b, a_0, a_1, \ldots, a_{m-2}: \mathbb{R} \to \mathbb{R}$ are continuous functions, then the differential equation

$$\left.\begin{aligned} f^{(m-1)}(t) + \sum_{j=0}^{m-2} a_j(t) f^{(j)}(t) &= b(t)\\ \text{with} \qquad f^{(j)}(0) &= y_j \quad [0 \leqslant j \leqslant m-1] \end{aligned}\right\}$$

has a unique continuously $(m-1)$ times differentiable solution.

20.5 (i) Use Question 20.4 to show that there are unique twice continuously differentiable functions $s, c: \mathbb{R} \to \mathbb{R}$ such that

$$s''(x) + s(x) = 0, \; s(0) = 0, \; s'(0) = 1$$
$$c''(x) + c(x) = 0, \; c(0) = 0, \; c'(0) = 0.$$

Show that s' satisfies the same equations as c and so $s'(x) = c(x)$. Show similarly that $c'(x) = -s(x)$, that $c(x) = c(-x)$ and $s(x) = -s(-x)$.

(ii) Show by imitating Question 20.3(i), or otherwise, that

$$s(x + y) = c(x)s(y) + s(x)c(y),$$
$$c(x + y) = c(x)c(y) - s(x)c(y),$$

$$c(x)^2+s(x)^2=1 \quad \text{for all } x,\, y\in\mathbb{R}.$$

(iii) Let $f(x)=s(x)-x+x^3/3!$. Show that $f'''(x)\geqslant 0$ for all x, and hence deduce that $f(x)\geqslant 0$ for all $x\geqslant 0$ i.e. $s(x)\geqslant x-x^3/3!$.

(iv) Hence show that $s(x)\neq 0$ for $0<x<6^{1/2}$. Deduce that if, $c(x)=c(y)=0$ and $x\neq y$, then $|x-y|\geqslant 6^{1/2}$.

(v) Show that

$$1-\frac{x^2}{2!}+\frac{x^4}{4!}\geqslant c(x)\geqslant 1-\frac{x^2}{2!}$$

for all $x\in\mathbb{R}$. Deduce that $c(1)>0>c(2)$ and thus that there is a unique $0<\omega<2$ such that $c(\omega)=0$.

(vi) Show that $c(x)=s(x+\omega)$.

(vii) Show that c is strictly decreasing from 1 to -1 on $[0,2\omega]$ and strictly increasing from -1 to 1 on $[2\omega,4\omega]$.

(viii) Deduce that if $(x,y)\in\mathbb{R}^2$ and $(x,y)\neq(0,0)$, there exist a unique $r\in\mathbb{R}$ with $r>0$ and a unique $\theta\in[0,4\omega)$ with $x=rc(\theta)$, $y=rs(\theta)$.

(ix) Generalise the inequality of the first sentence of (v) to higher powers of x. Deduce that

$$\left|c(x)-\sum_{r=0}^{n}(-1)^r\frac{x^{2r}}{(2r)!}\right|\leqslant\frac{x^{2n+2}}{(2n+2)!}$$

for all x and conclude that

$$c(x)=\sum_{r=0}^{\infty}(-1)^r\frac{x^{2r}}{(2r)!}\quad\text{for all } x.$$

Is the convergence uniform and why? (If you have easy access to a computer, it is instructive to graph $\sum_{r=0}^{n}(-1)^r x^{2r}/(2r)!$ for $n=1,2,3,4,5$.)

(x) Obtain the formula $s(x)=\sum_{r=0}^{\infty}(-1)^r x^{2r+1}/(2r+1)!$

(a) by imitating the argument of (ix),
(b) by differentiating the corresponding result of (ix) term by term (why can you do this?),
(c) by using the Taylor's theorem of 4.2(ii).

(The chain of argument ending in (viii) defines the angle θ made by the line joining $(0,0)$ to (x,y) with the positive real axis in terms of analysis. The formula 'sine = opposite/hypotenuse' becomes a lemma rather than a definition. Before the middle of the nineteenth century 'angle' was considered a primary quantity whose meaning was obvious to all except the general run of school children and the occasional deliberately obtuse philospher. Nowadays we replace 'all' by 'nobody' in the previous sentence.

In a very bleak moment of mathematical history, when the Jewish mathematician Landau had been driven out of Göttingen by a Nazi inspired student boycott, Bierberbach, himself a distinguished mathematician, justified this 'manly rejection' of 'teachers of alien type' by citing Landau's 'un-German' approach to the definition of sine, cosine and π. Landau's treatment is close to the one given here.)

20.6 (i) Suppose $F:\mathbb{R}^2 \to \mathbb{R}$ is a continuous function satisfying

$$|F(t, s_1) - F(t, s_2)| \leqslant K|s_1 - s_2| \quad \text{for all } t, s_1, s_2 \in \mathbb{R}. \qquad (*)$$

Suppose $f_0(t) = y_0$,

$$f_{n+1}(t) = y_0 + \int_0^t F(x, f_n(x))\, dx,$$

and $g_0(t) = y_1$

$$g_{n+1}(t) = y_1 + \int_0^t F(x, g_n(x))\, dx.$$

Show that, if $t \geqslant 0$, then

$$|f_{n+1}(t) - f_{n+1}(t)| \leqslant |y_0 - y_1| + Kt \sup_{0 \leqslant x \leqslant t} |f_n(x) - g_n(x)|$$

and deduce that

$$|f_n(t) - g_n(t)| \leqslant |y_0 - y_1| \sum_{r=0}^{n} (Kt)^r/r!.$$

Hence show that, if f, g are continuously differentiable functions with

$$f'(t) = F(t, f(t)), \quad f(0) = y_0,$$
$$g'(t) = G(t, g(t)), \quad g(0) = y_1,$$

then $|f(t) - g(t)| \leqslant |y_0 - y_1|e^{K|t|}$ for $t \in \mathbb{R}$.
(In other words initial errors give rise to errors which grow no faster than exponential.)

(ii) Suppose $F, G: \mathbb{R}^2 \to \mathbb{R}$ are continuous functions satisfying

$$|F(t, s_1) - F(t, s_2)| \leqslant K|s_1 - s_2|, \quad |G(t, s_1) - G(t, s_2)| \leqslant K|s_1 - s_2| \quad (*)$$

for all $t, s_1, s_2 \in \mathbb{R}$, and suppose further that $|F(t, s) - G(t, s)| \leqslant \varepsilon$ for all $t, s \in \mathbb{R}$. Let

$$f(t) = y_0 + \int_0^t F(x, f(x))\, dx,$$

$$g(t) = y_0 + \int_0^t G(x, g(x))\, dx.$$

Set $f_0(t) = g(t)$ and

$$f_{n+1}(t) = y_0 + \int_0^t F(x, f(x))\,dx$$

By using the formulae obtained in Question 20.1 (iii), or otherwise, show that

$$|f(t) - g(t)| \leqslant \sup_{0 \leqslant x \leqslant t} |f_1(t) - f_0(t)| \exp(Kt)$$

$$\leqslant \varepsilon t \exp(Kt)$$

for all $t \geqslant 0$. Write down a similar result for $t \leqslant 0$. (In other words errors due to inaccurate knowledge of F grow no faster than exponential.)

(iii) Suppose F satisfies the condition of (i) and in addition

$$|F(t, s)| \leqslant K(1 + |t|) \quad \text{for all } s, t \in \mathbb{R}.$$

Show, by taking $f_0(t) = y_0$, or otherwise, that, if

$$f(t) = y_0 + \int_0^t F(x, f(x))\,dx$$

then

$$|f(t) - y_0| \leqslant K(t + t^2/2) \exp(Kt) \quad \text{for all } t \geqslant 0.$$

(iv) Suppose that $\alpha, \beta: \mathbb{R} \to \mathbb{R}$ are continuous and $|\alpha(x)| \leqslant A(1 + x)$, $|\beta(x)| \leqslant A \exp Lx$ for $x \geqslant 0$. Show that, if y is a solution of

$$y'(x) = \alpha(x)y(x) + \beta(x), \quad y(0) = y_0$$

and $w(x) = \exp(Lx)y(x)$, then w is the solution of

$$w'(x) = \alpha_0(x)y(x) + \beta_0(x), \quad w(0) = y_0$$

with $|\alpha_0(x)|$, $|\beta_0(x)| \leqslant A_0$ for all $x \geqslant 0$ and some constant A_0. Deduce from (iii) that

$$|w(x) - y_0| \leqslant A_0(x + x^2/2) \exp(A_0 x)$$

and hence that

$$|y(x) - y_0| \leqslant A_1 \exp(A_2 x)$$

for all $x \geqslant 0$ and some constants A_1, A_2.

(v) Generalise (iv) to the context of Question 20.4 to show that, if all coefficients $a_j(t)$ in the equation

$$f^{(m)}(t) + \sum_{j=0}^{m-1} a_j(t) f^{(j)}(t) = b(t)$$

are of (at most) polynomial growth and $b(t)$ is of (at most) exponential growth the solutions are of (at most) exponential growth.

(This is important because it tells us that the Laplace transform methods of Chapters 74 to 78 can be applied to a large variety of differential equations. Note, however, that the Laplace transform of a nasty equation may be, and often is, a still nastier equation.)

(vi) Can the words 'polynomial growth' be replaced by 'exponential growth' in (v)? Give a proof or a counter-example.

20.7 As might be expected, the same ideas which are used to discuss the existence of solutions for differential equations reappear when we seek those solutions numerically. Suppose $F: \mathbb{R}^2 \to \mathbb{R}$ is an n times continuously differentiable function such that

$$|F(t, s_1) - F(t, s_2)| \leqslant K|s_1 - s_2| \quad \text{for all } t, s_1, s_2 \in \mathbb{R},$$

and suppose that there exists an M_1 such that

$$|F_{p,q}(t, s)| \leqslant M_1 \quad \text{for all } t, s \in \mathbb{R};\ p, q \in \mathbb{Z};\ p, q \geqslant 0;\ p + q \leqslant n.$$

(Here, as usual, $F_{p,q}(t, s) = \partial^{p+q} F(t, s)/\partial t^p \partial s^q$.) We wish to solve the equation

$$f'(t) = F(t, f(t)) \tag{*}$$

numerically. To be more specific, suppose we wish to find $f(a)$ for some $a > 0$, given $f(0)$.

(i) If $f'(t) = F(t, f(t))$ for all t, show that

$$f'(t) = G_1(t, f(t)), \quad f''(t) = G_2(t, f(t)), \quad f'''(t) = G_3(t, f(t))$$

where

$$G_1(x, y) = F_{0,0}(x, y)$$

$$G_2(x, y) = F_{1,0}(x, y) + F_{0,0}(x, y)F_{0,1}(x, y)$$

$$\begin{aligned} G_3(x, y) = {} & F_{2,0}(x, y) + 2F_{0,0}(x, y)F_{1,1}(x, y) + F_{1,0}(x, y)F_{0,1}(x, y) \\ & + F_{0,0}(x, y)(F_{0,1}(x, y))^2 + (F_{0,0}(x, y))^2 F_{0,2}(x, y). \end{aligned}$$

Show, more generally, that f is $(n + 1)$ times continuously differentiable and that, if $1 \leqslant r \leqslant n + 1$, then

$$f^{(r)}(t) = G_r(t, f(t))$$

where $G_r(x, y)$ is a multinomial involving only those $F_{p,q}(x, y)$ with $0 \leqslant p, q, (p + q) \leqslant r$. Show, in particular, that there exists an M such that

$$|f^{(n+1)}(t)| \leqslant M \quad \text{for all } t \in \mathbb{R}.$$

(ii) We shall consider the following scheme for estimating $f(a)$, given $f(0)$.

Choose an integer $N \geqslant 1$ and set $h = a/N$. Set $Y(0) = f(0)$ and define $Y(mh)$ inductively by

$$Y((m+1)h) = Y(mh) + \sum_{r=1}^{n} G_r(mh, Y(mh))\frac{h^r}{r!}.$$

Explain why you would expect $Y(mh)$ to remain close to $f(mh)$ for $m = 0, 1, \ldots, N$ and so $Y(a)$ to be close to $f(a)$.

(iii) We now seek to estimate the accuracy of our answer. Our method is high on notation (it may be helpful to draw diagrams), but contains an interesting basic idea. Let f_m be the exact solution of

$$f'_m(t) = F(t, f_m(t)) \quad \text{with } f_m(mh) = Y(mh).$$

Express $f_0(a)$ and $f_N(a)$ in terms of $Y(a)$ and $f(a)$. By using part (i) of Question 20.6 show that

$$|f_{m+1}(a) - f_m(a)| \leqslant |Y((m+1)h) - f_m((m+1)h)|e^{K(N-m-1)h}.$$

By using a simple form of Taylor's theorem (see Question 4.2 (iii) if necessary) show that

$$|Y((m+1)h) - f_m((m+1))h| \leqslant Mh^{n+1}/(n+1)!$$

so that

$$|f_{m+1}(a) - f_m(a)| \leqslant Mh^{n+1}e^{K(N-m-1)h}/(n+1)!.$$

Hence, by summing, show that

$$|Y(a) - f(a)| \leqslant \frac{Mh^{n+1}}{(n+1)!}\frac{e^{Ka}-1}{e^{Kh}-1}.$$

(Thus we have obtain a 'global' estimate for the error $|Y(a) - f(a)|$ in terms of a 'local' estimate for the error $(Y((m+1)h) - f_m((m+1)h)$.) Deduce that, if $Kh < 1/2$, then

$$|Y(a) - f(a)| \leqslant M_2 e^{Ka} h^n$$

for some constant M_2 independent of a and h.
(If we take $n = 1$ we obtain Euler's method

$$Y((m+1)h) = Y(m, h) + F(mh, Y(mh))h.$$

A major problem with the process given above is that, if $n \geqslant 2$, we need to be able to find the derivatives of F analytically. It is also the case, as we point out from time to time in these exercises (see e.g. Questions 45.1 to 45.5), that expansion in Taylor's series may not be the best way to use the information that a function f looks locally like a polynomial of low degree. Thus, although the 'Taylor method' is not an obviously bad method, it is rarely used in practice.)

21

THE DISTURBED DAMPED LINEAR OSCILLATOR DOES NOT EXPLODE

21.1 Consider a damped pendulum governed by the equation

$$\ddot{\theta} = -b \sin \theta - a\dot{\theta} \tag{*}$$

with a and b strictly positive constants.

(i) By constructing a suitable energy function show that $\dot{\theta}(t) \to 0$ as $t \to \infty$ and that there exists an integer n such that $\theta(t) \to n\pi$.

(ii) Give physical reasons why you would expect n to be even.

(This question is continued as Question 72.1.)

21.2 The following piece of algebra is often useful when discussing stability and other matters. Consider the quadratic form $Q: \mathbb{R}^n \to \mathbb{R}$ given by $Q(\mathbf{x}) = \sum_{i=1}^{n}\sum_{j=1}^{n} x_i a_{ij} x_j$ with $a_{ij} = a_{ji}[1 \leq i < j \leq n]$ and $a_{ij} \in \mathbb{R}[1 \leq i, j \leq n]$. We say that Q is positive definite, if $Q(\mathbf{x}) > 0$ whenever $\mathbf{x} \neq 0$.

(i) Show that, if Q is positive definite, then $a_{11} > 0$.

(ii) Show that, if $a_{11} > 0$, then

$$Q(\mathbf{x}) = a_{11}y_1^2 + a_{11}^{-1} \sum_{i=2}^{n}\sum_{j=2}^{n} x_i A_{ij} x_j$$

with $y_1 = x_1 + \sum_{j=2}^{n} a_{11}^{-1} a_{i1} x_j$ and $A_{ij} = a_{11}a_{i1} - a_{il}a_{1j}$.

(iii) Deduce that Q is positive definite if and only if both $a_{11} > 0$ and Q_1 is positive definite where $Q_1: \mathbb{R}^{n-1} \to \mathbb{R}$ is the quadratic form given by

$$Q(x_2, x_3, \ldots, x_n) = \sum_{i=2}^{n}\sum_{j=2}^{n} x_i A_{ij} x_j.$$

(iv) By induction, or otherwise, show that Q is positive definite if and only if

$$a_{11} > 0, \quad \det\begin{pmatrix} a_{11} & a_{12} \\ a_{21} & a_{22} \end{pmatrix} > 0, \quad \det\begin{pmatrix} a_{11} & a_{12} & a_{13} \\ a_{21} & a_{22} & a_{23} \\ a_{31} & a_{32} & a_{33} \end{pmatrix} > 0, \ldots \text{ etc.}$$

(v) It is a key result of multidimensional calculus (well concealed, though by different means, both in most old fashioned and many modern treatments) that, if $f: \mathbb{R}^n \to \mathbb{R}$ is sufficiently well behaved, then

$$f(\mathbf{x}_0 + \mathbf{h}) = f(\mathbf{x}_0) + \sum_{i=1}^{n} \frac{\partial f}{\partial x_i} h_i + \sum_{i=1}^{n} \sum_{j=1}^{n} h_i \frac{\partial^2 f}{\partial x_i \partial x_j} h_j + \varepsilon(\mathbf{h}) \sum_{i=1}^{n} h_i^2$$

with $\varepsilon(\mathbf{h}) \to 0$ as $\sum_{i=1}^{n} h_i^2 \to 0$ [cf. Exercise 4.2 (v)]. Give a sufficient condition along the lines of (iv) for f to have a minimum at $\mathbf{x}_0$. Explain briefly why it is not necessary.

(vi) By considering $f(x, y) = x^2 - 4xy + y^2$ show that the conditions $\partial f(0,0)/\partial x = \partial f(0,0)/\partial y = 0$, $\partial^2 f(0,0)/\partial x^2 > 0$, $\partial^2 f(0,0)/\partial y^2 > 0$ do not guarantee that $(0, 0)$ is a minimum.

21.3 (i) Consider the system of equations

$$x' = -x + f(z), \quad y' = -y + f(z), \quad z' = x + y - kf(z)$$

with $f: \mathbb{R} \to \mathbb{R}$ a continuous function. Suppose that

(a) $zf(z) > 0$ for $z \neq 0$,
(b) $k > 2$,
(c) $\int_0^\infty f(z)\,dz = \int_0^{-\infty} f(z)\,dz = \infty$.

By considering $\frac{1}{2}x^2 + \frac{1}{2}y^2 + \int_0^z f(t)\,dt$, or otherwise, show that

$$(x(t), y(t), z(t)) \to (0, 0, 0) \quad \text{as } t \to \infty.$$

(You may find Question 21.2 (iv) helpful.)

(ii) Show that, even if $f(z) = z$, the conclusion of (i) may fail if $k \leqslant 2$.

(iii) Show that the conclusion of (i) may fail if hypothesis (a) is omitted.

21.4 (i) Consider the system of differential equations

$$\dot{x} = x(1 - 4x^2 - y^2) - \tfrac{1}{2}y(1 + x)$$

$$\dot{y} = y(1 - 4x^2 - y^2) - 2x(1 + x).$$

By considering $V(x, y) = (1 - 4x^2 - y^2)^2$, or otherwise, show that either $(x(t), y(t)) = (0, 0)$ for all t or

$$(1 - 4x(t)^2 - y(t)^2)^2 \to 0 \quad \text{as } t \to \infty.$$

(ii) What happens as $t \to -\infty$? (There are three different possible types of behaviour.)

(iii) Show that there exists a $\delta > 0$ such that $\dot{x}^2 + \dot{y}^2 > \delta$ whenever $4x^2 + y^2 = 1$. Conclude that, if $4x(0)^2 + y(0)^2 = 1$, then $(x(t), y(t))$ cycles round the ellipse given by $4x^2 + y^2 = 1$.

23
THE LINEAR DAMPED OSCILLATOR WITH PERIODIC INPUT

23.1 (This pleasant computational exercise is a lightly modified version of a 1971 Tripos question.)

(i) A machine of mass m is cushioned from its foundations by an elastic spring and a viscous dashpot in parallel, so that the linearised equation of vertical motion $q(t)$ under an external force $F\cos pt$ is

$$m\ddot{q} + f\dot{q} + kq = F\cos pt$$

where $k = m\omega^2$ and $f > 0$. We wish to find the amplitude L of the load transmitted to the foundations disregarding transients. Explain why this can be done by seeking a solution Ae^{ipt} of the equation

$$m\ddot{q} + f\dot{q} + kq = Fe^{ipt}$$

and setting $L = \mathrm{Re}\,(Ae^{ipt})$ What is the physical meaning of $\arg A$?

(ii) Sketch the relation between L/F and p/ω for an arbitrary value of $f/m\omega$.

(iii) Deduce that $|L|$ can be kept less than F if and only if the spring is such that $\omega\sqrt{2}$ is below the working range of p.

(iv) Show also that in this range the damping coefficient f must be at least $m\omega\sqrt{(3/2)}$ if the amplitude of forced vibration of the machine is not to exceed one half its static displacement under a steady force F.

23.2 (This is based on a 1981 Tripos equation but since it looks so offputting I have added some hints.) Consider the following simplified model of the harpsichord. For each note there are two strings, each of which can be considered as a one-dimensional harmonic oscillator of frequency ω. However, the frame produces a resistive acceleration on each oscillator equal to k times the sum of the velocities of the oscillators. Thus the equations governing the displacements $x(t)$, $y(t)$ of the oscillators are

$$\ddot{x} + k(\dot{x} + \dot{y}) + \omega^2 x = 0, \quad \ddot{y} + k(\dot{x} + \dot{y}) + \omega^2 y = 0.$$

We shall suppose $k < \omega$.

(i) Find the equations for $X(t) = x(t) + y(t)$ and $Y(t) = x(t) - y(t)$.

(ii) In normal use, when a note is struck, the oscillators are excited equally (i.e. $x(0) = y(0)$, $\dot{x}(0) = \dot{y}(0)$). Show that $E(t) = \frac{1}{2}(\dot{x}(t)^2 + \dot{y}(t)^2)$ decays to zero at least as fast as $C_1 \exp(-2kt)$ where C_1 is some constant depending on the initial conditions. This is the usual percussive effect.

(iii) In order to sustain the note a pedal was added to arrange that only one string would be excited (i.e. $y(0) = \dot{y}(0) = 0$, say). How does E behave now?

(iv) It was finally discovered that by mistuning the strings, say to $\omega(1 \pm \varepsilon)$ with ε small, so that the equations become

$$\ddot{x} + k(\dot{x} + \dot{y}) + w^2(1 + \varepsilon)^2 x = 0, \quad \ddot{y} + k(\dot{x} + \dot{y}) + w^2(1 - \varepsilon)^2 y = 0,$$

a small controlled decay could be effected. Show that, if we seek solutions of the form $X(t) = \operatorname{Re}(X_0 e^{\Omega t})$ and $Y(t) = \operatorname{Re}(Y_0 e^{\Omega t})$, then Ω may take one of four possible values $\Omega_1, \Omega_2, \Omega_3, \Omega_4$, say, but do not seek to calculate these for the moment. Write down the general solution for X and Y.

(v) Show that two of the values of Ω_i $[i = 1, 2, 3, 4]$ correspond to rapidly decaying solutions and two to slowly decaying solutions. Show that, for t reasonably large, $E(t)$ decays to zero like $C_2 \exp(-2w^2\varepsilon^2 t/k)$ where C_2 is some constant depending on the initial conditions.

23.3 (The following is taken from a little book of home experiments entitled *Take Nobody's Word For It* by Auckland and Coates.) Stand two chairs back to back about 1 metre apart and tie a length of string between them. Tie two 0.5 metre-long pieces of string to two similar-sized potatoes and fix these to the string between the chairs in such a way that the potatoes are between 15 cm and 30 cm apart. Start one potato swinging like a pendulum in a direction parallel to the backs of the chairs. After a few seconds the other potato will start to swing and eventually take over the motion whilst the first potato stops dead. The process then goes into reverse.

A simple model of the process is given by

$$\ddot{x} + wx = f(y), \quad \ddot{y} + wy = f(x),$$

where x is the displacement of the first pendulum, y of the second and f represents the influence of one pendulum on the other. By considering the case of rest, $x(t) = y(t) = 0$ for all t, it is clear that $f(0) = 0$. If $f(u)$ is well behaved, its variation will be approximately linear for small changes in u. We thus replace our model by the still simpler one

$$\ddot{x} + \omega^2 x = \varepsilon y, \quad \ddot{y} + \omega^2 y = \varepsilon x.$$

Solve the equations exactly for the initial conditions $x(0) = A$, $y(0) = \dot{x}(0) = \dot{y}(0) = 0$. By using the addition formula for cosines, or otherwise, show that, if ε is small, the model exhibits the behaviour described at the end of the first paragraph. Roughly how long does it take the system to return to state where y hardly moves from 0 during a period of length $2\pi/\omega$?

In an even more spectacular version of the above experiment (which I, however, have never seen) a pendulum bob is suspended from an appropriately chosen spring. A vertical pull sets up vertical oscillation which slowly dies away as the bob begins to swing like a pendulum. The pendulum motion then dies away to be replaced by vertical oscillation and so on. Explain this.

27

POISSON SUMMATION

27.1 Substantiate the remarks at the end of Chapter 27 by using Theorem 27.1 to prove Theorems 2.4, 2.5 and 9.1.

27.2 Poisson summation is concerned with the behaviour $\sum_{n=-\infty}^{\infty} b_n r^{|n|} \exp in\theta$ as $r \to 1-$. If we are only concerned with behaviour at a single point θ, there is no loss in generality in taking $\theta = 0$. We then find ourselves dealing with the behaviour of $\sum_{n=0}^{\infty} a_n r^n$ as $r \to 1-$. If we write $s_n = \sum_{r=0}^{n} a_r$, we are then concerned with the behaviour of $\sum_{n=0}^{\infty}(1-r)r^n s_n$ as $r \to 1-$. Show that:

(A) if $s_n \to s$ as $n \to \infty$, then $\sum_{n=0}^{\infty}(1-r)r^n s_n \to s$ as $r \to 1-$;
(B) if $\sum_{n=0}^{\infty}(1-r)r^n s_n \to s$ as $r \to 1-$, then $\sum_{n=0}^{\infty}(1-r)r^n s_{n+1} \to s$ as $r \to 1-$;
(C) if s_n is real and non negative and $\sum_{n=0}^{\infty}(1-r)r^n s_n \to s$ as $r \to 1-$, then $s \geqslant 0$.

(These results link up with Question 1.11, but they can be readily obtained without using any of the questions on Chapter 1.)

27.3 Conclusion (A) of Question 27.2 shows that, if $\sum_{n=0}^{\infty} a_n$ converges, then $\sum_{n=0}^{\infty} a_n r^n \to \sum_{n=0}^{\infty} a_n$ as $r \to 1-$. Abel proved this result and then used it to prove a result which I still find surprising and which we obtain in part (iii) of this question.

Suppose $a_n, b_n \in \mathbb{C}$. We write $c_n = \sum_{r=0}^{n} a_{n-r} b_r$.

(i) Show that, if $a_n = b_n = (-1)^n (n+1)^{-1/2}$, then $\sum_{r=0}^{\infty} a_n$ and $\sum_{r=0}^{\infty} b_n$ converge but $\sum_{n=0}^{\infty} c_n$ diverges.

(ii) If $\sum_{n=0}^{\infty} a_n$, $\sum_{n=0}^{\infty} b_n$ converge, explain, quoting carefully any theorems that you use, why $\sum_{n=0}^{\infty} a_n z^n$, $\sum_{n=0}^{\infty} b_n z^n$ and $\sum_{n=0}^{\infty} c_n z^n$ exist with $\sum_{n=0}^{\infty} a_n z^n \sum_{n=0}^{\infty} b_n z^n = \sum_{n=0}^{\infty} c_n z^n$ for all $|z| < 1$.

(iii) Hence prove, using the result of the first paragraph, that, if $\sum_{n=0}^{\infty} a_n$, $\sum_{n=0}^{\infty} b_n$ and $\sum_{n=0}^{\infty} c_n$ all exist, then $\sum_{n=0}^{\infty} a_n \sum_{n=0}^{\infty} b_n = \sum_{n=0}^{\infty} c_n$.

27.4 Let us establish some notation. Let (a_n) be a bounded sequence of complex numbers. We write

$$s_n = \sum_{j=0}^{n} a_j,$$

$$\sigma_n = (n+1)^{-1}\sum_{j=0}^{n}(n+1-j)a_j,$$

$$A(r) = \sum_{j=0}^{n} a_j r^j \quad [0 \leqslant r < 1].$$

If $s_n \to s$ as $n \to \infty$, we say that (a_j) has sum s. If $\sigma_n \to \sigma$ as $n \to \infty$, we say that (a_j) has Cesàro sum σ. If $A(r) \to A$, we say that (a_j) has Abel sum A. We know from Lemma 1.4 that, if (a_j) has a sum s, then it has Cesàro sum s and that there exists a sequence which has a Cesàro sum but no sum. In this question we shall establish that, if (a_j) has a Cesàro sum σ, then it has Abel sum σ. We also observe that there exists a sequence which has an Abel sum but no Cesàro sum. (More briefly we say that Abel summation is strictly stronger than Cesàro summation.)

(i) By differentiating a power series term by term within its circle of convergence, or otherwise, evaluate $\sum_{j=0}^{\infty}(-1)^j(2j+1)r^{2j}$ and hence show that, if we set $b_{2j} = (-1)^j(2j+1)$, $b_{2j+1} = 0$ for $j \geqslant 0$, then (b_j) has Abel sum 0 but has no Cesàro sum.

(ii) Let (a_j) be a bounded sequence. By multiplying term by term within the circle of convergence, show that

$$(1-r)^{-2}A(r) = \sum_{j=0}^{\infty}(j+1)r^j \sum_{k=0}^{\infty} a_k r^k = \sum_{n=0}^{\infty}(n+1)\sigma_n r^n, \text{ and so}$$

$$A(r) = (1-r)^2 \sum_{n=0}^{\infty}(n+1)\sigma_n r^n \quad \text{for } 0 \leqslant r < 1.$$

Now observe, that if $\alpha \in \mathbb{C}$, $\alpha = (1-r)^2\sum_{n=0}^{\infty}(n+1)\alpha r^n$ and so

$$A(r) - \alpha = (1-r)^2 \sum_{n=0}^{\infty}(n+1)r^n(\sigma_n - \alpha) \quad \text{for } 0 \leqslant r < 1.$$

(iii) Suppose now that (a_j) has a Cesàro sum σ. Then, given $\varepsilon > 0$, we can find an M such that $|\sigma_n - \sigma| < \varepsilon$ for $n \geqslant M$. Taking $\alpha = \sigma$, show that

$$|A(r) - \sigma| \leqslant (1-r)^2 \sum_{n=0}^{M}(n+1)r^n|\sigma_n - \sigma| + (1-r)^2 \sum_{n=M+1}^{\infty}(n+1)r^n\varepsilon$$

$$\leqslant (1-r)^2 \sum_{n=0}^{M}(n+1)r^n|\sigma_n - \sigma| + \varepsilon \to \varepsilon \quad \text{as } r \to 1-.$$

Deduce that (a_j) has Abel sum σ as required.

27.5 (i) By trivial modifications of Theorem 15.3 show that, if (a_j) has a Cesàro sum and $a_j = O(j^{-1})$, then it has a sum.

A theorem which states that, if (a_j) has a sum in some sense and then a_j satisfy some further condition, then (a_j) has a sum in the ordinary sense is called a Tauberian theorem. Although Tauber was the first to prove such theorems, the first deep results along these lines are associated with Hardy and Littlewood. One of these will be the subject of the rest of the question.

(ii) In view of the final remarks for Chapter 27, it is natural to ask whether we can replace the words Cesàro sum by Abel sum in part (i). In view of Question 27.3, this must be harder to prove, but in fact the answer is yes. The proof that follows originated with Karamata. It echoes other proofs in this book in an interesting manner.

From now on we assume that

$$A(r) = \sum_{j=0}^{\infty} a_j r^j \to A \quad r \to 1-$$

and $|a_j| \leqslant Kj^{-1}$ for $j \geqslant 1$.

Prove first that

$$A(r^n) \to A \quad \text{as } r \to 1-,$$

and deduce that, if P is any polynomial with $P(0) = 0$,

$$\sum_{j=0}^{\infty} a_j P(r^j) \to AP(1).$$

(iii) It is natural to hope that we might have $\sum_{j=0}^{\infty} a_j \phi(r^j) \to A\phi(1)$ for any continuous function ϕ with $\phi(0) = 0$, but this is not the case.

Suppose, for example, that $a_j = (-1)^j (j(\log j)^{1/2})^{-1}$ for $j \geqslant 2$, $a_0 = a_1 = 0$. (Thus $j^{-1}a_j \to 0$ as $j \to \infty$ and $\sum_{j=0}^{\infty} a_j$ converges.) Choose primes $p(1) < p(2) < p(3) < \dots$ and integers $N(1)$, $N(2)$, $N(3), \dots$ such that $p(n)^{-1/N(n)} \to 1$ as $n \to \infty$. Show that there is a well-defined continuous function such that $\phi(0) = 0$, $\phi(t) = 0$ for $t \geqslant 1/2$, and $\phi(p(n)^{-k/N(n)}) = (-1)^k (\log k)^{-1/2}$ for $k \geqslant 10nN(n)$. Show that $\sum_{j=0}^{\infty} a_j \phi(p(n)^{-j/N(n)})$ diverges. In a similar way construct a continuous function ψ such that $\psi(t) = 0$ for $t \leqslant 1/2$, $\psi(1) = 0$ and yet $\sum_{j=0}^{\infty} a_j \psi(p(n)^{-j/N(n)}) \to \infty$ as $n \to \infty$.

(iv) None the less, we can obtain results on the behaviour of $\sum_{j=0}^{\infty} a_j \phi(r^j)$ as $r \to 1-$ provided that ϕ satisfies conditions which restrict its oscillation near 0 and 1. Suppose $|\phi(r)| \leqslant \varepsilon r(1 - r)$ for all $0 \leqslant r \leqslant 1$ and some $\varepsilon > 0$. Show that $1 - r^n \leqslant n(1 - r)$ for $0 \leqslant r \leqslant 1$ and deduce that

$$\sum_{j=0}^{\infty} |a_j \phi(r^j)| \leqslant K\varepsilon \quad \text{for all } 0 \leqslant r \leqslant 1.$$

(v) By applying the Weierstrass polynomial approximation theorem

(Theorem 4.3) together with parts (ii) and (iv) show that, if there exists a continuous function $f: [0, 1] \to \mathbb{C}$ with $\phi(r) = r(1-r)f(r)$ for $0 \leqslant r \leqslant 1$, then $\sum_{j=0}^{\infty} a_j \phi(r^j) \to 0$ as $r \to 1-$.

(vi) Deduce that, if $\phi: [0, 1] \to \mathbb{C}$ is continuous, has $\phi(0) = \phi(1) = 0$, is (right) differentiable at 0 and (left) differentiable at 1, then

$$\sum_{j=0}^{\infty} a_j \phi(r^j) \to 0 \quad r \to 1-.$$

By using part (ii) again, show that, if $\phi: [0, 1] \to \mathbb{C}$ is continuous and is (right) differentiable at 0 and (left) differentiable at 1 and $\phi(0) = 0$, then

$$\sum_{j=0}^{\infty} a_j \phi(r^j) \to A\phi(1) \quad \text{as } r \to 1-.$$

(vii) From now on the proof follows a familiar pattern (compare Step 6 of the proof of Theorem 3.1 and the method of Theorem 15.3). We start by observing that although good end-point behaviour is crucial, the behaviour of ϕ elsewhere is less important.

Let $1/2 > \delta > 0$ and $\psi: [0, 1] \to \mathbb{C}$ satisfy the conditions $\psi(t) = 0$ for $|t - 1/2| > \delta$, $|\psi(t)| \leqslant 1$ for all t. For any fixed $0 < r < 1$ write

$$B = \{j: \psi(r^j) \neq 0\}.$$

Observe that if $j \in B$ then $|a_j| \leqslant K\,|\log r / \log(\delta + 1/2)|$ and by estimating the number of points in B show that

$$\sum_{j=0}^{\infty} |a_j \psi(r^j)| \leqslant K(\log(1/2 - \delta) - \log(\delta + 1/2) - 1)/\log(\delta + 1/2).$$

(viii) Using (vi) and (vii), show that, if $\chi: [0, 1] \to \mathbb{R}$ is the step function given by $\chi(t) = 0$ for $0 \leqslant t < 1/2$, $\chi(t) = 1$ for $1/2 \leqslant t \leqslant 1$, then

$$\sum_{j=0}^{\infty} a_j \chi(r^j) \to A \quad \text{as } r \to 1-.$$

Conclude, by rewriting this formula, that

$$\sum_{j=0}^{m} a_j \to A \quad \text{as } m \to \infty$$

as required.

(ix) Show that the hypothesis is $\phi(0) = 0$ may be removed in the conclusion to part (vi).

27.6 (i) Use Question 27.5 and Theorem 27.1 to establish that, if $g: \mathbb{T} \to \mathbb{C}$ is continuous and $\hat{g}(n) = O(|n|^{-1})$ as $|n| \to \infty$, then $\sum_{j=-n}^{n} \hat{g}(j) \exp(ij\theta) \to g(\theta)$ as $n \to \infty$ for each $\theta \in \mathbb{T}$.

(ii) By modifying Theorem 27.1, show that, if $g: \mathbb{T} \to \mathbb{R}$ is Riemann integrable and $\hat{g}(n) = O(|n|^{-1})$ as $|n| \to \infty$, then $\sum_{j=-n}^{n} \hat{g}(j) \exp ij\theta \to g(0)$ as $n \to \infty$ at any point θ where g is continuous.

(iii) By strengthening the result of Question 27.5 in an appropriate way show that under the hypotheses of part (i), in fact $\sum_{j=-n}^{n} \hat{g}(j) \exp(ij\theta) \to g(\theta)$ uniformly on $\mathbb{T}$ as $n \to \infty$.

(Thus both parts of Theorem 15.3 can be obtained using the ideas of Chapter 27. The work involved is substantial, but if we were prepared to use slightly stronger hypotheses such as $n^{-1}\hat{g}(n) \to 0$ as $|n| \to \infty$, then the work would be much reduced.)

27.7 Poisson summation can be used to provide a slight improvement on the standard maximum principle for a disc. Suppose $f: \{z: |z| < 1\} \to \mathbb{C}$ is an analytic function with the following properties.

(a) There exists an M_1 such that $|f(z)| \leq M_1$ for all $|z| < 1$.

(b) We can find $\theta_1, \theta_2, \ldots, \theta_n$ such that $f(re^{i\theta})$ converges uniformly to $g(\theta)$, say, as $r \to 1-$ on each set of the form

$$\{\theta: |\theta - \theta_j| > \delta \text{ for } j = 1 \ldots n\} \; [\delta > 0].$$

(c) $|g(\theta)| \leq M$ for $\theta \notin \{\theta_1, \theta_2, \ldots, \theta_n\}$.

Then $|f(z)| \leq M$ for all $|z| < 1$.

Prove this results as follows.

(i) Explain why we can write $f(z) = \sum_{n=0}^{\infty} a_n z^n$ for $|z| < 1$. Explain why, if we define $g(\theta_j)$ arbitrarily (e.g. $g(\theta_j) = 0$), the function $g: \mathbb{T} \to \mathbb{C}$ is continuous except possibly at $\theta_1, \theta_2, \ldots, \theta_n$ and Riemann integrable.

(ii) Show carefully that

$$a_n r^n = \frac{1}{2\pi} \int_{\mathbb{T}} f(re^{i\theta}) e^{-in\theta} \, d\theta \to \frac{1}{2\pi} \int_{\mathbb{T}} g(\theta) e^{-in\theta} \, d\theta = \hat{g}(n)$$

as $r \to 1-$ and deduce that $\hat{g}(n) = a_n$ for $n \geq 0$. Show that $\hat{g}(n) = 0$ for $n < 0$.

(iii) Show that $f(re^{i\theta}) = (1/2\pi) \int_{\mathbb{T}} g(\theta - t) P_r(t) \, dt$ and use condition (c) and Lemma 27.3 (iii) to conclude that $|f(z)| \leq M$ for all $|z| < 1$.

27.8 The result of the previous question may not seem very impressive, but it does allow us to obtain a version of the maximum principle for certain unbounded regions in $\mathbb{C}$. Suppose $R = \{z: a < \operatorname{Re} z < b\}$, $\bar{R} = \{z: a \leq \operatorname{Re} z \leq b\}$. Let $f: \bar{R} \to \mathbb{C}$ be a continuous function on $\bar{R}$ which is analytic on R. Suppose that $|f(z)| \leq M_1$ for all $z \in \bar{R}$ and $|f(z)| \leq M$ for all z with $\operatorname{Re} z = a$ or $\operatorname{Re} z = b$.

(i) Construct a conformal map $F: R \to \{z: |z| < 1\}$.

(ii) Extend F in an obvious way to give a continuous bijective map $F: \bar{R} \to \{z: |z| \leq 1, z \neq z_1, z_2\}$.

(iii) Explain why f is uniformly continuous on $\{z: a \leq \operatorname{Re} z \leq b, |\operatorname{Im} z| \leq c\}$ for each c.

(iv) Set $z_1 = \exp i\theta_1$, $z_2 = \exp i\theta_2$. Show that $f \circ F^{-1}$ satisfies the hypotheses of Question 27.1 and conclude that $|f \circ F^{-1}(z)| \leqslant M$ for all $|z| < 1$. Deduce that $|f(z)| \leqslant M$ for all $z \in \bar{R}$.

(v) Is it true that, if $g: \bar{R} \to \mathbb{C}$ is continuous, g analytic on R and $|g(z)| \leqslant M$ when $\operatorname{Re} z = a$ or $\operatorname{Re} z = b$, that $|g(z)| \leqslant M$ for all $z \in \bar{R}$? (Give a proof or a counter example. You may find it helpful to consider the same question with R replaced by $H = \{z: \operatorname{Im} z > 0\}$.)

(We shall use the main result of this question in Question 34.2.)

27.9 (i) Lemma 27.2 tells us that, if $0 \leqslant r < 1$,

$$\text{(a)} \quad \frac{1 - r^2}{1 - 2r\cos\theta + r^2} = 1 + 2r\cos\theta + 2r^2\cos 2\theta + \ldots.$$

Show that, if $0 \leqslant r < 1$,

$$\text{(b)} \quad \frac{\sin\theta}{1 - 2r\cos\theta + r^2} = \sin\theta + r\sin 2\theta + r^2\sin 3\theta + \ldots.$$

By equating coefficients of r^n, or otherwise, use (b) to show that

$$\frac{\sin(n+1)\theta}{\sin\theta} = \sum_{0 \leqslant 2k \leqslant n} \frac{(-1)^k(n-k)!}{k!(n-2k)!}(2\cos\theta)^{n-2k}.$$

Give a proof along these lines that the Tchebychev polynomial, defined as in Chapter 5, by $\cos n\theta = T_n(\cos\theta)$ $[\theta \in \mathbb{T}]$, is a polynomial of degree n with $T_n(t) = \sum_{r=0}^{n} a_r t^r$, say. Obtain explicit expressions for the a_r.

(ii) By integrating (b) appropriately show that

$$\int_0^\pi \log(1 - 2r\cos\theta + r^2)\,d\theta = 0 \quad \text{if } |r| < 1.$$

Show that

$$\int_0^\pi \log(1 - 2r\cos\theta + r^2)\,d\theta = \pi\log r^2 \quad \text{if } |r| > 1.$$

Are either of these results obvious from general considerations? (This last question is a bit open ended but you should be able to say something if you have a reasonable background in complex variable or two-dimensional potential theory.)

28

DIRICHLET'S PROBLEM FOR THE DISC

28.1 (i) Show that, if we use polar coordinates, Laplace's equation $\nabla^2\phi = 0$ in two dimensions becomes

$$\frac{1}{r}\frac{\partial}{\partial r}\left(\frac{r\partial\phi}{\partial r}(r,\theta)\right) + \frac{1}{r^2}\frac{\partial^2\phi}{\partial\theta^2}(r,\theta) = 0.$$

(ii) Find the most general solution of the form $\phi(r,\theta) = e^{in\theta}f(r)$ to Laplace's equation valid in the punctured plane $\mathbb{R}^2\setminus\{0\} = \{(r,\theta): r > 0\}$ when $n \in \mathbb{Z}$. How many arbitrary constants does it contain? Why does it not make sense to seek solutions of the form $\phi(r,\theta) = e^{i\omega\theta}f(r)$ when ω is not an integer?

(iii) Suppose $0 < a < b$. Find a solution for the following set of conditions. The function $\phi: \{(r,\theta): a \leqslant r \leqslant b\} \to \mathbb{R}$ is continuous and is twice continuously differentiable with $\nabla^2\phi = 0$ on $\{(r,\theta): a < r < b\}$, $\phi(a,\theta) = |\theta|$ for $|\theta| \leqslant \pi$, $\phi(b,\theta) = 0$ for all θ.

(iv) More generally, if $G: \{(a,\theta): \theta \in \mathbb{R}\} \cup \{(b,\theta): \theta \in \mathbb{R}\} \to \mathbb{R}$ is continuous, can we always find $\phi: \{(r,\theta): a \leqslant r \leqslant b\} \to \mathbb{R}$ such that

(a) ϕ_{11} and ϕ_{22} exist with $\phi_{11} + \phi_{22} = 0$ at all points (r,θ) with $a < r < b$,

(b) ϕ is continuous at all points (r,θ) with $a \leqslant r \leqslant b$,

(c) $\phi = G$ at all points (r,θ) with $r = a$ or $r = b$?

You should support your answer with a rigorous proof.

29

POTENTIAL THEORY WITH SMOOTHNESS ASSUMPTIONS

29.1 (i) (The Neumann problem) Suppose Ω is a bounded open region in $\mathbb{R}^2$ with smooth boundary $\partial\Omega$ and $G\colon \partial\Omega \to \mathbb{R}$ is a well-behaved function. The Neumann problem asks for a function $\phi\colon \bar{\Omega} \to \mathbb{R}$ satisfying

(a) $\nabla^2\phi = 0$ on Ω,

(b) $\dfrac{\partial\phi}{\partial n} = G$ on $\partial\Omega$.

Discuss the extent of the uniqueness and possible existence of such ϕ in the manner of Theorem 29.3 and Lemmas 29.4 and 29.5. You should assume good behaviour throughout.

(ii) Let Ω be as in (i) and for each k suppose we have well-behaved fuction $\phi_k\colon \bar{\Omega} \to \mathbb{R}$ satisfying

(a) $\nabla^2\phi_k = k$ on Ω,

(b) $\phi_k = F$ on $\partial\Omega$

when $F\colon \partial\Omega \to \mathbb{R}$ is a well-behaved function. Show that

$$\int_\Omega (\nabla\phi_k)^2 - (\nabla\phi_0)^2 dA = -k\int_\Omega (\phi_k - \phi_0)dA$$

$$= \int (\nabla(\phi_k - \phi_0))^2 dA.$$

[Here, as usual, $(\nabla\phi)^2 = (\partial\phi/\partial x)^2 + (\partial\phi/\partial y)^2$.]

29.2 We work in polar coordinates (r, θ). If $k > 0$ and $0 < a < b$ show that the set of equations for $\phi(r, \theta)$

$$\nabla^2\phi - \frac{k^2}{r^2}\phi = 0 \quad (a < r < b)$$

$$\phi(a, \theta) = f(\theta)$$

$$\phi(b, \theta) = g(\theta)$$

can have at most one well-behaved solution.

By guesswork, along the lines of Question 28.1, or otherwise, find ϕ when

$$f(\theta) = -\cos 2\theta,$$
$$g(\theta) = \cos 2\theta.$$

29.3$^{(-)}$ (The Neumann problem again) Suppose, just as in Question 29.1, that Ω is a bounded open region in $\mathbb{R}^2$ with smooth boundary $\partial\Omega$ and $G: \partial\Omega \to \mathbb{R}$ is a well-behaved function. Suppose further that $F: \bar{\Omega} \to \mathbb{R}$ is a well-behaved function. Show that if $\phi: \bar{\Omega} \to \mathbb{R}$ satisfies

(a) $\nabla^2\phi = F$ on Ω,

(b) $\dfrac{\partial\phi}{\partial n} = G$ on $\partial\Omega$,

then

$$\iint_\Omega F dA = \int_{\partial\Omega} G ds. \qquad (*)$$

Comment on Question 29.1(i) in the light of this result.

(The reader may feel that it was unfair to place Question 29.3 after Question 29.1(i). If she carries away a belief that the Neumann problem is, in some sense, harder than the Dirichlet problem the effect would be unfortunate. *Provided the consistency condition corresponding to (*) is satisfied, and provided that we only demand uniqueness up to adding a constant* the Neumann problem runs entirely in parallel with the Dirichlet problem, at least for well behaved boundary conditions. The next question illustrates this.)

29.4 (This follows on from Questions 29.1 and 29.3.) Throughout this question we work with well-behaved functions associated with the unit disc $D = \{(x, y): x^2 + y^2 < 1\}$ in $\mathbb{R}^2$ but the arguments go through if D is replaced by any well-behaved (in particular, simply connected) region.

(i) Let ϕ, $\psi: \bar{D} \to \mathbb{R}$ be well-behaved functions such that $f(x + iy) = \phi(x, y) + i\psi(x, y)$ defines an analytic function on $\{z \in \mathbb{C}: |z| < 1\}$. Show that, on the boundary ∂D,

$$\frac{\partial\phi}{\partial n} = \frac{\partial\psi}{\partial s}, \quad \frac{\partial\psi}{\partial n} = -\frac{\partial\phi}{\partial s}$$

where $\partial/\partial n$ denotes differentiation in the direction of the external normal and $\partial/\partial s$ denotes differentiation in the counterclockwise tangential direction.

(The notion of 'good behaviour' is an elastic one. I suggest that you assume that f can be extended to an analytic function on $\{z \in \mathbb{C}: |z| < 1 + \varepsilon\}$ for some $\varepsilon > 0$. The result then reduces to the statement that the Cauchy–Riemann equations are rotation invariant. Weaker hypotheses produce the same result but Chapter 30 shows that some restriction is required.)

(ii) Let $H: \partial D \to \mathbb{R}$ be a well-behaved function satisfying the consistency condition (see Question 29.3)

$$\int_{\partial D} H\,ds = 0. \tag{*}$$

Suppose that we wish to solve the Neumann problem of finding ψ: $\bar{D} \to \mathbb{R}$ such that

(a) $\nabla^2 \psi = 0$ on D,

(b) $\dfrac{\partial \psi}{\partial n} = H$ on ∂D.

Choose an arc length parameterisation of ∂D and define

$$G(s) = \int_0^s H(\sigma)\,d\sigma$$

(thus G is the line integral of H along an arc of ∂D of length s starting at some fixed point). Explain why we need condition (*) to ensure that G is well-defined and continuous on all of ∂D. Show that, if we can solve the Dirichlet problem,

(a′) $\nabla^2 \phi = 0$ on D,
(b′) $\phi = G$ on ∂D

then we can solve our Neumann problem.

(iii) Show similarly how to use solutions of the Neumann problem to solve the Dirichlet problem.

29.5 (This continues Questions 29.4) (i) Let D, H and G be as in Question 29.4(ii). Set $h(\theta) = H(\cos\theta, \sin\theta)$. Assuming that H is well behaved, use the method of Question 29.4(ii) together with the solution of the Dirichlet problem for the disc given in Theorem 28.5 to obtain a solution for the Neumann problem

(a) $\nabla^2 \psi = 0$ on D,

(b) $\dfrac{\partial \psi}{\partial n} = H$ on ∂D

as an infinite series involving $\hat{h}(n)$.

(ii) Verify rigorously that, provided H is continuous, but with no further conditions on H, the ψ you have found in part (i) satisfies the following conditions:

(a) ψ is defined and continuous on $\bar{D}$,
(b) ψ is infinitely differentiable and $\nabla^2\psi = 0$ at all points of D,
(c) if $\tau = \partial\psi/\partial r$ on D and $\tau = H$ on ∂D then τ is continuous on $\bar{D}$.

30
AN EXAMPLE OF HADAMARD

30.1 Suppose $F(a, b) \to \mathbb{R}$ is continuous. We write

$$D_2F(x) = \lim_{h\to 0} \frac{F(x+h) - 2F(x) + F(x-h)}{h^2}$$

whenever the right hand side exists. (We call D_2F the Schwarz (or the Riemann) derivative.) In this question we seek to prove that, if $D_2F(x)$ exists and $D_2F(x) = 0$ for all $x \in (a, b)$, then F is linear. The resemblance of our proof to those of Theorems 31.2 and 31.3 is in no way accidental.

(i) Suppose $F: (a, b) \to \mathbb{R}$ is a continuous function such that $D_2F(x)$ exists for all $x \in (a, b)$. Show that, if $a < t < v < s < b$ and $F(v) \leqslant F(x)$ for all $x \in (t, s)$ (i.e. v is a local minimum), then $D_2F(v) \geqslant 0$.

(ii) Suppose $F: (a, b) \to \mathbb{R}$ is a continuous function such that $D_2F(x)$ exists and $D_2F(x) \geqslant 0$ for all $x \in (a, b)$. Show that, if $F(t_1) = F(s_1)$ and $a < t_1 < s_1 < b$, then $F(y) \leqslant F(t_1)\ \forall\ y \in [t_1, s_1]$. By considering $F(x) + Ax$ with suitable A, or otherwise, show that, if $a < t < s < b$, then $F(\lambda s + (1-\lambda)t) \leqslant \lambda F(s) + (1-\lambda)F(t)$ for all $0 \leqslant \lambda \leqslant 1$ (i.e. F is convex).

(iii) By considering F and $-F$, or otherwise, show that if $D_2F(x)$ exists and $D_2F(x) = 0$ for all $x \in (a, b)$ then F is linear.

30.2 (Cantor's uniqueness theorem) In this book we start with a function f and ask, for example, whether $f(t) = \sum_{n=-\infty}^{\infty} \hat{f}(n) \exp(int)$. In the paper Riemann offered for his Habilitationsschrift (examination for a university teaching post), he turned this question on its head and asked what we can say if we start with a sequence a_n and consider the function (if any) given by

$$\sum_{n=-\infty}^{\infty} a_n \exp(inx) = f(x).$$

In this context we may ask, for example, whether a function f can have two different representations, i.e. whether we can find an f and two distinct sequences (a_n), (b_n) such that

$$\sum_{n=-\infty}^{\infty} a_n \exp(inx) = \sum_{n=-\infty}^{\infty} b_n \exp(inx) = f(x)$$

for all x. By subtraction, we see that this amounts to asking whether, if $\sum_{n=-N}^{N} c_n \exp inx \to 0$ as $N \to \infty$ for all $x \in \mathbb{T}$, then $c_n = 0$ for all n? This question, in its full generality, was answered by Cantor using ideas of Riemann.

(i) In this question we consider $\sum_{n=-N}^{N} a_n \exp inx$ as a 2π periodic function on $\mathbb{R}$ so all our functions will map $\mathbb{R}$ into $\mathbb{C}$. Suppose that $\sum_{n=-N}^{N} a_n \exp inx \to f(x)$ pointwise on $\mathbb{R}$. Explain why $a_n \to 0$ as $|n| \to \infty$. Riemann considers the formal doubly integrated series

$$F(x) = (a_0/2)x^2 - \sum_{n \neq 0} (a_n/n^2) \exp(inx).$$

Explain why F is a well-defined continuous function, and why $F(x) - (a_0/2)x^2$ is 2π periodic. Riemann shows that

$$\frac{F(x+h) - 2F(x) + F(x-h)}{h^2} \to f(x) \quad \text{as } h \to 0 \text{ for all } x \in \mathbb{R}.$$

(ii) Assuming the final result of (i) use Question 30.1 to show that, if $\sum_{n=-N}^{N} c_n \exp inx \to 0$ as $N \to \infty$ for all x, then $c_n = 0$ for all $n \in \mathbb{Z}$.

(iii) Explain why the final result of (i) will follow if we can show that

$$\frac{F(h) - 2F(0) + F(-h)}{h^2} \to f(0) \quad \text{as } h \to 0,$$

and this in turn will follow if we can show that $\sum_{n=1}^{N} a_n \to A$ as $N \to \infty$ implies

$$\sum_{n=1}^{\infty} a_n \left(\frac{\sin nh}{nh}\right)^2 \to A \quad \text{as } h \to 0.$$

(iv) If you have done the questions for chapter 1, you will recognise that this is a question on summation methods. If you have not, you will still, I hope, find the mode of attack a familiar one. Write $u(h) = (\sin h/h)^2$ $[h \neq 0]$. Explain why it is sufficient to show that

$$\sum_{n=1}^{\infty} a_n u(nh_j) \to A \quad j \to \infty$$

for any fixed sequence of $h_j \neq 0$ with $h_j \to 0$. Writing $s_n = \sum_{m=1}^{n} a_m$ show that

$$\sum_{n=1}^{\infty} a_n u(nh_j) = \sum_{n=1}^{\infty} s_n(u(nh_j) - u((n+1)h_j))$$

and so our problem reduces to showing that

$$\sum_{n=1}^{\infty} s_n(u(nh_j) - u((n+1)h_j)) \to A \quad \text{as } j \to \infty$$

whenever $s_n \to A$.

(v) Show that

(a) $u(nh_j) - u((n+1)h_j) \to 0$ as $n \to \infty$ for each fixed j,
(b) $\sum_{n=1}^{\infty}(u(nh_j) - u((n+1)h_j)) = u(h_j) \to 1$ as $j \to \infty$.

Convince yourself graphically that $\sum_{n=1}^{\infty}|(u(nh_j) - u((n+1)h_j)| < K_1$ for some constant K_1 and all j. Now show that $\int_0^{\infty}|u(t)|\,dt$ converges to K, say, and justify the statements in (c) below.

(c) $\sum_{n=1}^{\infty}|u(nh_j) - u((n+1)h_j| = \sum_{n=1}^{\infty}|\int_{nh_j}^{(n+1)h_j} u'(t)\,dt| \leqslant K$.

(vi) Using (a), (b) and (c), prove that

$$\sum_{n=1}^{\infty} s_n(u(nh_j) - u((n+1)h_j) \to A.$$

(If you have done Question 1.6 you may quote it. If not you should not find it hard to prove the result directly.) Cantor's theorem now follows.
(Cantor went on to examine a more general problem. If $\sum_{n=-N}^{N} c_n \exp inx \to 0$ for all x not in some particular set E, can we deduce that $c_n = 0$ for all n? Sets E, for which this is true, he called sets of uniqueness; the result above shows that the empty set $\varnothing$ is a set of uniqueness. His work led him through the consideration of sets E of increasing complexity to lay the foundations of modern set theory, analytic topology and measure theory.)

31

POTENTIAL THEORY WITHOUT SMOOTHNESS ASSUMPTIONS

31.1 The following chain of reasoning is due to Jensen.

(i) Write $D(0, r) = \{z: |z| < r\}$. Suppose $r > 1$ and $g: D(0, r) \to \mathbb{C}$ is analytic with no zeros. Show that $\log|g|: D(0, r) \to \mathbb{R}$ is harmonic. (You can do this directly, but the simplest approach is to observe that $\log|g|$ is locally the real part of some analytic function.)

(ii) Deduce that $\log|g(0)| = (1/2\pi)\int_0^{2\pi} \log|g(e^{i\theta})|\, d\theta$.

(iii) In order to study what happens when g has zeros, we study a particular function with a single zero. Let $M_a(z) = (a - z)/(1 - a^*z)$ if $|a| < 1$. Show that M_a is analytic in $D(0, |a|^{-1})$, $|M_a(e^{i\theta})| = 1$ for all $\theta \in \mathbb{R}$ and M_a has a single simple zero at a.

(iv) Suppose $r > 1$ and $f: D(0, r) \to \mathbb{C}$ is analytic with only a finite number of zeros at $a_1, \ldots, a_m$ say (multiple zeros repeated according to multiplicity) and $0 < |a_1|, |a_2|, \ldots, |a_m| < 1$. Show that $f(z) = g(z)\prod_{j=1}^m (z - a_j)/(1 - \bar{a}_j z)$ where $g: D(0, r) \to \mathbb{C}$ is analytic with no zeros. Use (ii) to show that

$$\log|f(0)| = \frac{1}{2\pi}\int_0^{2\pi} \log|f(e^{i\theta})|\,d\theta + \sum_{j=1}^{m} \log|a_j|.$$

(v) Suppose now $f: D(0, 1) \to \mathbb{C}$ is analytic with $f(0) \neq 0$ and that f has an infinite number of zeros. Show that f can have only a finite number of zeros in each $D(0, \rho)$ with $\rho < 1$. (Otherwise we could choose distinct zeros ω_j with $|\omega_j| < \rho$. We may extract a subsequence $\omega_{j(k)}$ with $\omega_{j(k)} \to \omega$ as $k \to \infty$ for some ω with $|\omega| \leq \rho$.) Thus we may write the zeros as $a_1, a_2, \ldots$ (multiple zeros repeated according to multiplicity) so that $|a_1| \leq |a_2| \leq \ldots$.

(vi) Show by rescaling (iv) that, if $|a_N| < R < |a_{N+1}|$,

$$\log|f(0)| = \frac{1}{2\pi}\int_0^{2\pi} \log|f(\mathrm{Re}^{i\theta})|\, d\theta + \sum_{j=1}^{N} \log(|a_j|/R).$$

(vii) Deduce that, if f is as in (v) and $|f(z)| \leq K$ for all $|z| < 1$, then

$$\log|f(0)| \leqslant \log K + \sum_{j=1}^{N} \log|a_j| \quad \text{for all } N$$

and so $\sum_{j=1}^{\infty} \log|a_j|$ converges.

(viii) Show (by writing $f(z) = z^n g(z)$ with $g(0) \neq 0$, or otherwise) that, if $f: D(0,1) \to \mathbb{C}$ is analytic and non constant and $|f(z)| \leqslant M$ for $|z| < 1$, then either f has only finitely many zeros or, taking its zeros (apart from those at the origin) as $a_1, a_2, \ldots$, we have $\sum_{j=1}^{\infty} -\log|a_j|$ convergent.

(ix) Show that, if x is real and $|x| \leqslant \frac{1}{4}$, then $|\log(1+x) - x| \leqslant |x|/2$. Conclude that if $f: D \to \mathbb{C}$ is a bounded analytic function with zeros at $a_1, a_2, \ldots$ (multiple zeros noted multiply) then $\sum_{j=1}^{\infty}(1 - |a_j|)$ converges.

31.2 (i) Let $\bar{D} = \{(x, y): x^2 + y^2 \leqslant 1\}$ be the closed disc in $\mathbb{R}^2$ and $\partial D = \{(x, y): x^2 + y^2 = 1\}$ its boundary. Suppose that $\psi: \bar{D} \to \mathbb{R}$ is continuous and that ψ is twice continuously differentiable on $D\backslash\{0\}$ with $\nabla^2\psi = 0$ on $D\backslash\{0\}$. Explain why we can find $\chi: \bar{D} \to \mathbb{R}$ which is continuous on $\bar{D}$, twice continuously differentiable on D with $\nabla^2\chi = 0$ on D and such that $\chi(x, y) = \psi(x, y)$ for all $(x, y) \in \partial D$. By applying the arguments of Example 31.2 to $\phi = \chi - \psi$, show that $\chi = \psi$ on $\bar{D}$ and deduce that ψ is twice continuously differentiable on D with $\nabla^2\psi = 0$ on D.

(ii) Let Ω be a non empty open subset of $\mathbb{R}^2$ and let $a \in \Omega$. Show that, if $\phi: \Omega \to \mathbb{R}$ is continuous and ϕ is twice continuously differentiable on $\Omega\backslash\{a\}$ with $\nabla^2\phi = 0$ on $\Omega\backslash\{a\}$ then, in fact, ϕ is twice continuously differentiable with $\nabla^2\phi = 0$ on Ω. (Compare the case of removable singularities in complex variable theory.)

(iii) Let $T: \mathbb{R}^2\backslash\{0\} \to \mathbb{R}^2\backslash\{0\}$ be the inversion map given by $T(x, y) = (x/(x^2 + y^2), y/(x^2 + y^2))$. If Ω is an open bounded set containing 0 and $\phi: \Omega \to \mathbb{R}$ satisfies Laplace's equation on Ω, show that $\phi \circ T^{-1}: T\Omega \to \mathbb{R}(x, y)$ tends to a limit as $x^2 + y^2 \to \infty$.

(iv) If T and Ω are as in (iii) and $\psi: T\Omega \to \mathbb{R}$ satisfies Laplace's equation on $T\Omega$ whilst $\psi(x, y)$ tends to a limit a, say, as $x^2 + y^2 \to \infty$, show that, if $\tilde{\psi}(x, y) = \psi \circ T(x, y)$ for $(x, y) \in \Omega\backslash\{0\}$ and $\tilde{\psi}(0, 0) = a$, then $\tilde{\psi}: \Omega \to \mathbb{R}$ satisfies Laplace's equation on Ω.

31.3 (The Lebesgue thorn) In this question we consider the following three-dimensional version of the Dirichlet problem. Let Ω be an open set in $\mathbb{R}^3$ such that $\Omega \supseteq \{\mathbf{x} \in \mathbb{R}^3: \|\mathbf{x}\| > K\}$ for some K and let $\partial\Omega$ be its boundary and $\bar{\Omega}$ its closure. Given a continuous function $G: \partial\Omega \to \mathbb{R}$, can we find $\phi: \bar{\Omega} \to \mathbb{R}$ such that ϕ is continuous on $\bar{\Omega}$, ϕ satisfies Laplace's equation on Ω and $\phi(\mathbf{x}) \to 0$ as $\|\mathbf{x}\| \to \infty$? In the past, physicists have indulged in some friendly teasing of mathematicians for worrying about such a problem since, if we imagine putting a charge density G on the surface $\partial\Omega$, then the resulting electric field will be derived from a potential ϕ with precisely the properties required. It is thus, they said, physically obvious that the Dirichlet problem must always have a solution.

In 1913 Lebesgue produced an example, which will be the subject of this question.

(i) Suppose Ω is as above and $\psi: \Omega \to \mathbb{R}$ satisfies Laplace's equation. Suppose further that $\psi(\mathbf{x}) \to 0$ as $\|\mathbf{x}\| \to \infty$ and that, given $\varepsilon > 0$, we can find a $\delta > 0$ such that, if $\mathbf{x} \in \Omega$ and if we can find a $\mathbf{y} \in \partial\Omega$ with $\|\mathbf{y} - \mathbf{x}\| < \delta$, then $\psi(\mathbf{x}) \leq \varepsilon$. Show by modifying the proof of Theorem 31.2, or otherwise, that $\psi(\mathbf{x}) \leq 0$ for all $\mathbf{x} \in \Omega$. Deduce a uniqueness result along the lines of Theorem 31.3.

(ii) Lebesgue considers the electric potential due to a charge of line density t at $(t, 0, 0)$ $[0 \leq t \leq 1]$. More formally, he considers

$$u(x, y, z) = \int_0^1 \frac{t}{\sqrt{((x - t)^2 + y^2 + z^2)}}\, dt.$$

Verify that u is well defined on $X = \mathbb{R}^2 \backslash \{t, 0, 0): 0 \leq t \leq 1\}$. By differentiating under the integral sign (see Theorem 53.4 if necessary), show that $\nabla^2 u$ is well defined and $\nabla^2 u = 0$ on X. Show also that $u(\mathbf{x}) > 0$ for all $\mathbf{x} \in X$.

(iii) Show that, if $0 < x \leq 1$, then $u(x, y, z) \to \infty$ as $y^2 + z^2 \to 0$, but that $u(0, \lambda y, \lambda z) \to 1$ as $\lambda \to 0$, provided $(y, z) \neq (0, 0)$.

(iv) Suppose $\eta > 0$. Consider the equation

$$u(x, y(x), 0) = 1 + \eta$$

where we seek $y(x) \geq 0$. Show that there exists an $x(\eta) > 1$ such that the equation has a solution if and only if $0 < x \leq x(\eta)$. Show that, if $0 < x \leq x(\eta)$, the equation has a unique (positive) solution $y(x) = f_\eta(x)$, say.

(v) Explain why $f_\eta: (0, x(\eta)] \to \mathbb{R}$ is a continuous function with $f_\eta(x(\eta)) = 0$, a horizontal tangent at 0 and a vertical tangent at $x(\eta)$ (i.e. $h^{-1} f_\eta(h) \to 0$, $h^{-1} f_\eta(x(\eta) - h) \to \infty$ as $h \to 0+$). Sketch f_η. (If you find yourself bogged down in rigour, simply explain roughly why you expect f_η to behave as stated and continue.)

(vi) Sketch the surface

$$\partial\Omega_\eta = \{(x, y, z) \in X: u(x, y, z) = 1 + \eta\} \cup \{0\}.$$

It should look like a top, with axis the x axis, balancing with a thin pointed end on the origin. We take

$$\Omega_\eta = \{(x, y, z) \in X: u(x, y, z) < 1 + \eta\}$$

$$\bar{\Omega}_\eta = \{(x, y, z) \in X: u(x, y, z) \leq 1 + \eta\} \cup \{0\}.$$

Check that Ω_η is open with boundary $\partial\Omega_\eta$ and closure $\bar{\Omega}_\eta$. Check that, if $\pi > \eta' > 0$, then $\partial\Omega_{\eta'} \subseteq \Omega_\eta \cup \{0\}$. Explain why we can find a K_η such that $\Omega_\eta \supseteq \{\mathbf{x}: \|\mathbf{x}\| \geq K_\eta\}$.

(vii) Now fix $\eta > 0$ and set $\Omega = \Omega_\eta$, $K = K_\eta$. We claim that, if $G(\mathbf{x}) = 1 + \eta$, then the Dirichlet problem stated in the first paragraph of this question has no solution. For suppose ϕ were such a solution. By uniform continuity, or otherwise, show that given $\tau > 0$ we can find a $\delta > 0$ such that, whenever $\mathbf{x} \in \Omega$ and there exists a $\mathbf{y} \in \partial\Omega$ with $\|\mathbf{x} - \mathbf{y}\| < \delta$, we have $\phi(\mathbf{x}) < 1 + \eta + \tau/k$. Now set $\psi(\mathbf{x}) = \phi(\mathbf{x}) - u(\mathbf{x}) - \tau\|\mathbf{x}\|^{-1}$. By applying (i), show that $\phi(\mathbf{x}) \leqslant 0$ for $\mathbf{x} \in \Omega$ and deduce that

$$\phi(\mathbf{x}) \leqslant u(\mathbf{x}) + \tau\|\mathbf{x}\|^{-1} \quad \text{for } \mathbf{x} \in \Omega.$$

Conclude that $\phi(\mathbf{x}) \leqslant u(\mathbf{x})$ for $\mathbf{x} \in \Omega$ and, by using a similar argument, show that $u(\mathbf{x}) \leqslant \phi(\mathbf{x})$ and so $\phi(\mathbf{x}) = u(\mathbf{x})$ for all $\mathbf{x} \in \Omega$.

Thus, if our Dirichlet problem has a solution ϕ, it is given by $\phi(\mathbf{x}) = u(\mathbf{x})$ for $\mathbf{x} \in \Omega$. Explain why $u(\mathbf{x}) \not\rightarrow 1 + \eta$ as $\mathbf{x} \rightarrow \mathbf{0}$ with $\mathbf{x} \in \Omega$ and conclude that our Dirichlet problem has no solution.

(It is well known that, if we charge a pointed object like a needle, sparks fly from the point. This is because near sharp points the potential field has very high gradients. Very sharp points cannot hold a charge and so it is physically obvious that the Dirichlet problem need have no solution for very cusped regions. It is easy to modify Lebesgue's thorn to obtain a bounded open set Ω in $\mathbb{R}^3$ with closure $\bar{\Omega}$ whose boundary $\partial\Omega$ is smooth apart from a simple cusp and yet such that there exists no continuous function $\phi: \bar{\Omega} \rightarrow \mathbb{R}$ with $\nabla^2\phi = 0$ on Ω and $\phi(\mathbf{x}) = 1$ for all $\mathbf{x} \in \partial\Omega$.)

31.4$^{(-)}$ In Question 31.6 we shall require a more general version of Lemma 31.7.

(i) Let $\Omega = \{(x, y): (x - x_0)^2 + (y - y_0)^2 < R^2\}$ and suppose $\phi: \Omega \rightarrow \mathbb{R}$ satisfies Laplace's equation on Ω. Show that, if we set

$$\psi(x, y) = \int_{l((x_0,y_0),(x,y))} (-\phi_2 dx + \phi_1 dy)$$

(the integral being the line integral from (x_0, y_0) to (x, y)), then $\psi(x_0, y_0) = 0$ and $f(x + iy) = \phi(x, y) + i\psi(x, y)$ defines an analytic function on $\Omega_{\mathbb{C}}$.

(ii) Let Ω be an open subset of $\mathbb{R}^2$ let $(x_0, y_0) \in \Omega$ and suppose ϕ satisfies Laplace's equation on Ω. Show that, if the line integral

$$\psi(x, y) = \int_C (-\phi_2 dx + \phi_1 dy)$$

with C a well-behaved curve in Ω running from (x_0, y_0) to (x, y) is independent of the choice of C for all $(x, y) \in \Omega$, then $\psi(x_0, y_0) = 0$ and $f(x + iy) = \phi(x, y) + i\psi(x, y)$ defines an analytic function on $\Omega_{\mathbb{C}}$.

(iii) Under which circumstances is it true that the integral $\int_C(-\phi_2 dx + \phi_1 dy)$ is path independent? Show that, if we can apply Stokes' theorem, then the fact that ϕ satisfies Laplace's equation implies that $\int_C(-\phi_2 dx + \phi_1 dy) = 0$ for any closed curve in C and so we have path independence. Stokes' theorem applies to 'regions without holes'. Anyone who has done a second course in complex analysis will know that the 'regions without holes' are precisely the simply connected open sets. In a first course in complex analysis, we often only deal with convex or star shaped regions and the proof of Cauchy's theorem given there will also prove Stokes' theorem.

To summarize: if Ω is an open convex (or more generally simply connected) subset of $\mathbb{R}^2$ and ϕ satisfies Laplace's equation on Ω, then we can find $\psi: \Omega \to \mathbb{R}$ such that $f(x+iy) = \phi(x,y) + i\psi(x,y)$ defines an analytic function on $\Omega_{\mathbb{C}}$.

31.5 (The Riemann mapping theorem) Riemann's dissertation, presented when he was 25, is one of the most remarkable papers ever written. It contains the germ of a major part of the modern theory of analytic functions, it initiated the systematic study of topology, it revolutionised algebraic geometry and paved the way for Riemann's later foundation of modern differential geometry. His paper contains no hard calculations or complex formulae, but the ideas he uses are so profound that it took mathematicians sixty years to understand them and give them rigorous form. Klein spoke of Riemann's work as 'The book with seven seals'.

The most immediately striking result in Riemann's dissertation (though, in fact, only one consequence of more general considerations) is that any well-behaved proper open subset Ω of $\mathbb{C}$ can be conformally mapped to the unit disc $D = \{z: |z| < 1\}$. We now know that this result is true for any simply connected open subset Ω of $\mathbb{C}$ with $\Omega \neq \emptyset, \mathbb{C}$ but I do not think we do violence to Riemann's chain of thought if we restrict ourselves to much simpler Ω. The reader may, for the purpose of argument, suppose Ω bounded and convex.

Riemann's theorem was particularly remarkable because at the time no one knew of any explicit conformal map $f: \Omega \to D$ even when Ω was a polygon. In addition Riemann's proof made use of Dirichlet's principle – that is, it assumed that Dirichlet's problem can always be solved by a minimising argument of the type discussed in the paragraph after Lemma 29.5. However Weierstrass pointed out that such an assumption might not be valid (and, as Zaremba's and Lebesgue's examples show in Example 31.1 and Question 31.3, he was right).

Klein writes '[Riemann] fully recognised the justice and correctness of Weierstrass's critique; but he said, as Weierstrass once told me: "that he appealed to the Dirichlet Principle only as a convenient tool and that his existence theorems are still correct". Weierstrass probably agreed with this opinion for he induced his student H. A. Schwarz to undertake a thorough study of Riemann's existence theorems and to seek other proofs of them' (Klein, *Development of Mathematics in the 19th Century*).

We will return to Riemann's idea in the next question but first we give some useful standard background which the reader may well have met before.

(i) (Schwarz's lemma) Suppose $f: D \to D$ is analytic and $f(0) = 0$. Show that

$g(z) = f(z)/z$ for $1 > |z| > 0$, $g(0) = f'(z)$ defines an analytic function $g: D \to \mathbb{C}$. By applying the maximum principle to $\{z: |z| \leq r\}$, show that $|g(z)| < r^{-1}$ for all $|z| < r$ and all $r < 1$. Deduce that $|g(z)| \leq 1$ for all $|z| < 1$ and, by a further application of the maximum principle, that, if $|g(z_0)| = 1$ for any $|z_0| < 1$, then $g(z) = g(z_0)$ for all $|z| < 1$. Conclude that either

(a) $|f(z)| < z$ for all $0 < |z| < 1$, and $|f'(0)| < 1$,

or

(b) $f(z) = e^{i\theta} z$ for all $|z| < 1$ and some $\theta \in \mathbb{R}$.

(ii) Suppose that $f: D \to D$ is conformal (so f is bijective and f and f^{-1} are analytic) and that $f(0) = 0$. By differentiating both sides of the relation

$$z = f^{-1}(f(z)),$$

substituting $z = 0$ and using (i), show that $f(z) = e^{i\theta} z$ for all $|z| < 1$ and some $\theta \in \mathbb{R}$. (Thus the only conformal maps of D to itself which fix 0 are the rotations.)

(iii) Suppose $|a| < 1$. Show that, if we define $T_a: D \to D$ by $T_a z = (a - z)/(1 - a^* z)$ for $|z| < 1$, then T_a is a conformal map with $T_a(a) = 0$, $T_a(0) = a$ and $T_a^{-1} = T_a$. (Question 31.1 (iii) may be helpful.)

(iv) Let $f: D \to D$ by a conformal map. Write $a = f^{-1}(0)$. By considering $f \circ T_a$, or otherwise, show that $f(z) = e^{i\theta}(a - z)/(1 - a^* z)$. Conclude that the conformal maps $f: D \to D$ are precisely given by the maps $f(z) = e^{i\theta}(a - z)/(1 - a^* z)$ with $|a| < 1$, $\theta \in \mathbb{R}$.

(v) Let Ω be an open subset of $\mathbb{C}$ and T, $S: \Omega \to D$ conformal maps. Show that $T \circ S^{-1}(z) = e^{i\theta}(a - z)/(1 - a^* z)$ for all $z \in D$ and some $|a| < 1$, $\theta \in \mathbb{R}$. (Thus conformal maps $T: \Omega \to D$, if they exist at all, are unique up to the choice of two constants a and θ.)

31.6 Riemann's idea was to cut out a piece of copper in the shape of Ω, to place a source of heat at a point $\mathbf{a} \in \Omega$ and to keep the boundary $\partial\Omega$ at a fixed temperature 0, say. Under steady conditions, where the temperature U is independent of t, the heat equation $\nabla^2 u = K\partial U/\partial t$ becomes $\nabla^2 U = 0$. Near the source of heat we expect the temperature to depend essentially only on the distance from $\mathbf{a}$ (i.e. a radially symmetric distribution) so that $U(\mathbf{x}) \approx \psi(|\mathbf{x} - \mathbf{a}|)$. The only non trivial radially symmetric solution of $\nabla^2 \psi = 0$ is $\psi(\mathbf{x}) = A \log \|\mathbf{x} - \mathbf{a}\|$ (see the proof of Plausible Lemma 87.2). We thus have

$$\nabla^2 U = 0 \text{ on } \Omega \backslash \{\mathbf{a}\}, \quad U = 0 \text{ on } \partial\Omega$$

and

$$U(\mathbf{x}) \approx A \log \|\mathbf{x} - \mathbf{a}\| \quad \text{for } \|\mathbf{x} - \mathbf{a}\| \text{ small.}$$

If we look at the lines of equal temperature (i.e. U constant) and the lines of heat flow (lines whose tangents are ∇U), we see that they form an orthogonal net with the flow lines running from $\mathbf{a}$ and the zero line of equal temperature coincident with $\partial\Omega$. If we now do exactly the same for the disc D with $\mathbf{a}$ replaced by $\mathbf{0}$, we get a similar orthogonal pattern of lines of latitude (equal temperature) and longitude (flow lines) with pole (source of flow lines) at $\mathbf{0}$ and equator (zero temperature) the boundary ∂D. Mapping the one set of lines of longitude and latitude onto the other, we get the required conformal map. (If all this is double Dutch to the reader, she should probably ignore it. However, many first texts on complex variables contain a chapter on applications to two-dimensional steady heat flow, rotational fluid flow or electric currents. See, for example, Markushevich *Theory of Functions of a Complex Variable*, Volume 2, Chapter 6. If my discussion makes some sort of vague sense to her, she should find that the modern construction which follows corresponds quite closely to the programme outlined.)

Our object is to show that, if Ω is a non empty bounded convex open subset of $\mathbb{C}$ such that Dirichlet's problem is soluble for Ω (i.e. if $\Omega_{\mathbb{R}} = \{(x, y): x + iy \in \Omega\}$ is such that, given $g: \partial\Omega_{\mathbb{R}} \to \mathbb{R}$ continuous, we can find $\phi: \bar{\Omega}_{\mathbb{R}} \to \mathbb{R}$ continuous, satisfying Laplace's equation $\nabla^2\phi = 0$ on $\Omega_{\mathbb{R}}$ and with $\phi = g$ on $\partial\Omega_{\mathbb{R}}$), then we can find an analytic bijective $f: \Omega \to \mathbb{R}$.

(i) Show that, if $0 \in \Omega$, we can find $u: \bar{\Omega}_{\mathbb{R}} \to \mathbb{R}$ continuous such that $u(x, y) = \log(x^2 + y^2)^{1/2}$ for $(x, y) \in \partial\Omega_{\mathbb{R}}$ and $\nabla^2 u = 0$ on $\Omega_{\mathbb{R}}$. Setting $U(x, y) = \log(x^2 + y^2)^{1/2} - u(x, y)$ verify that $U: \bar{\Omega}_{\mathbb{R}}\backslash\{0\} \to \mathbb{R}$ is continuous,

$$\nabla^2 U = 0 \text{ on } \Omega_{\mathbb{R}}\backslash\{0\}, \quad U = 0 \text{ on } \partial\Omega_{\mathbb{R}},$$

and

$$U(x, y)/\log(x^2 + y^2)^{1/2} \to 1 \quad \text{as } x^2 + y^2 \to 0.$$

Such a U is called a Green's function. Use the maximum principle to show that $U(x, y) < 0$ for all $(x, y) \in \Omega_{\mathbb{R}}\backslash\{0\}$.

(ii) Explain why we can find $V: \Omega_{\mathbb{R}} \to \mathbb{R}$ such that $g(x + iy) = U(x, y) + iV(x, y)$ defines analytic function $g: \Omega \to \mathbb{C}$. Explain why, if we choose any $\alpha \in \mathbb{R}$, the function $\log_\alpha: \{re^{i\theta} \in \Omega: 2\pi + \alpha > \theta > \alpha\} \to \mathbb{C}$ given by $\log_\alpha(re^{i\theta}) = \log r + i\theta$ $[r > 0, 2\pi + \alpha > \theta > \alpha]$, is analytic. Show that

$$f(z) = e^{\log|z| + i\arg z - g(z)} = ze^{-g(z)}$$

is a well-defined analytic function $f: \Omega \to \mathbb{C}$. Show that

(I) f has a simple zero at 0 and nowhere else,
(II) $|f(z)| = e^{U(x,y)} < 1$ for $z = x + iy \in \Omega\backslash\{0\}$,
(III) if $z_0 \in \partial\Omega$ then $|f(z)| \to 1$ as $z \to z_0$ $[z \in \Omega]$.

More generally, conclude that, if $a \in \Omega$, we can find $f_a: \Omega \to D$ analytic such that

(I)$_a$ $f_a(z)$ has a simple zero at a and nowhere else,
(II)$_a$ if $z_0 \in \partial\Omega$ then $|f_a(z)| \to 1$ as $z \to z_0$ $[z \in \Omega]$.

We still have to show that f_a is injective and surjective and this is the purpose of the rest of the question.

(iii) If $a, b \in \Omega$ and $a \neq b$, set

$$f(z, a, b) = \frac{f_a(z) - f_a(b)}{1 - f_a(b)^* f_a(z)} \quad [z \in \Omega].$$

On the assumption that f_a and f_b are, in fact, conformal, use Question 31.5 to give a relation between $f(z, a, b)$ and $f_b(z)$. Of course, we do not yet know that f_a and f_b are conformal, so we proceed as follows. Show that $|f(z, a, b)| < 1$ for $z \in \Omega$ and use $(\mathrm{II})_b$ and the maximum principle to show that

$$\left|\frac{f(z, a, b)}{f_b(z)}\right| \leqslant 1 \quad \text{for all } z \in \Omega$$

with equality for all $z \in \Omega$ if there is equality at one point. By setting $z = a$, deduce that $|f_a(b)| \leqslant |f_b(a)|$, use symmetry to obtain $|f_a(b)| = |f_b(a)|$ and hence show that the inequality of the previous sentence is in fact an equality, i.e. $|f(z, a, b)| = |f_b(z)|$ for all $z \in \Omega$.

(iv) Use the final result of (iii), together with $(\mathrm{I})_b$, to show that $f_a(z) = f_a(b)$ has exactly one solution for each $b \in \Omega$, i.e. that f_a is injective.

(v) The proof that f_a is surjective is similar but considerably simpler. Suppose f_a is not surjective, so we can find a w with $|w| < 1$ but $f_a(z) \neq w$ for $z \neq w$. By applying the maximum principle to

$$h(z) = \frac{f_a(z) - w}{1 - w^* f_a(z)}$$

and to $1/h(z)$, show that $|h(z)| = 1$ for all $z \in \Omega$. By choosing a suitable value of z derive a contradiction.

(In his *Men of Mathematics*, Bell gives Gauss's report on Riemann's thesis:

> The dissertation submitted by Herr Riemann offers convincing evidence of the author's thorough and penetrating investigations in those parts of the subject treated in the dissertation, of a creative, active, truly mathematical mind, and of a gloriously fertile orginality. The presentation is perspicuous and concise and, in places, beautiful. The majority of readers would have preferred a greater clarity of arrangement. The whole is a substantial, valuable work, which not only satisfies the standards demanded for doctoral dissertations, but far exceeds them.)

32

MEAN SQUARE APPROXIMATION I

32.1$^{(-)}$ Geometric intuition can be pushed a great deal further to give results useful in analysis. An example is given by the following sequence of lemmas. Let us work for the moment in $\mathbb{R}^n$ rather than $\mathbb{C}^n$. (This enables the reader to consider the meaning of our results in the context of the known spaces $\mathbb{R}^2$ and $\mathbb{R}^3$.) If $\mathbf{x} = (x_1, x_2, \ldots, x_n) \in \mathbb{R}^n$ and $\mathbf{y} \in (y_1, y_2, \ldots, y_n) \in \mathbb{R}^n$, define $(x, y) = \sum_{j=1}^n x_j y_j$.

(i) Verify that, if $\mathbf{x}, \mathbf{y}, \mathbf{z} \in \mathbb{R}^n$, $\lambda, \mu \in \mathbb{R}$,

(a) $(\mathbf{x}, \mathbf{y}) = (\mathbf{y}, \mathbf{x})$,
(b) $(\lambda\mathbf{x} + \mu\mathbf{y}, \mathbf{z}) = \lambda(\mathbf{x}, \mathbf{z}) + \mu(\mathbf{y}, \mathbf{z})$,
(c) $(\mathbf{x}, \mathbf{x}) \geq 0$,
(d) if $(\mathbf{x}, \mathbf{x}) = 0$, then $\mathbf{x} = 0$.

(ii) Prove the Cauchy, Schwarz, Buniakowski inequality to the effect that if $\mathbf{x}, \mathbf{y} \in \mathbb{R}^n$ then $|(\mathbf{x}, \mathbf{y})|^2 \leq (\mathbf{x}, \mathbf{x})(\mathbf{y}, \mathbf{y})$, with equality if and only if $\lambda\mathbf{x} + \mu\mathbf{y} = 0$ for some $\lambda, \mu \in \mathbb{R}$ not both zero.

(iii) If $\mathbf{x} \in \mathbb{R}^n$, let us set $|\mathbf{x}| = (\mathbf{x}, \mathbf{x})^{1/2}$ (where the positive square root is taken). Show that, if $\mathbf{x}, \mathbf{y} \in \mathbb{R}^n$, $\lambda \in \mathbb{R}$,

(a) $|\lambda\mathbf{x}| = |\lambda|\,|\mathbf{x}|$,
(b) $|\mathbf{x}| \geq 0$, with equality if and only if $\mathbf{x} = 0$,
(c) $|\mathbf{x} + \mathbf{y}| \leq |\mathbf{x}| + |\mathbf{y}|$.

(iv) Observe that, if $\mathbf{x}, \mathbf{y} \in \mathbb{R}^n$, then $|\mathbf{x} - \mathbf{y}|$ is the usual Euclidean distance.

(v) Show that, if $\mathbf{x}, \mathbf{y} \in \mathbb{R}^n$, then $|\mathbf{x} + \mathbf{y}|^2 + |\mathbf{x} - \mathbf{y}|^2 = 2|\mathbf{x}|^2 + 2|\mathbf{y}|^2$.

(vi) Use (v) to show that, if *ABCD* is a parellelogram, then the sum of the squares of the lengths of the two diagonals is equal to the sum of the squares of the lengths of the sides. (This explains why (v) is called the parallelogram law.) By applying this result to a rectangle, obtain the theorem of Pythagoras.

32.2 We call a subset E of $\mathbb{R}^n$ *convex* if, whenever $\mathbf{x}, \mathbf{y} \in E$, then $\lambda\mathbf{x} + (1 - \lambda)\mathbf{y} \in E$ for all $0 \leq \lambda \leq 1$ (i.e. the line segment joining $\mathbf{x}$ and $\mathbf{y}$ lies in E).

(i) Show that any vector subspace of $\mathbb{R}^n$ is convex. Show that any ball $B(\mathbf{z}, r) = \{\mathbf{x}: |\mathbf{x} - \mathbf{z}| < r\}$ is convex $[r \geqslant 0]$.

(ii) Show that the intersection of two convex sets is convex. Is the same true for their union? (Give a proof or a counter example.)

(iii) Show that the closure of a convex set is convex.

In this question we obtain, at least for $\mathbb{R}^n$, a substantial generalisation of the type of result conjectured in Hypothesis C on page 147 of the textbook, and proved in Theorem 32.5.

(iv) Let E be a non empty closed convex subset of $\mathbb{R}^n$ and suppose $\mathbf{a} \notin E$. Choose $\mathbf{e} \in E$ and let $d = |\mathbf{a} - \mathbf{e}|$. Observe that $E \cap \{\mathbf{x}: |\mathbf{x} - \mathbf{a}| \leqslant d\}$ is a closed bounded non empty set. Show that the map $f: \mathbb{R}^n \to \mathbb{R}$ given by $f(\mathbf{x}) = |\mathbf{x} - \mathbf{a}|$ is continuous and hence, setting out your reasoning carefully, show that there is an $\mathbf{e}_0 \in E$ such that $|\mathbf{a} - \mathbf{e}_0| \leqslant |\mathbf{a} - \mathbf{e}|$ for all $\mathbf{e} \in E$.

(v) Suppose $\mathbf{e}_1 \in E$ and $|\mathbf{a} - \mathbf{e}_1| = |\mathbf{a} - \mathbf{e}_0|$. Since E is convex $(\mathbf{e}_0 + \mathbf{e}_1)/2 \in E$. By considering $|\mathbf{a} - (\mathbf{e}_0 + \mathbf{e}_1)/2|$ and applying the parallelogram law, or otherwise, show that $\mathbf{e}_0 = \mathbf{e}_1$. Thus $|\mathbf{a} - \mathbf{e}_0| < |\mathbf{a} - \mathbf{e}|$ for all $\mathbf{e} \in E$ with $\mathbf{e} \neq \mathbf{e}_0$.

(vi) Suppose for the moment that $\mathbf{e}_0 = \mathbf{0}$. If $\mathbf{e} \in E$, use the fact that every point $\lambda\mathbf{e} + (1 - \lambda)\mathbf{e}_0$ with $0 \leqslant \lambda \leqslant 1$ lies in E, together with the (algebraic expression of the) fact that the nearest point to $\mathbf{a}$ on the line $\{\lambda\mathbf{e} + (1 - \lambda)\mathbf{e}_0: \lambda \in \mathbb{R}\}$ is the foot of the perpendicular from $\mathbf{a}$ to the line (essentially Theorem 32.5), to show that $(\mathbf{a}, \mathbf{e}) \leqslant 0$.

(vii) By translating the result of (vi), show that in general $(\mathbf{a} - \mathbf{e}_0, \mathbf{e} - \mathbf{e}_0) \leqslant 0$ for all $\mathbf{e} \in E$. Is it necessarily true that $(\mathbf{a} - \mathbf{e}_0, \mathbf{e} - \mathbf{e}_0) < 0$ for all $\mathbf{e} \in E$ with $\mathbf{e} \neq \mathbf{e}_0$?

(viii) Draw diagrams to illustrate this result for the convex sets of part (i). (Take $n = 2$, $n = 3$.)

32.3 (i) Suppose E is a closed convex set in $\mathbb{R}^m$ and $\mathbf{a}_0 \in E$, but we can find $\mathbf{a}_n \notin E$ with $\mathbf{a}_n \to \mathbf{a}_0$ (we say that $\mathbf{a}_0$ is in the *frontier* of E). Write $\mathbf{b}_n = |\mathbf{a}_n - \mathbf{a}_0|^{-1}(\mathbf{a}_n - \mathbf{a}_0)$ and show that we can find a $\mathbf{b} \in \mathbb{R}^n$ and a subsequence $n(j) \to \infty$ with $\mathbf{b}_{n(j)} \to \mathbf{b}$. Show that $|\mathbf{b}| = 1$ and, using part (vii) of Question 32.2, that $(\mathbf{b}, \mathbf{e} - \mathbf{a}_0) \leqslant 0$ for all $\mathbf{e} \in E$.

(ii) By considering the case $n = 2$, $E = \{(x, y): 0 \leqslant x, y \leqslant 1\}$, $\mathbf{a}_0 = (0, 0)$ or otherwise show that $\mathbf{b}$ need not be unique.

(iii) Suppose H is an open convex set in $\mathbb{R}^n$. Show that the closure E of H is also convex. Show that, if $\mathbf{a} \notin H$, we can find $\mathbf{b} \in \mathbb{R}^n$ with $|\mathbf{b}| = 1$ such that $(\mathbf{b}, \mathbf{e} - \mathbf{a}) < 0$ for all $\mathbf{e} \in H$. (Note that $\mathbf{a}$ may or may not be a frontier point of E, but no point of H is a frontier point of E.)

(iv) Conclude that if H is an open convex subset of $\mathbb{R}^m$ and $\mathbf{a} \notin H$, we can find $\alpha: \mathbb{R}^m \to \mathbb{R}$ linear such that $\alpha(\mathbf{h}) < \alpha(\mathbf{a})$ for all $\mathbf{h} \in H$.

(Many readers will recognise (iv) as a finite-dimension version of a powerful infinite-dimensional result known as the Hahn–Banach theorem.)

(v) If $A, B \in \mathbb{R}^m$ let us write

$$A + B = \{\mathbf{a} + \mathbf{b}: \mathbf{a} \in A, \mathbf{b} \in B\}.$$

Show that, if A and B are convex, then $A + B$ is convex. Show that, if A is arbitrary and B is open, then $A + B$ is open. Use (iv) to conclude that, if K is a closed convex subset of $\mathbb{R}^m$ and $\mathbf{a} \notin K$, then we can find $\alpha: \mathbb{R}^m \to \mathbb{R}$ linear such that $\alpha(\boldsymbol{k}) < \alpha(\boldsymbol{a})$ for all $\boldsymbol{k} \in K$.

32.4 We now restrict ourselves still further. Let us call a function $f: (a, b) \to \mathbb{R}$ concave, if $\lambda f(t) + (1 - \lambda)f(s) \leqslant f(\lambda t + (1 - \lambda)s)$ for all $a < s,t < b$, $0 \leqslant \lambda \leqslant 1$.

(i) Show that, if $F: (a, b) \to \mathbb{R}$ is twice differentiable with $F''(x) < 0$ for all $x \in a < x < b$, then, if $a < s < t < b$, the function $G: [0, 1] \to \mathbb{R}$, given by $G(\lambda) = F(\lambda t + (1 - \lambda)s) - \lambda F(t) - (1 - \lambda)F(s)$, increases as λ runs from 0 to some point c and then decreases. Hence show that F is concave. Prove in particular that $x \geqslant \sin x \geqslant 2x/\pi$ for $0 \leqslant x \leqslant \pi/2$.

(ii) If f satisfies the conditions set out at the beginning of the question, show that $H = \{(x, y) \in \mathbb{R}^2: y < f(x), a < x < b\}$ is an open convex subset of $\mathbb{R}^2$.

(iii) By using the results of the previous questions, deduce that, if $a < t < b$, there exists a $K \in \mathbb{R}$ such that $f(t) + K(s - t) \geqslant f(s)$ for all $a < s < b$.

(iv) Now suppose that $h: (a, b) \to \mathbb{R}$ is continuous, $h(s) > 0$ for $a < s < b$, $b > t(1) \geqslant t \geqslant t(2) > a$, $\int_{t(1)}^{t(2)} h(s)\, ds = 1$ and $\int_{t(1)}^{t(2)} sh(s)\, ds = t$. By multiplying the inequality of (iii) by $h(s)$ and then integrating, show that $f(t) \geqslant \int_{t(1)}^{t(2)} f(s)h(s)\, ds$.

(v) Deduce that, if $g: (a, b) \to \mathbb{R}$ is continuous, $g(s) > 0$ for $a < s < b$, and $b > t(1) > t(2) > a$, then

$$f\left(\int_{t(1)}^{t(2)} sg(s)\, ds \Bigg/ \int_{t(1)}^{t(2)} g(s)\, ds\right) \geqslant \int_{t(1)}^{t(2)} f(s)g(s)\, ds \Bigg/ \int_{t(1)}^{t(2)} g(s)\, ds.$$

(vi) In the same way, show that if $\lambda_1, \lambda_2, \ldots, \lambda_n \geqslant 0$, $\sum_{j=1}^n \lambda_j = 1$ and $t_1, t_2, \ldots, t_n \in (a, b)$, then

$$f\left(\sum_{j=1}^{n} \lambda_j t_j\right) \geqslant \sum_{j=1}^{n} \lambda_j f(t_j).$$

(vii) By taking $f(t) = \log t$, $\lambda_j = n^{-1}$, deduce the famous arithmetic–geometric mean inequality of Cauchy

$$n^{-1} \sum_{j=1}^{n} t_j \geqslant \left(\prod_{j=1}^{n} t_j\right)^{1/n}$$

for all $t_1, t_2, \ldots, t_n > 0$.

(viii) Use (vii) to show that $(n + 1)/2 \geqslant (n!)^{1/n}$.

(ix) By taking $f(t) = t \log t$ and choosing $\lambda_1, \lambda_2, t_1, t_2$ suitably show that $x \log(x/a) + y \log(y/b) \geqslant (x + y) \log((x + y)/(a + b))$ for all $a, b, x, y > 0$.

(The credit for observing the existence and utility of inequality (vi) belongs to Jensen.)

32.5 In what follows, p, q are positive real numbers with $p^{-1} + q^{-1} = 1$.

(i) By applying Jensen's inequality (Question 32.4 (vi)) with $f(t) = \log t$, $\lambda_1 = p^{-1}$, $\lambda_2 = q^{-1}$, show that

$$x_1^{1/p} x_2^{1/q} \leqslant p^{-1} x_1 + q^{-1} x_2.$$

(ii) Suppose $f, g\colon \mathbb{T} \to \mathbb{C}$ are continuous functions which are not identically zero. Write $\|f\|_p = ((1/2\pi)\int_{\mathbb{T}} |f(t)|^p\, dt)^{1/p}$, $\|g\|_q = ((1/2\pi)\int_{\mathbb{T}} |g(t)|^q\, dt)^{1/q}$. By replacing x_1 by $|f(t)|^p/\|f\|^p$, x_2 by $|g(t)|^q/\|g\|_q^q$ in the inequality of (i) and integrating with respect to t, prove Hölder's inequality

$$\frac{1}{2\pi}\int_{\mathbb{T}} |f(t)g(t)|\, dt \leqslant \|f\|_p \|g\|_q.$$

(iii) Suppose $f, g\colon \mathbb{T} \to \mathbb{C}$ are continuous functions. Observe that

$$\frac{1}{2\pi}\int_{\mathbb{T}} |f(t) + g(t)|^p\, dt \leqslant \frac{1}{2\pi}\int_{\mathbb{T}} |f(t)|\, |f(t) + g(t)|^{p-1}\, dt$$

$$+ \frac{1}{2\pi}\int_{\mathbb{T}} |g(t)|\, |f(t) + g(t)|^{p-1}\, dt$$

and, by applying Hölder's inequality to each of the two terms on the right hand side of the inequality, prove Minkowski's inequality

$$\|f + g\|_p \leqslant \|f\|_p + \|g\|_p.$$

(iv) Verify that $\|\ \|_p$ is a norm of $C(\mathbb{T})$, i.e.

(a) $\|f\|_p \geqslant 0$ with equality only if $f = 0$,
(b) $\|\lambda f\|_p = |\lambda|\, \|f\|_p$ for all $\lambda \in \mathbb{C}$,
(c) $\|f + g\|_p \leqslant \|f\|_p + \|g\|_p$.

(v) In this context we write $\|f\|_1 = (1/2\pi)\int_{\mathbb{T}} |f(t)|\, dt$. By applying Hölder's inequality with $g = 1$, show that $\|f\|_1 \leqslant \|f\|_p$ for $f \in C(\mathbb{T})$. By replacing f by $|f|^r$ in this inequality show, more generally, that if $1 \leqslant r \leqslant u$ then $\|f\|_r \leqslant \|f\|_u$. (The norm $\|\ \|_p$ is a clear generalisation of the norms $\|\ \|_2$ and $\|\ \|_1$ of great importance in more advanced work.)

(vi) Let $0 < r < 1$. Show that, if we write $\|f\| = ((1/2\pi)\int_{\mathbb{T}} |f(t)|^r\, dt)^{1/r}$, then, if $f, g\colon \mathbb{T} \to \mathbb{C}$ are continuous and f, g are not identically zero but $f(t)g(t) = 0$ for all t, then $\|f\|_r + \|g\|_r < \|f + g\|_r$.

(vii) Suppose $p(1), p(2), \ldots, p(n)$ are positive real numbers with $\sum_{j=1}^{n} p(j)^{-1} = 1$. Show that, if $f_1, f_2, \ldots, f_n \in C(\mathbb{T})$,

$$\frac{1}{2\pi}\int_{\mathbb{T}} \left|\prod_{j=1}^{n} f_j(t)\right| dt \leqslant \prod_{j=1}^{n} \|f_j\|_{p(j)}.$$

32.6 In what follows p and q are positive real numbers.

(i) By imitating the proofs of Question 32.5, show that, if $a_1, a_2, \ldots, a_n$, $b_1, b_2, \ldots, b_n \in \mathbb{C}$, then, if $p^{-1} + q^{-1} = 1$,

$$\sum_{j=1}^{n} |a_j b_j| \leq \left(\sum_{j=1}^{n} |a_j|^p\right)^{1/p} \left(\sum_{j=1}^{n} |b_j|^q\right)^{1/q}$$

and, if $p \geq 1$, $(\sum_{j=1}^{n} |a_j + b_j|^p)^{1/p} \leq (\sum_{j=1}^{n} |a_j|^p)^{1/p} + (\sum_{j=1}^{n} |b_j|^p)^{1/p}$.

(ii) Let us write l^p for the space of infinite sequences $\mathbf{a} = (a_1, a_2, \ldots)$ such that $\sum_{j=1}^{\infty} |a_j|^p$ converges. Show that, if $\mathbf{a} \in l^p$, $\mathbf{b} \in l^q$, then

$$\sum_{j=1}^{\infty} |a_j b_j| \leq \|\mathbf{a}\|_p \|\mathbf{b}\|_q.$$

Show that, if we use the standard definition of addition and scalar multiplication ($(\mathbf{a} + \mathbf{b})$ is the sequence $(a_1 + b_1, a_2 + b_2, \ldots)$, $\lambda \mathbf{a}$ is the sequence $(\lambda a_1, \lambda a_2, \ldots)$), then if, $\mathbf{a}, \mathbf{b} \in l^p$ and $\lambda \in \mathbb{C}$, if follows that $\mathbf{a} + \mathbf{b} \in l^p$, $\lambda \mathbf{a} \in l^p$ and

$$\|\mathbf{a} + \mathbf{b}\|_p \leq \|\mathbf{a}\|_p + \|b\|_p$$

$$\|\lambda \mathbf{a}\|_p = |\lambda| \, \|\mathbf{a}\|_p.$$

Observe also that $\|\mathbf{a}\|_p \geq 0$ and that, if $\|\mathbf{a}\|_p = 0$, then $\mathbf{a} = \mathbf{0}$.

(iii) Show that, if $1 \leq r \leq u$, then $l^u \supseteq l^r$. By considering $a_n = ((n+1)(\log(n+1))^2)^{-1/2}$, or otherwise, show that

$$l^u \neq \bigcup_{u > r \geq 1} l^r.$$

(iv) (Converse to Hölder's inequality) Suppose that $\mathbf{a}$ is a sequence such that $\sum_{j=1}^{\infty} |a_j b_j|$ converges and $\sum_{j=1}^{\infty} |a_j b_j| \leq A \|\mathbf{b}\|_q$ whenever $\mathbf{b} \in l^q$. By considering $\mathbf{b}$ defined by $b_j = |a_j|^{p-1}$ for $1 \leq j \leq N$, $b_j = 0$ for $j \geq N + 1$, show that $\mathbf{a} \in l^p$ and $\|\mathbf{a}\|_p \leq A$.

(v) State and prove a version of (iv) which provides a converse to part (ii) of Question 32.5.

32.7 In this question we show that l^p is complete i.e. that, if $\mathbf{a}(n) \in l^p$ and $\|\mathbf{a}(n) - \mathbf{a}(m)\|_p \to 0$ as $n, m \to \infty$, then there exists an $\mathbf{a} \in l^p$ with $\|\mathbf{a}(n) - \mathbf{a}\|_p \to 0$.

(i) More formally, suppose $\mathbf{a}(n) \in l^p$ for each n and, given $\varepsilon > 0$, we can find an $N(\varepsilon)$ such that $\|\mathbf{a}(n) - \mathbf{a}(m)\|_p < \varepsilon$ whenever $n, m \geq N(\varepsilon)$. Show that for each fixed j, $|a_j(n) - a_j(m)| < \varepsilon$ whenever $n, m \geq N(\varepsilon)$ and explain why this implies the existence of an $a_j \in \mathbb{C}$ with $|a_j(n) - a_j| \to 0$ as $n \to \infty$.

(ii) If $m \geq N(1)$, observe that, whenever $n \geq N(1)$, we have

$$\left(\sum_{j=1}^{M}|a_j(n)|^p\right)^{1/p} \leq \left(\sum_{1}^{M}|a_j(n)-a_j(m)|^p\right)^{1/p} + \left(\sum_{1}^{M}|a_j(n)|^p\right)^{1/p}$$
$$\leq 1 + \|\mathbf{a}(m)\|$$

and deduce, by first allowing $n \to \infty$ and then allowing M to vary, that $\mathbf{a} \in l^p$.

(iii) Now note that, if $n, m \geq N(\varepsilon)$, we have

$$\left(\sum_{j=1}^{M}|a_j(m)-a_j|^p\right)^{1/p} \leq \left(\sum_{j=1}^{M}|a_j(m)-a_j(n)|^p\right)^{1/p} + \left(\sum_{j=1}^{M}|a_j(n)-a_j|^p\right)^{1/p}$$
$$\leq \varepsilon + \left(\sum_{j=1}^{M}|a_j(n)-a_j|^p\right)^{1/p}$$

and deduce, by first allowing $n \to \infty$ and then allowing M to vary, that $\|\mathbf{a}(m)-\mathbf{a}\|_p \leq \varepsilon$ for all $m \geq N(\varepsilon)$. Thus $\|\mathbf{a}(m)-\mathbf{a}\|_p \to 0$ as $m \to \infty$.

(iv) Sometimes instead of considering l^p we consider $l^p(\mathbb{Z})$ consisting of all two-sided sequences $\mathbf{a} = (\ldots, a_{-2}, a_{-1}, a_0, a_1, a_2, \ldots)$ such that $\sum_{j=-\infty}^{\infty}|a_j|^p < \infty$. We define

$$\|\mathbf{a}\|_p = \left(\sum_{j=-\infty}^{\infty}|a_j|^p\right)^{1/p},$$

$$\lambda\mathbf{a} + \mu\mathbf{b} = (\ldots, \lambda a_{-2} + \mu b_{-2}, \lambda a_{-1} + \mu b_{-1}, \lambda a_0 + \mu b_0, \lambda a_1 + \mu b_1, \lambda a_2 + \mu b_2, \ldots).$$

Explain briefly why l^p and $l^p(\mathbb{Z})$ have essentially the same properties.

32.8 In the last question we saw that l^p is complete for all real $p \geq 1$. In this question we shall show that there is no way of obtaining a similar result for integrals within the context of Riemann integration by constructing $g_n \in C(\mathbb{T})$ with $\|g_n - g_m\|_p \to 0$ as $n, m \to \infty$ for all real $p \geq 1$ but such that there does not exist an Riemann integrable (and so, trivially, no continuous) function $g: \mathbb{T} \to \mathbb{R}$ such that $(1/2\pi)\int_{\mathbb{T}}|g(t)-g_n(t)|^p\,dt \to 0$ as $n \to \infty$ for any $p \geq 1$.

(i) Let $g: \mathbb{T} \to \mathbb{R}$ be a continuous function such that $|g(t)| \leq 1$ for all $t \in \mathbb{T}$. Suppose $x \in \mathbb{T}$, $\pi/2 > \delta > 0$ and $k \in \mathbb{R}$ with $|k| = 1$ are given. Show that we can find $f: \mathbb{T} \to \mathbb{R}$ a continuous real valued function such that $|f(t)| \leq 1$ for all $t \in \mathbb{T}$, $f(t) = g(t)$ for $|t-x| > \delta$, $f(t) = k$ for $|t-x| < \delta/2$.

(ii) Explain why we can find a sequence $y_n \in \mathbb{T}$ such that, given $y \in \mathbb{T}$, there exist $n(j) \to \infty$ such that $y_{n(j)} \to y$. Let us set $x_{2n-1} = x_{2n} = y_n$. Now construct inductively a sequence $g_n: \mathbb{T} \to \mathbb{R}$ of continuous functions such that

(a) $|g_n(t)| \leq 1$ for all $t \in \mathbb{T}$, $n \geq 1$,
(b) $g_n(t) = g_{n-1}(t)$ for all $|t - x_n| > 10^{-n}\pi$, $n \geq 2$,
(c) $g_n(t) = (-1)^n$ for all $|t - x_n| < 2^{-1}10^{-n}\pi$, $n \geq 1$.

(iii) If $m > n$ show that we can find a set $J(n, m)$ which is the union of finitely many intervals of total length less than $3^{-1}10^{-n}\pi$ such that $g_n(t) = g_m(t)$ for all $t \notin J(n, m)$. Deduce that

$$\|g_n - g_m\|_p \leq (2^p 3^{-1} 10^{-n})^{1/p}$$

and so $\sup_{m,n \geq M} \|g_n - g_m\|_p \to 0$ as $M \to \infty$ for all real $p \geq 1$.

(iv) Suppose now that g is Riemann integrable. Then, in particular, given any $\varepsilon > 0$, we can find $0 \leq a_0 < a_1 < a_2 < \ldots < a_M < 2\pi$ and $\lambda_0, \lambda_1, \ldots, \lambda_M$, $\mu_0, \mu_1, \ldots, \mu_M$ real with $\lambda_j \geq \mu_j$ $[0 \leq j \leq M]$ such that, writing $a_{M+1} = 2\pi + a_0$, we have

(d) $\lambda_j \geq g(t) \geq \mu_j$ for $t \in [a_j, a_{j+1}), 0 \leq j \leq M$,
(e) $\sum(\lambda_j - \mu_j)(a_{j+1} - a_j) < 2\pi\varepsilon$.

Show that this means we can find a $j(0)$ with $0 \leq j(0) \leq M$ such that $\lambda_{j(0)} - \mu_{j(0)} < \varepsilon$. Conclude that, if g is Riemann integrable, we can find an interval $I = [a, b)$, say, and a $\lambda \in \mathbb{R}$ such that $|g(t) - \lambda| < 2^{-1}$ for all $t \in I$.

(v) We now have two cases according as $\lambda > 0$ or $\lambda \leq 0$. Suppose that $\lambda \leq 0$. Show that we can find an integer $N \geq 1$ such that $(x_{2N} - 10^{-2N}\pi, x_{2N} + 10^{-2N}\pi) \subseteq I$. Show that, if $m \geq 2N$ and $J(2N, m)$ is defined as in part (iii), then $g_m(t) = 1$ for all $t \in \mathbb{T}$ with $|t - x_{2N}| < 2^{-1}10^{-2N}\pi$ and $t \in J(2N, m)$. Deduce that, if $m \geq 2N$, then $|g_m(t) - g(t)| > 2^{-1}$ on a finite set of disjoint intervals of length at least $2^{-1}10^{-2N}\pi$ and so $\|g_m - g\|_p \geq 2^{-1}(4^{-1}10^{-2N})^{1/p}$. Thus $\|g_m - g\|_p \not\to 0$ as $m \to \infty$. Explain briefly how to obtain the same final result if $\lambda > 0$.

(We shall return to this example in Questin 34.1. From a modern point of view this question exposes a fundamental defect in the Riemann integral.)

32.9 The following elegant proof of two inequalities of Hilbert is due to Schur.

(i) Let h be the (translated) sawtooth function given by $h(t) = t - \pi$ for $0 < t < 2\pi$, $h(0) = 0$. Compute $\hat{h}(r)$ for each r.

(ii) Let $P = \sum_{j=0}^{n} a_j e_{-j}$, $Q = \sum_{j=0}^{n} b_j e_{-j}$ (where $e_n(t) = \exp(int)$). Show that

$$\frac{1}{2\pi i} \int_{\mathbb{T}} P(t)Q(t)h(t)e_{-1}(t)\,dt = \sum_{j=0}^{n} \sum_{j=0}^{n} \frac{a_j b_k}{j + k + 1}.$$

(iii) Use the fact that $|h(t)e_{-1}(t)| \leq \pi$ for all t together with the Cauchy–Schwarz–Buniakowski inequality to show that

$$\left|\frac{1}{2\pi} \int_{\mathbb{T}} P(t)Q(t)h(t)e_{-1}(t)\,dt\right| \leq \pi\left(\frac{1}{2\pi} \int_{\mathbb{T}} |P(t)|^2\,dt \,\frac{1}{2\pi} \int_{\mathbb{T}} |Q(t)|^2\,dt\right)^{1/2}$$

$$= \pi\left(\sum_{j=0}^{n} |a_j|^2 \sum_{k=0}^{n} |b_j|^2\right)^{1/2}.$$

(iv) Deduce that $\sum_{j=0}^{n}\sum_{k=0}^{n} a_j b_k/(j+k+1) \leqslant \pi(\sum_{j=0}^{n}|a_j|^2)^{1/2}(\sum_{j=0}^{n}|b_j|^2)^{1/2}$ and so $\sum_{j=0}^{n}\sum_{k=0}^{n}|a_j|\,|b_k|/(j+k+1) \leqslant \pi(\sum_{j=0}^{n}|a_j|^2)^{1/2}(\sum_{j=0}^{n}|b_j|^2)^{1/2}$.

(v) Conclude that, if $\sum_{j=0}^{\infty}|a_j|^2$ and $\sum_{j=0}^{\infty}|b_j|^2$ converge, then $\lim_{n\to\infty}\sum_{j=0}^{n}\sum_{k=0}^{n}|a_j|\,|b_k|/(j+k+1)$ exists and is no greater than $\pi(\sum_{j=0}^{\infty}|a_j|^2)^{1/2}(\sum_{j=0}^{\infty}|b_j|^2)^{1/2}$.

(vi) Prove similarly that

$$\left|\sum_{|j|,|k|\leqslant n, j\neq k} a_j b_k/(j-k)\right| \leqslant \pi\left(\sum_{j=0}^{n}|a_j|^2\right)^{1/2}\left(\sum_{j=0}^{n}|b_j|^2\right)^{1/2}.$$

32.10 Here is a corollary of the result of Question 32.9 due to Hardy and Littlewood. (It is taken from the treasure house of Hardy, Littlewood and Polya *Inequalities*.)
Suppose $f: [0,1]\to\mathbb{R}$ is continuous and let $a_j = \int_0^1 x^j f(x)\,dx$ be the jth moment of f (see Chapter 6).

(i) Show that, if $b_j \in \mathbb{R}$, then $\int_0^1(\sum_{j=0}^{n} b_j x^j)^2\,dx \leqslant \pi\sum_{j=0}^{n} b_j^2$.

(ii) Show that $\sum_{j=0}^{n} a_j b_j = \int_0^1(\sum_{j=0}^{n} b_j x^j) f(x)\,dx$ and, by applying the Cauchy–Schwarz–Bunikowski inequality, deduce that $(\sum_{j=0}^{n} a_j b_j)^2 \leqslant \pi\sum_{j=0}^{n} b_j^2 \int_0^1 (f(x))^2\,dx$.

(iii) By choosing b_j suitably, deduce that $\sum_{j=0}^{n} a_j^2 \leqslant \pi\int_0^1 (f(x))^2\,dx$ and conclude that $\sum_{j=0}^{\infty} a_j^2$ converges with $\sum_{j=0}^{\infty} a_j^2 \leqslant \pi\int_0^1 f(x)^2\,dx$.

32.11 (Van der Corput's inequalities) These inequalities are simple to prove but of considerable utility both in number theory and Fourier analysis. We shall need the second mean value theorem for integrals (proved in Question 52.5) which states that, if $F: [a,b]\to\mathbb{R}$ is continuous and $G: [a,b]\to\mathbb{R}$ is continuous and increasing, then there exists a $c\in[a,b]$ with

$$\int_a^b F(x)G(x)\,dx = G(a)\int_a^c F(x)\,dx + G(b)\int_c^b F(x)\,dx$$

(i) Show that, if $F: [a,b]\to\mathbb{R}$ is continuous and $G: [a,b]\to\mathbb{R}$ continuous and monotone (i.e. everywhere increasing or everywhere decreasing), then

$$\left|\int_a^b F(x)G(x)\,dx\right| \leqslant 2\sup_{x\in[a,b]}|G(x)| \sup_{a\leqslant\alpha\leqslant\beta\leqslant b}\left|\int_\alpha^\beta F(x)\,dx\right|.$$

If G is as before and $F: [a,b]\to\mathbb{C}$ is continuous, show that

$$\left|\int_a^b F(x)G(x)\,dx\right| \leqslant 4\sup_{x\in[a,b]}|G(x)| \sup_{a\leqslant\alpha\leqslant\beta\leqslant b}\left|\int_\alpha^\beta F(x)\,dx\right|.$$

(ii) (Van der Corput's first inequality) Suppose that $f: [a,b]\to\mathbb{R}$ is continuously differentiable with monotone derivative f' and that $|f'(x)|\geqslant\lambda$ for all $x\in[a,b]$ and some $\lambda>0$. Show that, if we write $G(x)=1/f'(x)$, $F(x)=\exp(2\pi i f(x))$, then

$$\int_a^b e^{2\pi i f(x)}\,dx = \frac{1}{2\pi i}\int_a^b F'(x)G(x)\,dx$$

and deduce that $|\int_a^b e^{2\pi i f(x)}\,dx| \leqslant 4/\pi\lambda$.

(iii) Suppose that $f\colon [a, b] \to \mathbb{R}$ is twice differentiable with $f''(x) \geqslant \rho$ for all $x \in [a, b]$ and some $\rho > 0$. Show that, if $f'(x) \geqslant 0$ for all $x \in [a, b]$, then

$$\left|\int_a^b e^{2\pi i f(x)}\,dx\right| \leqslant \left|\int_a^c e^{2\pi i f(x)}\,dx\right| + \left|\int_c^b e^{2\pi i f(x)}\,dx\right|$$

$$\leqslant (c - a) + 4/(\pi(c - a)\rho)$$

for all $a < c < b$. By choosing c appropriately, show that $|\int_a^b e^{2\pi i f(x)}\,dx| \leqslant 3\rho^{-1/2}$. (Here and elsewhere the constants can be improved.)

(iv) (Van der Corput's second inequality) More generally show that, if $f\colon [a, b] \to \mathbb{R}$ is twice differentiable with $f''(x) \geqslant \rho$ for all $x \in [a, b]$ or $f''(x) \leqslant -\rho$ for all $x \in [a, b]$ and some $\rho > 0$, then

$$\left|\int_a^b e^{2\pi i f(x)}\,dx\right| \leqslant 6\rho^{-1/2}.$$

32.12 The inequality proved in this question is also due to Van der Corput.

(i) Let $h(x) = x - [x] - 1/2$ (here as usual $[x]$ is the integer part of x) and let $h_m(x) = -\sum_{n=1}^m (\sin 2\pi nx)/(\pi n)$ for all $x \in \mathbb{R}$. Graph $h\colon \mathbb{R} \to \mathbb{R}$ and, by using Question 18.1, show that $\int_a^b h_m(x)F(x)dx \to \int_a^b h(x)F(x)dx$ whenever $F\colon [a, b] \to \mathbb{C}$ is continuous.

(ii) Suppose $f\colon [a, b] \to \mathbb{R}$ is once continuously differentiable with monotone derivative f' satisfying $|f'(x)| \leqslant 1/2$ for all $x \in [a, b]$. Show that if $F(x) = \exp(2\pi i f(x))$ and $F_n(x) = \exp(2\pi i(f(x) - nx))$, then

$$\int_a^b h_m(x)F'(x)\,dx =$$

$$\frac{1}{2\pi i}\sum_{n=1}^m \frac{1}{n}\left(\int_a^b \frac{f'(x)}{f'(x) - n} F'_n(x)\,dx - \int_a^b \frac{f'(x)}{f'(x) + n} F'_{-n}(x)\,dx\right).$$

(iii) Show that $G_n(x) = f'(x)/(f'(x) - n)$ defines a monotone function $G_n\colon [a, b] \to \mathbb{R}$. By applying the second inequality of Question 32.11 (i) show that

$$\left|\int_a^b \frac{f'(x)}{f'(x) - n} F'_n(x)\,dx\right| \leqslant \frac{4}{|n| - 1/2}$$

and deduce that

$$\left|\int_a^b h_m(x)F'(x)\,dx\right| \leqslant \frac{4}{\pi}\sum_{n=1}^m \frac{1}{n(n - 1/2)} \leqslant \frac{4}{\pi}\sum_{n=2}^{2m} \frac{2}{n(n - 1)} \leqslant \frac{8}{\pi}.$$

32.13 We now combine the results of Question 3.2 (iii), and Questions 32.11 and 32.12. Suppose N and M are integers with $M < N$ and $f\colon [a, b] \to \mathbb{R}$ is twice differentiable with $f''(x) \geqslant \rho$ for all $x \in [M, N]$ (or $f''(x) \leqslant -\rho$ for all $x \in [M, N]$) and some $\rho > 0$. Our object is to bound $\sum_{n=m}^{N} e^{2\pi i f(n)}$.

(i) For each integer p, let us write

$$\Lambda(p) = \{M \leqslant n \leqslant N\colon p - 1/2 \leqslant f'(n) < p + 1/2\}.$$

Explain why either $\Lambda(p) = \varnothing$ or $\Lambda(p) = \{n \in \mathbb{Z}\colon M(p) \leqslant n \leqslant N(p)\}$ for some $M \leqslant M(p) \leqslant N(p) \leqslant N$. If $\Phi = \{p \in \mathbb{Z}\colon \Lambda(p) \neq \varnothing\}$, show that Φ has at most $|f'(N) - f'(M)| + 2$ distinct members.

(ii) Let $F_p(x) = \exp(2\pi i(f(x) - px))$. Check that, if $\Lambda(p) \neq \varnothing$, then the arguments of Question 3.2 give us

$$\sum_{n\in\Lambda(p)} e^{2\pi i f(n)} = \sum_{n=M(p)}^{N(p)} F_p(n) = \int_{M(P)}^{N(p)} F_p(x)\,dx + \frac{1}{2}(F_p(N(p)) + F_p(M(p)) + \int_{M(p)}^{N(p)} h(x)F'_p(x)\,dx.$$

(This is also, of course, a very simple case of the Euler–MacLaurin summation formula discussed, for example, in Question 18.7.) By using Question 32.11, trivial estimation and Question 32.12, deduce that

$$|\sum_{n\in\Lambda(p)} e^{2\pi i f(n)}| \leqslant 4 + 6\rho^{-1/2}.$$

(iii) Conclude that

$$\left|\sum_{n=M}^{N} e^{2\pi i f(n)}\right| \leqslant (|f'(N) - f'(M)| + 2)(4 + 6\rho^{-1/2}).$$

(iv) Let $\alpha, \theta \in \mathbb{R}$ with $\alpha \neq 0$. By setting $f(x) = (2\pi)^{-1}(\alpha x \log x + \theta x)$, show that

$$\left|\sum_{n=M}^{N} e^{i\alpha n \log n} e^{in\theta}\right| \leqslant C_1(\alpha)2^{k/2} \text{ for all } 2^k \leqslant M < N \leqslant 2^{k+1}$$

where $C_1(\alpha)$ depends only on α. By observing that

$$\sum_{n=2}^{2^l} \beta_n = \sum_{p=1}^{l}\left(\sum_{n=2^{p-1}+1}^{2^p} \beta_n\right),$$

or otherwise, deduce that there exists a constant $C(\alpha)$ depending only on α such that

$$\sup_{\theta\in\mathbb{T}}\left|\sum_{n=1}^{N} e^{i\alpha n \log n} e^{in\theta}\right| \leqslant C(\alpha)N^{1/2}.$$

(v) We have thus found $a_1, a_2, \ldots \in \mathbb{C}$ with $|a_n| = 1$ for all n together with a constant C such that $\sup_{\theta\in\mathbb{T}}|\sum_{n=1}^N a_n e^{in\theta}| \leqslant CN^{1/2}$. Show that, if $b_1, b_2, \ldots, b_N \in \mathbb{C}$ with $|b_n| = 1$, for all n then $(1/2\pi)\int_{\mathbb{T}}|\sum_{n=1}^N b_n e^{in\theta}|^2 d\theta = N$ and so $\sup_{\theta\in\mathbb{T}}|\sum_{n=1}^N b_n e^{in\theta}| \geqslant N^{1/2}$. Thus (apart from the constant C) our result cannot be improved. The reader might like to consider whether we could find $c_1, c_2, \ldots \in \{-1, 1\}$ and a constant C_3 such that $|\sum_{n=1}^N c_n e^{in\theta}| \leqslant C_3 N^{1/2}$. We shall consider this problem in Question 34.5.

(vi) Use Weyl's criterion (see Question 3.1) and part (iii) of this question to establish that, if $\alpha, \tau \in \mathbb{R}$, $\alpha \neq 0$, $\tau > 0$, then $\alpha n(\log n)^\tau$ is an equidistributed sequence.

33

MEAN SQUARE APPROXIMATION II

33.1 (Bessel's inequality for almost periodic functions) This question continues on from Question 8.7 and uses its notations.

(i) Suppose f is an almost periodic function, $\zeta(1), \zeta(2), \ldots, \zeta(n)$, are distinct real numbers and $a_1, a_2, \ldots, a_n \in \mathbb{C}$. If $\zeta \in \mathbb{R}$ we write $e_\zeta(t) = \exp(i\zeta t)$ $[t \in \mathbb{R}]$. Show that

$$M|f|^2 + \sum_{j=1}^{n} |(\mathscr{F}_M f)(\zeta(j)) - a_j|^2 = M\left|f - \sum_{j=1}^{n} a_j e_{\zeta(j)}\right|^2 + \sum_{j=1}^{n} |(\mathscr{F}_M f)(\zeta(j))|^2.$$

(ii) Deduce that, if f and $\zeta(1), \zeta(2), \ldots, \zeta(n)$ are fixed, the quantity

$$M|f - \sum_{j=1}^{n} a_j e_{\zeta(j)}|^2$$

is minimised uniquely by taking $a_j = (\mathscr{F}_M f)(\zeta(j))$ $[1 \leqslant j \leqslant n]$.

(iii) Show also that $M|f|^2 \geqslant \sum_{j=1}^{n} |(\mathscr{F}_M f)(\zeta(j))|^2$.

(iv) Let us write $E_0 = \{\lambda \in \mathbb{R}: |(\mathscr{F}_M f)(\lambda)| > 1\}$ and

$$E_n = \{\lambda \in \mathbb{R}: n^{-1} \geqslant |(\mathscr{F}_M f)(\lambda)| > (n+1)^{-1}\} \ [n \in \mathbb{Z}, n \geqslant 1].$$

Deduce from (iii) that E_n is finite for each $n \in \mathbb{Z}$, $n \geqslant 0$ and conclude that, if we write $E = \{\lambda: (\mathscr{F}_M f)(\lambda) \neq 0\}$, then either E is finite or we can find a sequence of distinct real numbers $\lambda(1), \lambda(2), \ldots$ such that $|(\mathscr{F}_M f)(\lambda(1))| \geqslant |(\mathscr{F}_M f)(\lambda(2))| \geqslant \ldots$, $|(\mathscr{F}_M f)(\lambda(j))| \to 0$ as $j \to \infty$ and $E = \{\lambda(j): j \geqslant 1\}$. (In more sophisticated terms, we have shown that $(\mathscr{F}_M f)(\lambda) = 0$ except at a countable set of points.) We write $E = \sigma(f)$ and refer to $\sigma(f)$ as the spectrum of f.

33.2$^{(-)}$ Let $a_1, a_2, \ldots, a_n$ be non negative real numbers such that $\sum_{k=1}^{n} a_k = nc$. Prove that

$$nc^2 \leqslant \sum_{k=1}^{n} a_k^2 \leqslant n^2c^2,$$

giving necessary and sufficient conditions for equality to hold in each of the two relations.

34
MEAN SQUARE CONVERGENCE

34.1 To see how unwelcome the result of 32.8 is, let us specialise to the case $p = 2$. We then have a sequence $g_n \in C(\mathbb{T})$ with $|g_n(t)| \leqslant 1$ for all $t \in \mathbb{T}$, $\|g_n - g_m\|_2 \to 0$ as $m, n \to \infty$ but such that there exits no Riemann integrable g with $(1/2\pi)\int_{\mathbb{T}}|g_n(t) - g(t)|^2\,dt \to 0$.

(i) Let us write $\mathbf{a}(n)$ for the two-sided sequence given by $a_j(n) = \hat{g}_n(j)$. Show that $\mathbf{a}(n) \in l^2(\mathbb{Z})$ and, using Bessel's inequality, that the $\mathbf{a}(n)$ form a Cauchy sequence in $l^2(\mathbb{Z})$. Since $l^2(\mathbb{Z})$ is complete, deduce that there exists an $\mathbf{a} \in l^2(\mathbb{Z})$ with $\|\mathbf{a}(n) - \mathbf{a}\|_2 \to 0$.

(ii) Consider $P_m(t) = \sum_{r=-m}^{m} a_r \exp(irt)$. Show, using Parseval's equality, that, if $\mathbf{b}(m)$ is defined by $b_j(m) = a_j$ for $|j| \leqslant m$, $b_j(m) = 0$ otherwise, then

$$\begin{aligned}\|P_m - g_m\|_2 &= \|\mathbf{b}(m) - \mathbf{a}(m)\|_2 \\ &\leqslant \|\mathbf{b}(m) - \mathbf{a}\|_2 + \|\mathbf{a} - \mathbf{a}(m)\|_2 \to 0\end{aligned}$$

as $m \to \infty$. Deduce that there is no Reimann integrable g such that $\|P_m - g\|_2 \to 0$. (Thus, although $\sum_{r=-\infty}^{\infty} |a_r|^2$ converges, there is no Reimann integrable sum $\sum_{r=-\infty}^{\infty} a_r \exp(irt)$. This defect is removed if we use the Lebesgue integral.)

34.2 The following results will be needed in the next question.

(i) Show that, if $|z| \leqslant 1$,

$$|e^z - 1 - z| \leqslant \sum_{r=2}^{\infty} |z|^r/r! \leqslant (|z|^2/2) \sum_{r=0}^{\infty} (|z|/2)^r \leqslant |z|^2.$$

(ii) Deduce that, if $f, g\colon \mathbb{T} \to \mathbb{C}$ are continuous (and so bounded), we have

$$\left| h^{-1}\left(\frac{1}{2\pi}\int_{\mathbb{T}} f(t)\exp((z+h)g(t))\,dt - \frac{1}{2\pi}\int_{\mathbb{T}} f(t)\exp(zg(t))\,dt\right)\right.$$
$$\left. - \frac{1}{2\pi}\int_{\mathbb{T}} f(t)g(t)\exp(zg(t))\,dt\right| \to 0 \quad \text{as } |h| \to 0$$

and so $(1/2\pi)\int_{\mathbb{T}} f(t)\exp(zg(t))\,dt$ is an analytic function of z. (Another result of this type is given in Theorem 75.2.)

(iii) Explain why a trigonometric polynomial can have only a finite number of zeros. Show that, if P is a trigonometric polynomial and $\varepsilon > 0$, we can find a nowhere zero continuous function g with $|g(t) - P(t)| < \varepsilon/2$ for all $t \in \mathbb{T}$. Hence, using Fejér's theorem, or otherwise, show that we can find a nowhere zero trigonometric polynomial A with $|A(t) - P(t)| < \varepsilon$ for all t.

(iv) If $f: \mathbb{T} \to \mathbb{C}$ is continuous and $\varepsilon > 0$, show that we can find a nowhere zero trigonometric polynomial P with $|P(t) - f(t)| < \varepsilon$.

34.3 (The Hausdorff–Young inequality. This question requires the result of Question 27.8.) Parseval's formula tells us that, if $f: \mathbb{T} \to \mathbb{C}$ is continuous, then

$$\left(\sum_{n=-\infty}^{\infty} |\hat{f}(n)|^2\right)^{1/2} = \left(\frac{1}{2\pi}\int_{\mathbb{T}} |f(t)|^2\,dt\right)^{1/2}$$

or, more briefly, that

$$\|\hat{f}\|_2 = \|f\|_2. \tag{A$_2$}$$

It is trivial that

$$\sup|\hat{f}(n)| \leqslant \frac{1}{2\pi}\int_{\mathbb{T}} |f(t)|\,dt$$

or, more briefly, that

$$\|\hat{f}\|_\infty \leqslant \|f\|_1. \tag{A$_1$}$$

Can we find similar inequalities for other norms?

The following result was proved for certain p by Young and extended to all $2 \geqslant p \geqslant 1$ by Hausdorff. If $2 \geqslant p \geqslant 1$ and $p^{-1} + q^{-1} = 1$ (when $p = 1$ we take $q = \infty$ by convention), then, if $f: \mathbb{T} \to \mathbb{C}$ is continuous, we have (for $1 < p \leqslant 2$).

$$\left(\sum_{n=-\infty}^{\infty} |\hat{f}(n)|^q\right)^{1/q} \leqslant \left(\frac{1}{2\pi}\int_{\mathbb{T}} |f(t)|^p\,dt\right)^{1/p}$$

or, more briefly, (for all $1 \leqslant p \leqslant 2$)

$$\|\hat{f}\|_q \leqslant \|f\|_p. \tag{A$_p$}$$

We say that the result (A$_p$) 'interpolates between result (A$_1$) and result (A$_2$)'. There now exist several powerful techniques for proving such results 'by interpolation'. In this question we give a proof by Calderon and Zygmund which shows in this simple special case something of what can be done in general.

(i) Explain why, if $1 < p \leqslant 2$, the result (A$_p$) will follow if we can show that $(\sum_{r=-N}^{N} |\hat{f}(n)|^q)^{1/q} \leqslant \|f\|_p$ for all N and, using the previous question, why it is thus sufficient to prove (A_p) when f is a nowhere zero trigonometric polynomial P, say. Henceforth we take $f(t) = P(t) = \sum_{r=-N}^{N} \hat{P}(r)\exp(irt)$ with $\|P\|_p = 1$.

(ii) Using Question 32.6 (iv), or proceeding directly, show that it is sufficient to show that

$$\left|\sum_{r=-N}^{N} \hat{P}(r)a_r\right| \leqslant 1 \quad \text{whenever} \quad \sum_{r=-N}^{N} |a_r|^p = 1.$$

(iii) Show that we can write $P(t) = g(t)^{1/p}u(t)$ where g and u are continuous $|u(t)| = 1$ for all t, $g(t) \geqslant 0$ for all t (so we can take $g(t)^{1/p}$ to be the positive pth root of $g(t)$). Write $a_r = b_r^{1/p}v_r$ where $|v_r| = 1$, $b_r \geqslant 0$ for all r.

(iv) Set

$$F(z) = \sum_{r=-N}^{N} v_r \exp(z\log b_r)\frac{1}{2\pi}\int_{\mathbb{T}} u(t)\exp(z\log g(t))\exp(-irt)\,dt$$

(more concisely

$$F(z) = \sum_{r=-N}^{N} v_r b_r^z = \frac{1}{2\pi}\int_{\mathbb{T}} u(t)g(t)^z \exp(-irt)\,dt.)$$

(a) Show, using the previous question, that F is analytic.
(b) Observe that $F(1/p) = \sum_{r=-N}^{N} \hat{P}(r)a_r$.
(c) Show by trivial estimation that there exists an M_1 with $|F(z)| \leqslant M_1$ for all z with $\frac{1}{2} \leqslant \operatorname{Re} z \leqslant 1$.
(d) Use the result of Question 27.8 to show that the result of (ii), and so the Hausdorff–Young inequality, will follow if we can show $|F(z)| \leqslant 1$ if $\operatorname{Re} z = 1$ or if $\operatorname{Re} z = 1/2$.

(v) Write $G_z(t) = u(t)g(t)^2$ and using the trivial (A_1) show that, if $\operatorname{Re} z = 1$,

$$|F(z)| \leqslant \sum_{r=-N}^{N} |b_r| \frac{1}{2\pi}\int_{\mathbb{T}} |G_1(t)|\,dt = \sum_{r=-N}^{N} |a_r|^p \frac{1}{2\pi}\int_{\mathbb{T}} |P(t)|^p\,dt = 1.$$

(vi) Using the Cauchy–Schwarz–Buniakowski inequality and Parseval's equality (A_2) show that, if $\operatorname{Re} z = 1/2$,

$$\begin{aligned}|F(z)| &\leqslant \sum_{r=-N}^{N} |b_r| \sum_{r=-N}^{N} |\hat{G}_2(r)|^2 \leqslant \sum_{r=-N}^{N} |a_r|^p \frac{1}{2\pi}\int_{\mathbb{T}} |G_z(t)|^2\,dt \\ &= \sum_{r=-N}^{N} |a_r|^p \frac{1}{2\pi}\int_{\mathbb{T}} |P(t)|^p\,dt = 1.\end{aligned}$$

Combine (iv)(d), (v) and (vi) to obtain the Hausdorff–Young inequality.

34.4 We can get a similar result in a reverse direction as follows. As before, $p, q > 0$ with $p^{-1} + q^{-1} = 1$ (if $p = 1$, $q = \infty$ and vice versa, by convention).

(i) Let P be a trigonometric polynomial of degree N say. Using the reverse Hölder inequality (question 32.6 (v)) show that given any $\varepsilon > 0$ we can find a trigonometric polynomial with $\|Q\|_p = 1$ and

$$\left|\frac{1}{2\pi}\int_{\mathbb{T}} P(t)Q(t)\,dt\right| \geq (1-\varepsilon)\|P\|_q.$$

(ii) Show that $(1/2\pi)\int_{\mathbb{T}} P(t)Q(t)\,dt = \sum_{r=-N}^{N} \hat{P}(r)\hat{Q}(-r)$ and use Hölder's inequality (Question 32.6 (i)) to show that $\|\hat{P}\|_p\|\hat{Q}\|_q \geq (1-\varepsilon)\|P\|_q$. Hence conclude using Question 34.3 that $\|\hat{P}\|_p \geq \|P\|_q$ whenever $2 \geq p \geq 1$.

(iii) Suppose more generally that $u, v > 0$ with $u^{-1} + v^{-1} = 1$. Show that if $\|\hat{Q}\|_q \leq A\|Q\|_u$ for all trigonometric polynomials Q then $A\|\hat{P}\|_p \geq \|P\|_v$ for all trigonometric polynomials P.

(iv) Show by a similar argument that if $\|Q\|_q \leq B\|\hat{Q}\|_u$ for all trigonometric polynomials Q then $B\|P\|_p \geq \|\hat{P}\|_v$ for all trigonometric polynomials P.

(The knowledgeable reader will recognise an argument involving adjoint mappings. She is invited, as a useful exercise, to recast the argument in a form which involves adjoint mapping explicitly, but reminded that historically it was this kind of special argument which inspired the general notion of an adjoint.)

34.5 (The Rudin Shapiro polynomials)

(i) Show by using Parseval's equality that if $|a_1| = |a_2| = \ldots = |a_N| = 1$ then $\sup_{t\in\mathbb{T}}|\sum_{n=1}^{N} a_n \exp(int)| \geq N^{1/2}$.

It is therefore of interest to find $a_n \in \mathbb{C}$ with $|a_n| = 1$ $[n = 1, \ldots, N]$ such that $\sum_{n=1}^{N} a_n \exp(int)| \leq AN^{1/2}$ for all $t \in \mathbb{T}$ where A is some constant independent of N. It was known for a long time that $a_n = \exp(in\log n)$ gave such coefficients (see Question 32.13) but not whether the a_n could be taken real. To quote Littlewood 'As a matter of cold fact many had doubted its truth'. However, Shapiro in his MSc thesis gave the following happy argument. (Since no one looks for brilliant ideas in a MSc thesis the argument had to be rediscovered by Rudin before the mathematical community took notice.)

(ii) Let $P_m(z) = \sum_{n=1}^{2m} \varepsilon_{nm} z^n$, $Q_m(z) = \sum_{n=1}^{2m} \eta_{nm} z^n$ with $\varepsilon_{nm}, \eta_{nm} \in \{-1, 1\}$. Show that, if we set

$$P_{m+1}(z) = P_m(z) + z^{2^m} Q_m(z), \quad Q_{m+1} = P_m(z) - z^{2^m} Q_m(z),$$

then

(a) $P_{m+1}(z) = \sum_{n=1}^{2m+1} \varepsilon_{nm+1} z^n$, $Q_{m+1}(z) = \sum_{n=1}^{2m+1} \eta_{nm+1} z^n$ with ε_{nm+1}, $\eta_{nm+1} \in \{-1, 1\}$;

(b) if $|z| = 1$, then

$$|P_{m+1}(z)|^2 + |Q_{m+1}(z)|^2 = 2(|P_m(z)|^2 + |Q_m(z)|^2)$$

(recall $|a|^2 = aa^*$).

(iii) By taking $P_0(z) = Q_0(z) = z$ deduce that we can find $P_m(z) = \sum_{n=1}^{2^m} \varepsilon_{nm} z^n$ with $\varepsilon_{nm} \in \{-1, 1\}$ such that $|P_m(\exp it)| \leqslant 2^{(m+1)/2}$.

(iv) If $r \geqslant 1$ and $2 > q \geqslant 1$, show that $\|\hat{P}_m\|_q / \|P_m\|_r \to \infty$ as $m \to \infty$. Deduce that no inequality of the form $A\|f\|_r \geqslant \|\hat{f}\|_q$ can hold if $2 > q \geqslant 1$. Using Question 34.4 (iv) show also that no inequality of the form $A\|\hat{P}\|_r \geqslant \|P\|_q$ can hold for trigonometric polynomials if $2 > q > 1$. (Thus the range of q obtained in the Hausdorff–Young inequality if the largest possible.) Does there exist an inequality of the form $A\|\hat{P}\|_\infty \geqslant \|P\|_1$?

(v) If N is a positive integer, then we can write $N = \sum_{k=0}^{m} 2^{n(k)}$ with $0 \leqslant n(0) < n(1) < \ldots < n(k)$. By considering polynomials $P(\exp it) = \sum_{k=0}^{m} \exp(2^{n(k)} it) P_{n(k)}(\exp it)$, show that there exists a constant A independent of N such that, given $N \geqslant 1$, we can find $a_1, a_2, \ldots, a_N \in \{-1, 1\}$ with

$$\left| \sum_{r=1}^{N} a_r \exp(irt) \right| \leqslant AN^{1/2} \quad \text{for all } t.$$

(vi) In fact a stronger result along the lines of Question 32.13 is true. Show that, if $\varepsilon_{n,m}$ are the coefficients of P_m as in (ii), then $\varepsilon_{n,m} = \varepsilon_{n,M}$ for all $1 \leqslant n \leqslant 2^m$ and $1 \leqslant m \leqslant M$. Thus we may define a sequence (ε_n) by $\varepsilon_n = \varepsilon_{n,M}$ for $1 \leqslant n \leqslant 2^m$. Define the sequence (η_n) similarly. Show that $\varepsilon_{n+2^m} = \eta_n$ for $1 \leqslant n \leqslant 2^m$. Show by induction, or otherwise, that $|\sum_{n=1}^{N} \varepsilon_n \exp(in\theta)| \leqslant \sum_{m=1}^{M} 2^{(m+1)/2}$ for all $1 \leqslant N \leqslant 2^M$. Conclude that there exists a constant A independent of N such that $|\sum_{n=1}^{N} \varepsilon_n \exp(in\theta)| \leqslant AN^{1/2}$ for all $N \geqslant 1$.
(It remain an open question whether, given $\varepsilon > 0$, we can find an $N \geqslant 2$ and $a_1, a_2, \ldots, a_N \in \{-1, 1\}$ such that $|\sum_{r=1}^{N} a_r \exp(irt)| \leqslant (1+\varepsilon)N^{1/2}$ for all t. Admittedly the problem has no deep mathematical significance, but I will happily send the cost of a bottle of good champagne to the first person to answer it.)

35

THE ISOPERIMETRIC PROBLEM I

35.1 (i) Consider the collection E of continuously once differentiable functions $f: [-1, 1] \to \mathbb{R}$ with $f(-1) = -1$, $f(1) = 1$. Show that

$$\inf_{f \in E} \int_{-1}^{1} f'(x)^2 x^4 \, dx = 0,$$

but that this infimum is not attained.

(ii) Consider the collection E of continuously once differentiable functions $f: [-1, 1] \to \mathbb{R}$. Show that

$$\inf_{f \in E} \int_{-1}^{1} ((f'(x)^2 - 1)^2 + f(x)^2) \, dx = 0,$$

but that this infimum is not attained.

(Part (i) gives one of the examples used by Weierstrass. Energy functions resembling that in (ii) occur physically in the theory of crystallisation. Electron microscope photography reveals a crinkly pattern that the reader will now have no difficulty in explaining.)

36

THE ISOPERIMETRIC PROBLEM II

36.1 (i) Prove Wirtinger's inequality. If $u\colon \mathbb{T} \to \mathbb{R}$ is once continuously differentiable and $\int_{\mathbb{T}} u(t)\,dt = 0$, then $\int_{\mathbb{T}}(u(t))^2\,dt \leqslant \int_{\mathbb{T}}(u'(t))^2\,dt$ with equality if and only if $u(t) = C\cos(t+\psi)$ for some $C, \psi \in \mathbb{R}$.

(ii) By extending v in an appropriate manner use (i) to show that if $v\colon [0, \pi/2] \to \mathbb{R}$ is once continuously differentiable with $v(0) = 0$ and $v'(\pi/2) = 0$ then

$$\int_0^{\pi/2} (v(t))^2\,dt \leqslant \int_0^{\pi/2} (v'(t))^2\,dt$$

with equality if and only if $v(t) = C\sin t$ for some $C \in \mathbb{R}$.

(iii) By approximating w by functions v of the type considered in (ii) show that if $w\colon [0, \pi/2] \to \mathbb{R}$ is once continuously differentiable with $w(0) = 0$ then

$$\int_0^{\pi/2} (w(t))^2\,dt \leqslant \int_0^{\pi/2} (w'(t))^2\,dt.$$

37

THE STURM–LIOUVILLE EQUATION I

37.1$^{(-)}$ A simple animal absorbs oxygen from its surroundings by diffusion. Its cells consume oxygen at a fixed rate per unit volume of k mole cm^{-3} s^{-1}, and the animal can not survive if the oxygen concentration C anywhere falls below C_0mole cm^{-3}. The oxygen concentration in the surroundings is C_1. If the reader has been exposed to the discipline of considering what goes in and what comes out of small cubes side length δx, she should be able to show that the governing equation for the steady state concentration distribution C is

$$K\nabla^2 C = k \qquad (*)$$

where K cm^2 s^{-1} is the diffusivity of oxygen in the animal's tissue. If not she should just assume $(*)$.

(i) Suppose that the animal is a long worm of radius a and $C = C(r)$ depends only on r the radial distance from the worm's axis. Show that $(*)$ becomes

$$K\frac{d}{dr}\left(r\frac{dC}{dr}(r)\right) = kr,$$

and by solving this equation with appropriate boundary conditions show that the maximum radius of such a worm is

$$a = \left[\frac{4K}{k}(C_1 - C_0)\right]^{1/2}.$$

(ii) Investigate the same problem for a spherical animal.

(iii) Evaluate the maxima of (i) and (ii) when

$$C_1 = 8 \times 10^{-6} \text{ mole cm}^{-3},\ C_0 = 2 \times 10^{-6} \text{ mole cm}^{-3},$$

$$k = 7 \times 10^{-8} \text{ mole cm}^{-3}\text{ s}^{-1},\ K = 2 \times 10^{-5} \text{ cm}^2\text{ s}^{-1}.$$

(Roughly speaking the more complex an animal the greater its volume has to be. Thus a simple spherical model is ruled out and, although a worm-shaped model is not ruled out directly (why?), complex animals employ different mechanisms to supply oxygen to their cells. It is no accident that the structure of our lungs resembles the curves of infinite length constructed as counter examples by mathematicians at the end of the nineteenth century.)

38

LIOUVILLE

38.1 Cantor gave a very different proof of the existence of transcendentals.

(i) Let $\mathcal{P}$ be the set of polynomials $\sum_{r=0}^{n} a_r x^r$ with integer coefficients. Let $\mathcal{P}_m$ be the set of polynomials $\sum_{r=0}^{n} a_r x^r$ with integer coefficients and such that $m \geqslant n + \sum_{r=0}^{n} |a_r|$. Show that $\mathcal{P}_m$ is finite for each m and $\bigcup_{m=1}^{\infty} \mathcal{P}_m = \mathcal{P}$. Deduce that we can find $g_1, g_2, \ldots$ (not necessarily distinct) such that $\mathcal{P} = \{g_r : r \geqslant 1\}$.

(ii) Let $x_0 \in \mathbb{R}$, $\varepsilon_0 > 0$. Construct $x_n \in \mathbb{R}$, $\varepsilon_n > 0$ inductively as follows. Suppose x_{n-1}, ε_{n-1} given. Since $g_n(t) = 0$ has only a finite set of solutions, we can find x_n with $|x_n - x_{n-1}| < \varepsilon_{n-1}/2$ such that $g_n(x_n) \neq 0$. Since g_n is continuous, we can find ε_n with $0 < \varepsilon_n < \varepsilon_{n-1}/2$ such that $g_n(t) \neq 0$ for $|t - x_n| < 2\,\varepsilon_n$.

(iii) Show by induction that $|x_{n+r} - x_n| < \varepsilon_n$ for all $r \geqslant 0$ and all $n \geqslant 0$ and deduce firstly that x_m tends to a limit x as $m \to \infty$ and secondly that $|x - x_n| \leqslant \varepsilon_n$ for each $n \geqslant 0$. Conclude that $g_n(x) \neq 0$ for all $n \geqslant 1$ and so x is not algebraic. Deduce that the transcendental numbers are dense in the reals.

(iv) Use the same sort of argument to show that if $y_1, y_2, \ldots$ is any sequence of real numbers then $\mathbb{R} \neq \{y_j : j \geqslant 1\}$ (i.e. the real numbers are uncountable).

(v) Which argument (Cantor's or Liouville's) do you prefer and why? Most texts give a different Cantorian argument that the reals are uncountable based on decimal expansions. If you know it, do you prefer it to be the one presented here? Why?

38.2 (i) Let $g_1, g_2, \ldots$ be as in Question 38.1. Show that we can find sets U_n such that each U_n is the union of a finite set of open intervals of total length 2^{-10-n} and U_n contains all the real roots of g_n.

(ii) Let $V_q = \bigcup_{p=0}^{q} \{x : |x - p/q| < 2^{-10} q^{-3}\}$ for $q \geqslant 1$. Show by considering lengths, or otherwise, that there exists an $x_n \in [0, 1]$ with $x_n \notin \bigcup_{r=1}^{n} U_r$ and $x_n \notin \bigcup_{r=1}^{n} V_r$. By extracting a convergent subsequence, or otherwise, show that there exists an $x \in [0, 1]$ which is transcendental but such that

$|x - p/q| \geq 2^{-10}q^{-3}$ for all integers p and q with $q \geq 1$. (Some transcendental numbers cannot be well approximated by rational numbers.)

(iii) By modifying the argument above show that there exists a transcendental number x such that, if $\alpha > 2$, the inequality $|x - p/q| < q^{-\alpha}$ has only finitely many solutions with p and q integers and $q \geq 1$.

38.3 (i) Prove by induction, or otherwise, that, if P and Q are polynomials which are not identically zero, then we can find a (possibly zero) polynomial R, distinct complex numbers $\lambda_1, \lambda_2, \ldots, \lambda_m$, integers $n(1), n(2), \ldots, n(m) \geq 1$ and complex numbers a_{rs} $[1 \leq r \leq n(s), 1 \leq s \leq m]$ such that

$$\frac{P(z)}{Q(z)} = R(z) + \sum_{s=1}^{m} \sum_{r=1}^{n(s)} \frac{a_{rs}}{(z - \lambda_s)^r}$$

(See Question 74.1 if you need a hint. For completeness we must allow $m = 0$ and $P(z)/Q(z) = R(z)$.)

(ii) Suppose that $\dfrac{d}{dz}\dfrac{P(z)}{Q(z)} = \dfrac{P_1(z)}{Q_1(z)}$ where P_1, Q_1 are polynomials, Q_1 is not identically zero and P_1, Q_1 have no common factor. Show that (if P_1 is not identically zero) either z^2 is a factor of $Q_1(z)$ or z is not a factor of $Q(z)$.

(iii) Deduce that we can not find polynomials P and Q such that

$$\frac{d}{dz}\frac{P(z)}{Q(z)} = \frac{1}{z}.$$

(The object of this question was to show how it is possible to prove a result of the form 'there does not exist a function of type A which, when differentiated, gives a function f'. The proof of the theorem of Liouville on the integration of $e^{-x^2/2}$ mentioned in the paragraph before Theorem 38.2 runs along similar (but more complicated) lines. In particular, problems of this type seem to be of an *algebraic* rather than an *analytic* nature. As an introduction to the topic, together with an indication as to why yet another cobweb covered piece of nineteenth-century mathematics has suddenly become fashionable and of some practical importance, consult Chapter 5 of *Computer Algebra* by Davenport, Suet and Tournier.)

38.4 (i) Recall that $\exp z = e^z$. Consider the map $\exp: \mathbb{C} \to \mathbb{C}\backslash\{0\}$. If $z(0) \in \mathbb{C}$ write down all the solutions of $\exp z = \exp z(0)$. If $U = \{z: |z - z(0)| < 1/4\}$ show that the restriction map $\exp|_U: U \to \exp U$ is conformal.

(ii) Let $f: \mathbb{C} \to \mathbb{C}$ be an analytic function with period $2\pi i$ (i.e. $f(z + 2\pi i) = f(z)$ for all $z \in \mathbb{C}$). Show carefully that $g(\exp w) = f(w)$ defines an analytic function $g: \mathbb{C}\backslash\{0\} \to \mathbb{C}$. By applying Laurent's theorem, show that $f(z) = \sum_{n=-\infty}^{\infty} c_n \exp(nz)$ where $c_n = (1/2\pi i) \int_{\Gamma(a,a+2\pi i)} f(z) \exp(-nz)\, dz$ and $\Gamma(a, a + 2\pi i)$ is a straight line from some $a \in \mathbb{C}$ to $a + 2\pi i$.

(iii) If we define $F: \mathbb{T} \to \mathbb{C}$ by $F(t) = f(it)$, what is the relation between $\hat{F}(n)$ and c_n?

(iv) Generalise the result of (ii) to the case when f has period ω.

(v) Let $f: \mathbb{C} \to \mathbb{C}$ be analytic and periodic with period ω. Show that we can write

$$f(z) = \sum_{n=-N}^{N} C_n \exp((2\pi inz)/\omega)$$

if and only if $\limsup_{|z|\to\infty} (\log|f(z)|)/|z| < \infty$.

(vi) Give, with proof, an example of a non constant analytic function $f: \mathbb{C} \to \mathbb{C}$ which is not periodic and satisfies $\limsup_{|z|\to\infty} (\log|f(z)|)/|z| < \infty$.

38.5 (Recall that a meromorphic function can only have a finite set of poles in any closed bounded set (see e.g. the second paragraph of the proof of Lemma 76.14).)

(i) Let f be a meromorphic function on $\mathbb{C}$ which is periodic with period $2\pi i$. Show that we can find real a and b such that f has no poles in the strip $S = \{z: a < \operatorname{Re} z < b\}$. Show further that $f(z) = \sum_{n=-\infty}^{\infty} c_n \exp(nz)$ where $c_n = (1/2\pi i) \int_{\Gamma(w,w+2\pi i)} f(z) \exp(-nz)\, dz$ and $\Gamma(w, w + 2\pi i)$ is a straight line from w to $w + 2\pi i$ with $w \in \mathbb{C}$ such that $a < \operatorname{Re} w < b$.

(ii) Does c_n depend on the choice of S and why?

(iii) Show that there is a meromorphic function γ_1 with period $2\pi i$ such that

(a) $\gamma_1(z) = \sum_{n=0}^{\infty} \exp -nz$ for $\operatorname{Re} z > 0$, $z \notin \{2\pi ir: r \in \mathbb{Z}\}$,
(b) γ_1 has a simple pole at $2r\pi i$ $[r \in \mathbb{Z}]$ but nowhere else.

Give an expression of the same kind as (a) for $\gamma_1(z)$ when $\operatorname{Re} z < 0$, $z \notin \{2\pi ir: r \in \mathbb{Z}\}$. What is the residue of γ_1 at 0?

(iv) For each integer $m \geq 1$ find a meromorphic function γ_m with period $2\pi i$ such that $\gamma_m(z) = \sum_{n=0}^{\infty} c_{mn} \exp(-nz)$ for $\operatorname{Re} z \geq 0$, $z \notin \{2\pi ir: r \in \mathbb{Z}\}$ and γ_m has a pole of order m exactly at $2\pi ir$ $[r \in \mathbb{Z}]$ but no other poles. Give the values of c_{mn} explicitly.

(v) Show that, if f is a meromorphic function on $\mathbb{C}$ with period $2\pi i$, then, given any R, we can write f in the form

$$f(z) = \sum_{n=-\infty}^{\infty} c_n \exp(nz) + \sum_{k=1}^{N} \lambda_k \gamma_{n(k)}(z - \alpha_k)$$

for all z with $|\operatorname{Re} z| < R$.

38.6 (A doubly periodic meromorphic function)

(i) If $N \geq 1$ is an integer, let us write

$$\Gamma(N) = \{n + mi: n, m \in \mathbb{Z} \text{ and } |n|, |m| \leq N\}.$$

Sketch $\Gamma(N)$. Sketch $\Gamma(M) \setminus \Gamma(N)$ when $M \geq N + 1$.

(ii) Let $R>0$. Show that, if $n, m \in \mathbb{Z}$, $n^2+m^2 \geqslant 4R^2$ and $z \in \mathbb{C}$, $|z| \leqslant R$, we have

$$|(n+mi+z)^{-3}| \leqslant 8(n^2+m^2)^{-3/2}.$$

(iii) Let $R>4$ be fixed for the time being and choose $N(0)$ an integer with $N(0)>2R$. If $N>N(0)$, set

$$f_N(z) = \sum_{n+im \in \Gamma(N)\setminus\Gamma(N(0))} (n+mi+z)^{-3}.$$

By showing that, if $|z| \leqslant R$, then

$$\sum_{\Gamma(N)\setminus(N-1)} |(n+mi+z)|^{-3} \leqslant 64N^{-2},$$

(of course, you may replace 64 by any other suitable constant), or otherwise, show that, if $M>N>N(0)$ and $|z| \leqslant R$,

$$|f_M(z) - f_N(z)| \leqslant 64N^{-1}.$$

Conclude that $f_N(z)$ converges uniformly to a function $f(z)$, say, on the region $\{z: |z| \leqslant R\}$. Quote a result (see e.g. Theorem 75.3 (iv)) which tells us that f is analytic in $\{z: |z|<R\}$.

(iv) Deduce that, if $R>4$ is fixed,

$$g(z) = \lim_{N\to\infty} \sum_{n+im\in\Gamma(N)} (n+im+z)^{-3}$$

defines a function meromorphic on $\{z: |z|<R\}$ with poles of order 3 at points of the form $n+mi$ with $n, m \in \mathbb{Z}$. By allowing R to vary, conclude that

$$g(z) = \lim_{N\to\infty} \sum_{n+im\in\Gamma(N)} (n+im+z)^{-3}$$

defines a meromorphic function on $\mathbb{C}$ with poles of order 3 at points of the form $n+mi$ with $n, m \in \mathbb{Z}$.

(v) Set $g_N(z) = \sum_{n+im\in\Gamma(N)} (n+im+z)^{-3}$. If z is fixed and $N>2|z|$, show that

$$|g_N(z+1) - g_N(z)| \leqslant 64N^{-2}$$

and conclude that $g(z+1) = g(z)$. Show similarly that $g(z+i) = g(z)$ for all $z \in \mathbb{C}$. Thus g is doubly periodic with periods 1 and i.

(vi) Suppose $\omega_1, \omega_2 \in \mathbb{C}\setminus\{0\}$ and $\omega_1/\omega_2 \notin \mathbb{R}$. Show that there exists a $\delta>0$ such that

$$\{z: |z|<\delta\} \cap \{n\omega_1 + m\omega_2: n, m \in \mathbb{Z}\} = \{0\}.$$

Show that, if $N \geqslant 1$ is an integer,

$$\{z: |z| < N\delta\} \cap \{n\omega_1 + n\omega_2: n, m \in \mathbb{Z} \text{ and } |n|, |m| \geqslant N\} = \{0\}.$$

Conclude that, if R is fixed, we can find a $K > 0$ and an $N(0) \geqslant 1$ such that

$$|(n\omega_1 + m\omega_2 + z)^{-3}| \leqslant K(m^2 + n^2)^{-3/2}$$

whenever $|z| < R$, $m^2 + n^2 \geqslant N(0)^2[m, n \in \mathbb{Z}]$. Sketch a proof that there exists a non trivial doubly periodic function with periods ω_1, ω_2.

(vii) Let $\omega_1, \omega_2, \omega_3 \in \mathbb{C}\backslash\{0\}$ be such that $\omega_1/\omega_2, \omega_2/\omega_3, \omega_3/\omega_1 \notin \mathbb{R}$ and $n_1\omega_1 + n_2\omega_2 + n_3\omega_3 = 0$ has no integer solution apart from $n_1 = n_2 = n_3 = 0$. What is wrong with the following argument? 'Let us define

$$G(z) = \lim_{N\to\infty} \sum_{|n_1|,|n_2|,|n_3|\leqslant N, n_1,n_2,n_3\in\mathbb{Z}} (n_1\omega_1 + n_2\omega_2 + n_3\omega_3 + z)^{-3}.$$

The argument of (v) shows that G has periods ω_1, ω_2 and ω_3. Thus G is a triply periodic meromorphic function.'

38.7 (Non trivial triply periodic meromorphic functions do not exist.)

(i) Let f be a meromorphic function on $\mathbb{C}$. Let us write

$$\Lambda = \{\omega \in \mathbb{C}: f(z + \omega) = f(z) \text{ for all } z \in \mathbb{C}\}.$$

Show that

(a) if $\omega_1, \omega_2 \in \Lambda$ and $n_1, n_2 \in \mathbb{Z}$, then $n_1\omega_1 + n_2\omega_2 \in \Lambda$,
(b) there exists a $\delta > 0$ such that, if $|\omega| < \delta$ and $\omega \in \Lambda$, then $\omega = 0$.

(Look at Theorem 75.6 if you need a hint.)

(ii) We now forget about f and simply look at a Λ satisfying (a) and (b). (Λ is called a lattice; in the case (i) we say that Λ is the period lattice of f.) Show that either $\Lambda = \{0\}$ of $\inf\{|\omega|: \omega \in \Lambda\backslash\{0\}\}$ exists and has value $\delta_1 > 0$. Show that, if $\omega_1, \omega_2 \in \Lambda$, either $\omega_1 = \omega_2$ or $|\omega_1 - \omega_2| \geqslant \delta_1$. Conclude that $\{\omega \in \Lambda: 0 < |\omega| < 2\delta_1\}$ is a finite and non empty and so there exists an $\omega_1 \in \Lambda$ such that $|\omega_1| = \delta_1$.

(iii) Suppose ω_1 is as in (ii) and that $\omega \in \{\lambda\omega_1: \lambda \in \mathbb{R}\} \cap \Lambda$. Show that we can find an $n \in \mathbb{Z}$ such that $|\omega - n\omega_1| \leqslant |\omega_1|/2$. By observing that $\omega - n\omega_1 \in \Lambda$, deduce that $\omega = n\omega_1$.

(iv) Suppose that $\Lambda \neq \{0\}$ and ω_1 has been obtained as above. By adapting the arguments already employed in (ii) and (iii) show that either $\Lambda = \{n\omega_1: n \in \mathbb{Z}\}$ or we can find an $\omega_2 \in \Lambda$ such that $\omega_2 \notin \{n\omega_1: n \in \mathbb{Z}\}$ and, if $\omega \notin \Lambda\backslash\{n\omega_1: n \in \mathbb{Z}\}$, then

$$|\omega - n\omega_1| \geqslant |\omega_2| \quad \text{for all } n \in \mathbb{Z}.$$

Show further that, if $\omega \in \Lambda$, then we can find $n_1, n_2 \in \mathbb{Z}$ such that $\omega = n_1\omega_1 + n_2\omega_2$.

(v) Conclude that, if Λ satisfies (a) and (b), then either $\Lambda = \{0\}$ or $\Lambda = \{n_1\omega_1 : n_1 \in \mathbb{Z}\}$ for some $\omega_1 \neq 0$ or $\Lambda = \{n_1\omega_1 + n_2\omega_2 : n_1, n_2 \in \mathbb{Z}\}$ for some $\omega_1, \omega_2 \neq 0$ with $\omega_1/\omega_2 \in \mathbb{R}$. Returning to part (i), classify the possible sets of periods for a meromorphic function.

38.8 Are there any further restrictions on doubly periodic meromorphic functions? Suppose f is a non constant meromorphic function on $\mathbb{C}$ with periods ω_1, ω_2 $[\omega_1, \omega_2 \in \mathbb{C}\backslash\{0\}, \omega_1/\omega_2 \notin \mathbb{R}]$. If $a \in \mathbb{C}$, let us write

$$P_a = \{a + \lambda_1\omega_1 + \lambda_2\omega_2 : 0 \leqslant \lambda_1, \lambda_2 < 1\}.$$

(P_a is called a period parallelogram.)

(i) If $b \in \mathbb{C}$, explain why there are only finitely many $z \in P_a$ with $f(z) = b$. Let us write $N_a(b)$ for the number of solutions of $f(z) = b$ with $z \in P_a$, multiple roots being counted multiply. (If you do not understand the meaning of the last phrase, your knowledge of complex variables is probably insufficient for the rest of the question.) Explain also why $N_a(\infty)$, the number of poles in P_a (multiple poles being counted multiply), is a well-defined integer. Using periodicity show that $N_a(b)$ and $N_a(\infty)$ are independent of a, so we may write $N_a(b) = N(b)$, $N_a(\infty) = N(\infty)$.

(ii) Let C be the contour consisting of the straight lines a to $a + \omega_1$, $a + \omega_1$ be $a + \omega_1 + \omega_2$, $a + \omega_1 + \omega_2$ to $a + \omega_2$, $a + \omega_2$ to a. Show that, if g is a meromorphic function, with no poles on C having periods ω_1 and ω_2, then $\int_C g(z)\, dz = 0$.

(iii) Explain why we can choose a so that f has neither zeros nor poles on C. By considering $\int_C (f'(z)/f(z))\, dz$, show that $N(0) = N(\infty)$. (If you need a hint look at Theorem E.1.) By looking at $f - b$ deduce that $N(b) = N(\infty)$ for all $b \in \mathbb{C}$. In view of this, let us write $N(\infty) = N(f)$.

(iv) Suppose that f has poles at $\gamma_1, \gamma_2, \ldots, \gamma_M$ with residues $\rho_1, \rho_2, \ldots, \rho_M$ within P_a. Show, by considering $\int_C f(z)\, dz$, that $\sum_{j=1}^{M} \rho_j = 0$. Deduce in particular that $N(\infty) \geqslant 2$.

(v) Now suppose that f has poles $\alpha_1, \alpha_2, \ldots, \alpha_m$ (multiple poles repeated multiply) and zeros $\beta_1, \beta_2, \ldots, \beta_m$ (multiple zeros repeated multiply) within P_a. By considering $\int_C (zf'(z)/f(z))\, dz$, show that

$$\sum_{j=1}^{m} \alpha_j - \sum_{j=1}^{m} \beta_j \in \{n_1\omega_1 + n_2\omega_2 : n_1, n_2 \in \mathbb{Z}\}.$$

In particular, remark that, once $\alpha_1, \alpha_2, \ldots, \alpha_m, \beta_1, \beta_2, \ldots, \beta_{m-1}$ are known, β_m is uniquely determined.

(In the next questions we shall see that the restrictions obtained in (iv) and (v) are the only ones on the poles and zeros of a doubly periodic meromorphic function.)

38.9 (i) Let α_1, α_2 lie in $\{x+iy: 0 \leqslant x, y<1\}$ with $\alpha_1 \neq \alpha_2$. Show that, for a unique choice of $a, b \in \mathbb{C}$ and some suitable choice of K, we have

$$\left|\frac{1}{(z-\alpha_1+n+im)}+\frac{a}{(z-\alpha_2+n+im)}+\frac{b}{(n+im)^2}\right| \leqslant K(n^2+m^2)^{-3/2}$$

whenever $n^2+m^2 \geqslant 4|z|^2$ and $n, m \in \mathbb{Z}$.

(ii) By imitating Question 38.6, show that

$$g(z)=\lim_{n+im \in \Lambda(N)}\left(\frac{1}{(z-\alpha_1+n+im)}+\frac{a}{(z-\alpha_2+n+im)}+\frac{b}{(n+im)^2}\right)$$

defines a meromorphic function which is doubly periodic with periods 1 and i and has simple poles at α_1+n+im, α_2+n+im $[n, m \in \mathbb{Z}]$.

(iii) Let $\beta_1 \in \{x+iy: 0 \leqslant x, y<1\}$, $\beta_1 \neq \alpha_1, \alpha_2$. Show that $f=g-g(\beta_1)$ is a doubly periodic meromorphic function with periods 1 and i such that the only poles of f in $\{x+iy: 0 \leqslant x, y<1\}$ are at α_1 and α_2 and the only zeros at β_1, β_2 where $\alpha_1+\alpha_2-\beta_1-\beta_2 \in \{n+mi: n, m \in \mathbb{Z}\}$.

(iv) By considering

$$g(z)=\lim_{n \rightarrow \infty} \sum_{n+mi \in \Lambda(N)}\left(\frac{1}{(z-\alpha+n+im)^2}-\frac{b}{(n+im)^2}\right),$$

or otherwise, show that, given $\alpha, \beta_1, \beta_2 \in \{x+iy: 0 \leqslant x, y<1\}$ with

$$2\alpha-\beta_1-\beta_2 \in \{n+mi: n, m \in \mathbb{Z}\}$$

and $\beta_1, \beta_2 \neq \alpha$, we can find a doubly periodic meromorphic function with periods 1 and i such that the only pole of f in $\{x+iy: 0 \leqslant x, y<1\}$ is a double pole at α and the only zeros are at β_1, β_2.

(v) Suppose $\alpha_1, \alpha_2, \alpha_3 \in S=\{x+iy: 0 \leqslant x, y<1\}$ are distinct and $\beta_1, \beta_2, \beta_3 \in S$ are such that $\beta_1, \beta_2, \beta_3 \notin \{\alpha_1, \alpha_2, \alpha_3\}$ and

$$\alpha_1+\alpha_2+\alpha_3-\beta_1-\beta_2-\beta_3 \in \{n+mi: n, m \in \mathbb{Z}\}.$$

By (iii), we can find meromorphic doubly periodic functions F, G with periods 1 and i such that F has simple poles in S at α_1, α_2 and a zero at β_1, (together with some other zero) and G has simple poles in S at α_2, α_3 and a zero at β_1 (together with some other zero). Explain why there is no doubly periodic meromorphic function with periods 1 and i whose poles in S are α_1, α_3 and whose zeros are β_1, β_3. Show that $F(\beta_3)$, $G(\beta_3) \neq 0$ and that, for a suitable choice of λ, $\mu \in \mathbb{C}$, we have $f=\lambda F+\mu G$ a doubly periodic function with periods 1 and i having simple poles in S at $\alpha_1, \alpha_2, \alpha_3$ and zeros at $\beta_1, \beta_2, \beta_3$.

(vi) Prove more generally that, if $\alpha_1, \alpha_2, \alpha_3, \beta_1, \beta_2, \beta_3 \in S$ with $\beta_1, \beta_2, \beta_3 \notin \{\alpha_1, \alpha_2, \alpha_3\}$ and

$$\alpha_1+\alpha_2+\alpha_3-\beta_1-\beta_2-\beta_3 \in \{n+mi: n, m \in \mathbb{Z}\},$$

then there exists a doubly periodic function with periods 1 and i, with poles in S given by α_1, α_2, α_3 and zeros in S given by β_1, β_2, β_3 (multiple zeros and poles represented multiply). (You may wish to recall Question 38.7 at some point in the argument.)

(vii) Discuss very briefly what changes, if any, would be required in your proofs if we replace the words 'periods 1 and i' by 'periods ω_1 and ω_2' (with $\omega_1, \omega_2 \neq 0$, $\omega_1/\omega_2 \notin \mathbb{R}$).

38.10 This question uses a lot of notation but is otherwise rather easy. Let $\omega_1, \omega_2 \in \mathbb{C}$ be such that $\omega_1, \omega_2 \neq 0$ and $\omega_1/\omega_2 \notin \mathbb{R}$. We shall be dealing with the set $\mathcal{F}$ of doubly periodic meromorphic functions with periods ω_1 and ω_2. We shall take P to be a fundamental parallelogram and write

$$\Lambda = \{n_1\omega_1 + n_2\omega_2 : n_1, n_2 \in \mathbb{Z}\}.$$

We say that $((\alpha_1, \alpha_2, \ldots, \alpha_m), (\beta_1, \beta_2, \ldots, \beta_m))$ are specified poles and zeros if

(a) $m \geqslant 2$;
(b) $\alpha_j, \beta_j \in P\ [1 \leqslant j \leqslant m]$; $\alpha_i \neq \beta_j\ [1 \leqslant i, j \leqslant m]$;
(c) $\sum_{j=1}^{m} \alpha_j - \sum_{j=1}^{m} \beta_j \in \Lambda$.

We say that $f \in \mathcal{F}$ has specified poles and zeros $((\alpha_1, \alpha_2, \ldots, \alpha_m), (\beta_1, \beta_2, \ldots, \beta_m))$ if the poles of f in P are $\alpha_1, \alpha_2, \ldots, \alpha_m$ (multiple poles being given multiply) and the zeros of f in P are $\beta_1, \beta_2, \ldots, \beta_m$ (multiple zeros being given multiply).

(i) If $f, g \in \mathcal{F}$, explain how the poles and zeros of f and g in P are related to the poles and zeros of fg in P. (Pay particular attention to the possibility of cancellation.) By using Question 38.9, or otherwise, show that, given any specified poles and zeros, we can find an $f \in \mathcal{F}$ with those specified poles and zeros.

(ii) Suppose that $f, g \in \mathcal{F}$ have the same specified poles and zeros. Show that there exist $\lambda, \mu \neq 0$ such that $\lambda f + \mu g = 0$.

(The results of this question are not only interesting in themselves, but indicate that the subject of doubly periodic functions is likely to be rich in algebraic identities. Just as today's students of analytic topology may feel themselves to be drowning in a flood of more and more peculiar topological spaces, so the late nineteenth-century student, faced with the 65 pages of formulae at the end of Volume 2 of Molk and Tannery's *Eléments de la Théorie des Fonctions Elliptiques*, could be forgiven a certain lowering of the spirits. In the next question we shall derive a few of these identities and, I hope, see how pretty and useful they are. What I shall not be able to show the reader is the fact that they are an indication of a more profound underlying order, the glint in a rock, as it were, which indicates to the experienced prospector the presence of a rich seam of ore deep underground.)

38.11 (The Weierstrass $\mathcal{P}$ function)

(i) By referring to earlier questions or by sketching the outlines of a proof, show that

$$\mathcal{P}(z) = \frac{1}{z^2} + \lim_{N\to\infty} \sum_{(n,m)\neq(0,0);|n|,|m|\leq N} \left\{\frac{1}{(z+n+mi)^2} - \frac{1}{(n+mi)^2}\right\}$$

defines a doubly periodic meromorphic function with periods 1 and i having a double pole at points of the form $n + mi$ $[n, m \in \mathbb{Z}]$. (If we replace 1 and i by more general periods ω_1, ω_2, the more general $\mathcal{P}$ will have similar properties but some of the algebraic expressions below take a more complicated form.)

(ii) Show that $\mathcal{P}(z)$ can be written as

$$\mathcal{P}(z) = \frac{1}{z^2} + f(z),$$

where f is analytic in $\{z: |z| < 1/2\}$. By direct computation, exploiting the appropriate symmetries, find $f^{(r)}(0)$ for $0 \leq r \leq 5$ and show that

$$\mathcal{P}(z) = \frac{1}{z^2} + a_2 z^2 + \sum_{r=3}^{\infty} a_{2r} z^{2r},$$

where the sum $\sum_{r=3}^{\infty} a_{2r} z^{2r}$ has radius of convergence at least 1/2 and $a_2 = 3\sum_{(n,m)\neq(0,0)}(n+mi)^{-4}$. Can you prove any others of the a_{2r} to be 0?

(iii) Show that $\mathcal{P}'(z)^2 - 4\mathcal{P}(z)^3$ is a doubly periodic meromorphic function with periods 1 and i having a double pole (at worst) at points $n + mi$ $[n, m \in \mathbb{Z}]$. Show that for a suitable a (to be found) the function $h(z) = \mathcal{P}'(z)^2 - 4\mathcal{P}(z)^3 - a\mathcal{P}(z)$ is a doubly periodic analytic function (if we silently remove all removable singularities) with $h(0) = 0$. Deduce, giving your reason, that

$$\mathcal{P}'(z)^2 = 4\mathcal{P}(z)^3 - g_2\mathcal{P}(z)$$

where $g_2 = 60\sum_{(n,m)\neq(0,0)}(n+mi)^{-4}$.

38.12 (In this question we obtain a fairly simple relation between $\mathcal{P}(2z)$ and $\mathcal{P}(z)$. The early investigators of doubly periodic functions were guided to some extent by analogy with simple, singly periodic functions like cos and sin. Since

$$\cos 2\theta = 2(\cos\theta)^2 - 1 \text{ and } (\sin 2\theta)^2 = 4(\sin\theta)^2(1 - (\sin\theta)^2),$$

it is natural to seek similar formulae for doubly periodic functions.)

(i) Fix $u \notin \{n + mi: n, m \in \mathbb{Z}\}$. Explain briefly why we can find a $\delta(u) > 0$ such that, if $0 < |u - v| < \delta(u)$, then v is not a pole of $\mathcal{P}$ and $\mathcal{P}(v) \neq \mathcal{P}(u)$. In what follows we shall take $0 < |u - v| < \delta(u)$.

(ii) Show that we can find $A = A(u, v)$, $B = B(u, v)$ such that, writing

$$f(z) = \mathcal{P}'(z) - A\mathcal{P}(z) - B,$$

we have $f(u) = f(v) = 0$. You need not calculate B explicitly, but you should show that $A(u, v) = (\mathcal{P}'(v) - \mathcal{P}'(u))/(\mathcal{P}(v) - \mathcal{P}(u))$.

(iii) Observe that f is a doubly periodic function, find the location and order of its poles and hence deduce that $-(u + v)$ is a zero of f. Explain why we can find a $0 < \delta'(u) \leqslant \delta(u)$ such that $\mathcal{P}(u)$, $\mathcal{P}(v)$ and $\mathcal{P}(-(u + v))$ are all distinct for $0 < |u - v| < \delta'(u)$. In what follows we shall take $0 < |u - v| < \delta'(u)$.

(iv) Show, using Question 38.11 (iii), that

$$(f(z) + A\mathcal{P}(z) + B)^2 = 4\mathcal{P}(z)^3 - g_2\mathcal{P}(z)$$

and deduce that $\mathcal{P}(u)$, $\mathcal{P}(v)$, $\mathcal{P}(-(u + v))$ are the roots of the equation

$$(Az + B)^2 = 4z^3 - g_2 z.$$

Conclude that

$$\mathcal{P}(u) + \mathcal{P}(v) + \mathcal{P}(-(u + v)) = A^2/4$$

and so, using the symmetry of $\mathcal{P}$,

$$\mathcal{P}(u) + \mathcal{P}(v) + \mathcal{P}(-(u + v)) = \frac{1}{4}\left(\frac{\mathcal{P}'(v) - \mathcal{P}'(u)}{\mathcal{P}(v) - \mathcal{P}(u)}\right)^2.$$

Are restrictions on u and v actually needed for the truth of this equation, or does it hold for all $u, v \in \mathbb{C}$ provided suitable conventions are made about poles? Why?

(v) By allowing $v \to u$, show that

$$2\mathcal{P}(u) + \mathcal{P}(2u) = (\mathcal{P}''(u)/\mathcal{P}'(u))^2/4$$

and hence, using Question 38.11 (iii) again, show that

$$\mathcal{P}(2z) = R(\mathcal{P}(z)),$$

where $R(z) = (16z^4 + 8g_2z^2 + g_2^2)/16z(4z^2 - g_2)$.

(vi) We have not actually shown that $g_2 \neq 0$. Do this by using the formula $\mathcal{P}(2z) = R(\mathcal{P}(z))$ proved above, or otherwise.

38.13 (An example of Lattès) In this question we shall find a rational function P/Q (with P, Q polynomials) such that, writing $T(z) = P(z)/Q(z)$ and T^n for the nth iterate $T^n(z) = T(T \ldots (T(z)))$, the following is true. Given any $z_0 \in \mathbb{C}$ and any $\delta > 0$, we can find an n such that (with the usual conventions $1/0 = \infty$, $1/\infty = 0$)

$$T^n(\{z\colon |z - z_0| < \delta\}) = \{T^n z\colon |z - z_0| < \delta\} = \mathbb{C} \cup \{\infty\}.$$

Before starting, the reader should spend some time thinking about what this result says. If after some thought it appears obvious, or simply likely, that such a T exists, she should make a serious effort to construct it and prove the stated property.

(i) Let R be the rational function defined in Question 38.12. Let $\delta > 0$ and z_0 be given. We write

$$D = \{z: |z - z_0| < \delta\},$$
$$U = \mathcal{P}^{-1}(D) = \{w: \mathcal{P}w \in D\}$$

and define ϕ by $\phi(z) = 2z$ $[z \in \mathbb{C}]$. Explain why U must contain an open disc $\{w: |w - w_1| < \delta_1\}$, say, and deduce that, if we take n sufficiently large, the set $\phi^n(U) = \{2^n z: z \in U\}$ must contain a period parallelogram $P = \{z_1 + \lambda + \mu i: 0 \leq \lambda, \mu < 1\}$.

(ii) Use Question 38.12 (v) and standard properties of doubly periodic meromorphic functions to deduce that

$$R^n(D) = R^n(\mathcal{P}(U)) \supseteq \mathcal{P}(\phi^n(U)) \supseteq \mathcal{P}(P) = \mathbb{C} \cup \{\infty\}$$

as required.

(iii) We can produce a somewhat prettier rational function as follows. Let $h(z) = 2z/g_2^{1/2} - 2$ where $g_2^{1/2}$ is either square root of g_2 and set $S(z) = h \circ R \circ h^{-1}(z)$. Show that

$$S(z) = \frac{(z^2 + 1)^2}{4z(z^2 - 1)}$$

and that, given any disc $D = \{|z: z_0| < \delta\}$, we can find an m such that $S^m(D) = \mathbb{C} \cup \{\infty\}$.

(This example is taken from Beardon's *Iteration of Rational Functions* which I expect to become the standard introduction for the serious student.)

38.14 (This question and the next are more in the nature of remarks bringing out the analogies between the study of some singly periodic and some doubly periodic functions. The reader may well choose to omit them or, better, simply glance through their content. If she wishes to know more about a subject central to the mathematical lives of Legendre, Gauss, Abel, Jacobi, Weierstrass and Hermite, she could begin with the appropriate parts of Jones and Singerman *Complex Functions* which provides an introduction from a modern standpoint.)

(i) Let us look again at the formula

$$\mathcal{P}'(z)^2 = 4\mathcal{P}(z)^3 - g_2\mathcal{P}(z)$$

proved in Question 38.12. Taking a square root always requires care and we proceed as follows. By taking conjugates and using symmetry, we see that the constant g_2 of Question 38.11 (iii) is real and we know from Question 38.12 (vi) that g_2 is not zero. It is possible, by taking sufficiently many terms in the defining sum $60 \sum_{(n,m) \neq (0,0)} (n + mi)^{-4}$, to establish that $g_2 > 0$. (There is no need for the reader to do this, since the other possibility could be dealt with by an argument of the same type as that set out below.) Let $g_2^{1/2}$ be the positive square root of g_2 and let

$$\Omega = \mathbb{C}\setminus\{x \in \mathbb{R}: x \leqslant 2^{-1}g^{1/2}\}.$$

The reader should convince herself that $4z^3 - g_2 z$ has an analytic square root in Ω or, more precisely, that there exists an analytic function $f: \Omega \to \mathbb{C}$ such that $f(z)^2 = 4z^3 - g_2 z$ and $f(x)$ is real and positive when x is real and $x > 2^{-1}g_2^{1/2}$. (The reader may very well know general theorems which give this result. If not, an ad hoc proof on the lines of, or using, the results of Appendix E can easily be produced. In any case she should not worry too much about rigour in this question.)

Show that, if $z_0 \in \Omega$ and $\Gamma(z_0, z)$ is a smooth path within Ω from z_0 to z, the value

$$G(z) = \int_{\Gamma(z_0,z)} \frac{1}{f(\zeta)}\, d\zeta$$

is independent of Γ. Show by the usual arguments that $G: \Omega \to \mathbb{C}$ is an analytic function with $G'(z) = 1/f(z)$.

(ii) Now show that $((G \circ \mathcal{P})'(z))^2 = 1$ and use continuity to deduce that $G(\mathcal{P}(z)) = az + b$ for all $z \in \Omega$ where $a \in \{-1, 1\}$, $b \in \mathbb{C}$. Write $\surd(4z^3 - g_2 z) = af(z)$ and show that, if we choose w_0 correctly, then setting

$$H(z) = w_0 + \int_{\Gamma(z_0,z)} \frac{1}{\surd(4\zeta^3 - g_2\zeta)}\, d\zeta$$

we have $H(\mathcal{P}(z)) = z$.

(Thus $\mathcal{P}$ is an inverse function associated with the integral $\int(1/\surd(4\zeta^3 - g_2\zeta))\, d\zeta$. Integrals of forms similar to $\int_{x_0}^{t}(1/\surd(x^3 - kx))\, dx$ occur in the evaluation of the lengths of curves such as ellipses and lemniscates and are called elliptic integrals. It was Abel who spotted that, whilst the integral $\sin^{-1} t = \int_0^t(1/\surd(1 - x^2))\, dx$ is hard to work with, the inverse function $\sin x$ is much nicer and that, similarly, instead of working with *elliptic integrals* it is better to work with their inverses, the so-called *elliptic functions*.)

38.15 (i) Establish, in the manner of Questions 38.6 and 38.9, that

$$G_1(z) = \lim_{N\to\infty} \sum_{n=-N}^{N} (z + n)^{-2}$$

and

$$G_2(z) = \lim_{N\to\infty} \left(z^{-1} + \sum_{1\leqslant|n|\leqslant N} ((z + n)^{-1} - n^{-1})\right)$$

are well-defined meromorphic functions. Establish the position and the order of their poles. Show also that G_1 and G_2 are periodic with period 1.

(ii) Let C_N be the square contour consisting of line segments joining $(N+1/2)+(N+1/2)i$, $-(N+1/2)+(N+1/2)i$, $-(N+1/2)-(N+1/2)i$, $(N+1/2)-(N+1/2)i$ in the order given. Show carefully that there exists a K independent of N such that $|\cot \pi w| \leqslant K$ for all $w \in C_N$ and all integers $N \geqslant 1$. By considering

$$\int_{C_N} \frac{\cot \pi w}{(z+w)^2}\, dw \quad \text{and} \quad \int_{C_N} \frac{\cot \pi w}{(z+w)}\, dw$$

show that

$$G_1(z) = (\pi/\sin \pi z)^2 \quad \text{and} \quad G_2(z) = \pi \cot \pi z.$$

(iii) By re-examining the arguments above, or otherwise, show that, in particular, if $1/2 > \delta > 0$ is fixed, then

$$x^{-1} - \sum_{1 \leqslant n \leqslant N} 2x/(n^2 - x^2) \to \pi \cot \pi x$$

uniformly as $N \to \infty$ for all real $x \in [\delta, 1-\delta]$. (We shall make use of this result in Question 76.2.)

(iv) By considering $\int_{C_N} ((\cot \pi w)/w^{2n})\, dw$, show that $\sum_{r=1}^{\infty} r^{-2n}$ is a rational multiple of π^{2n} whenever n is a strictly positive integer.
(This gives an alternative proof for the result of Question 18.6(iv). However, a glance at Question 18.8 shows that the two proofs are fairly closely related.)

39
THE STURM–LIOUVILLE EQUATION II

39.1 We consider sequences $\mathbf{y} = (y_1, y_2, \ldots, y_N)$. We write $\Delta y_k = y_k - y_{k-1}$ and $\Delta(p_{k-1}\,\Delta y_{k-1}) = p_k(y_{k+1} - y_k) - p_{k-1}(y_k - y_{k-1})$. If $\mathbf{y}(\lambda) = (y_1(\lambda),\ y_2(\lambda),\ \ldots, y_N(\lambda))$ is a solution of the system of equations

$$\Delta(p_{k-1}\,\Delta y_{k-1}(\lambda)) - (\lambda\, r_k + q_k)y_k(\lambda) = 0 \text{ for } 1 \leqslant k \leqslant N,$$

where $y_0(\lambda)$ and $y_{N+1}(\lambda)$ are defined by

$$\alpha_0 y_0(\lambda) + \alpha_1 y_1(\lambda) = 0,\ \alpha_N y_N(\lambda) + \alpha_{N+1} y_{N+1}(\lambda) = 0,$$

we say that $\mathbf{y}(\lambda)$ is an eigensolution with associated eigenvalue λ. (Here the α_0, α_1, α_N, α_{N+1} and q_k are fixed real numbers, $r_k > 0$ and $\alpha_0 \neq 0$, $\alpha_{N+1} \neq 0$.)

(i) Show that the eigenvalues are real.

(ii) Show that we can choose real eigensolutions associated to each eigenvalue. We do this henceforth.

(iii) Show that, if λ, μ are distinct eigenvalues and $\mathbf{y}(\lambda)$, $\mathbf{y}(\mu)$ are eigensolutions, then $\sum_{k=1}^{N} r_k y_k(\lambda) y_k(\mu) = 0$.

(iv) Show that there are at most N distinct eigenvalues.

(v) Show that, if there are N distinct eigenvalues $\lambda(1), \lambda(2), \ldots, \lambda(N)$, then we can find associated eigensolutions $\mathbf{u}(1), \mathbf{u}(2), \ldots, \mathbf{u}(N)$ with $\sum_{k=1}^{N} r_k u_k(j)^2 = 1$ for each j with $1 \leqslant j \leqslant N$. Show that, if $\mathbf{x} = (x_1, x_2, \ldots, x_N) \in \mathbb{R}^N$, then $\mathbf{x} = \sum_{j=1}^{N} (\sum_{k=1}^{N} r_k x_k u_k(j))\mathbf{u}(j)$.

39.2 Let us look deeper at the system of equations in Question 39.1 using ideas from a second course in linear algebra.

(i) Initially we take $r_1 = r_2 = \ldots = r_N = 1$. Show that the defining equations have the form

$$\sum_{j=1}^{N} a_{ij} y_j(\lambda) = \lambda y_i(\lambda), \quad 1 \leqslant i \leqslant N,$$

with $a_{ij} \in \mathbb{R}$ and $a_{ij} = a_{ji}$.

We are thus concerned with the eigenvalues and eigenvectors of the linear map $\alpha: \mathbb{R}^N \to \mathbb{R}^N$ associated with the symmetric matrix (a_{ij}). In what follows $(\mathbf{x}, \mathbf{y}) = \sum_{i=1}^N x_i y_i$ will be the standard inner product on $\mathbb{R}^N$ and $\|x\| = \sqrt{(\mathbf{x}, \mathbf{x})}$ the standard Euclidean norm.

(ii) Show that a linear map $\beta: \mathbb{R}^N \to \mathbb{R}^N$ is associated with a symmetric matrix (b_{ij}) if and only if $(\mathbf{x}, \beta\mathbf{y}) = (\beta\mathbf{x}, \mathbf{y})$ for all $\mathbf{x}, \mathbf{y} \in \mathbb{R}^N$. More briefly we call β a symmetric linear map.

(iii) Our first concern is to show that α does indeed have a real eigenvalue. To this end consider the function $f: \mathbb{R}^N \to \mathbb{R}$ given by $f(\mathbf{x}) = (\mathbf{x}, \alpha\mathbf{x})$. Explain why we can find a vector $\mathbf{u} \in \mathbb{R}^N$ with $\|\mathbf{u}\| \leqslant 1$ and such that $|f(\mathbf{u})| \geqslant |f(\boldsymbol{v})|$ whenever $\|\boldsymbol{v}\| \leqslant 1$. Show that (unless $\alpha = 0$, in which case our results are trivial) $\|\mathbf{u}\| = 1$.

(iv) Now let $\mathbf{w}$ be any vector with $\|\mathbf{w}\| = 1$ and $\mathbf{w}$ perpendicular to $\mathbf{u}$. Explain why if δ is any real number $\|\mathbf{u} + \delta\mathbf{w}\|^2 = (1 + \delta^2)$ and why it follows that $|f(\mathbf{u} + \delta\mathbf{w})| \leqslant (1 + \delta^2)|f(\mathbf{u})|$. Deduce that

$$|f(\mathbf{u}) + 2\delta(\mathbf{w}, \alpha\mathbf{u}) + \delta^2 f(\mathbf{w})| \leqslant (1 + \delta^2)\, |f(\mathbf{u})|$$

and, by considering what happens when $|\delta|$ is small, deduce that $(\mathbf{w}, \alpha\mathbf{u}) = 0$. By allowing $\mathbf{w}$ to vary show that $\alpha\mathbf{u} = \lambda\mathbf{u}$ for some $\lambda \in \mathbb{R}$.

(v) Write $\mathbf{u}_1 = \mathbf{u}$ and let $M = \mathbf{u}_1^\perp$ i.e. let $M = \{\mathbf{w}: (\mathbf{w}, \mathbf{u}_1) = 0\}$. By using (ii), or otherwise, show that $\alpha\mathbf{w} \in M$ whenever $\mathbf{w} \in M$. Thus we may define $\beta: M \to M$ by taking β to be the restriction of α to M. Observe that $(\mathbf{x}, \beta\mathbf{y}) = (\beta\mathbf{x}, \mathbf{y})$ for all $\mathbf{x}, \mathbf{y} \in M$ and M has dimension $N - 1$.

(vi) Now use induction to show that, if $\alpha: \mathbb{R}^N \to \mathbb{R}^N$ is a symmetric linear map, we can find $\mathbf{u}_1, \mathbf{u}_2, \ldots, \mathbf{u}_N$ orthogonal eigenvectors with $\|\mathbf{u}_j\| = 1$ $[1 \leqslant j \leqslant N]$.

(vii) Conclude that, if $r_1 = r_2 = \ldots = r_n = 1$, we can find N orthogonal eigensolutions $\mathbf{u}(1), \mathbf{u}(2), \ldots, \mathbf{u}(N)$ for our system with $\sum_{k=1}^N u_k(j)^2 = 1$.

(viii) Now suppose merely that $r_1, r_2, \ldots, r_n > 0$. By considering the system of equations satisfied by $\mathbf{z}(\lambda)$ where $z_k(\lambda) = r_k^{1/2} y_k(\lambda)$ show that we can find eigensolutions $\mathbf{u}(1), \mathbf{u}(2), \ldots, \mathbf{u}(N)$ to our original system such that

$$\sum_{k=1}^N r_k u_k(i) u_k(j) = 0 \text{ if } i \neq j, \quad \sum_{k=1}^N r_k u_k(i) u_k(i) = 1$$

for all $1 \leqslant i, j \leqslant N$. Show that, if $\mathbf{x} = (x_1, x_2, \ldots, x_N) \in \mathbb{R}^N$, then

$$\mathbf{x} = \sum_{j=1}^N \left(\sum_{k=1}^N r_k x_k u_k(j) \right) \mathbf{u}(j).$$

(ix) By considering the system with $N = 2$, $q_1 = q_2 = 0$, $r_1 = r_2 = 1$, $\alpha_0 = \alpha_3 = 1$, $\alpha_1 = \alpha_2 = 0$, $p_1 = 0$, $p_0 = p_2 = 1$ show that cases can arise in which there are fewer than N distinct eigenvalues.

39.3 (Readers who have a little knowledge of difference equations such as is found in Question 3.5 will be able to shorten this question drastically.) It is natural to try to understand the Sturm–Liouville differential equations of Chapter 39 by analogy with the, in some ways more tractable, Sturm–Liouville difference equations of the previous two questions. Well before the time of Fourier, mathematicians sought, in effect, to understand the system

$$y''(t) - \lambda y(t) = 0 \quad \text{for } 0 < t < \pi,\ y(0) = y(\pi) = 0$$

by looking at the system

$$\Delta(\Delta y_{k-1}) - \lambda y_k = 0 \quad \text{for } 1 \leqslant k \leqslant N,\ y_0 = y_{N+1} = 0$$

i.e. by looking at the system

$$y_{k+1} - (2 + \lambda)y_k + y_{k-1} = 0 \quad \text{for } 1 \leqslant k \leqslant N,\ y_0 = y_{N+1} = 0.$$

(i) Show that, if $\lambda \neq 0$, the equation $z^2 - (2 + \lambda)z + 1 = 0$ has two distinct roots α, β with $\alpha\beta = 1$. What happens if $\lambda = 0$?

(ii) Show that, if $\lambda \neq 0$, the sequence $y_k = A\alpha^k + B\beta^k$ satisfies the system $y_{k+1} - (2 + \lambda)y_k + y_{k-1} = 0$ for $1 \leqslant k \leqslant N$, and that, if $\lambda = 0$, the sequence $y_k = A + Bk$ does.

(iii) By observing that y_{k-1} and y_k determine y_{k+1}, show that the system

$$y_{k+1} - (2 + \lambda)\, y_k + y_{k-1} = 0 \quad \text{for } 1 \leqslant k \leqslant N,\ y_0 = 0,\ y_1 = c_1$$

has one and only one solution for each c_1.

(iv) Deduce that, if

$$y_{k+1} - (2 + \lambda)y_k + y_{k-1} = 0 \quad \text{for } 1 \leqslant k \leqslant N,\ y_0 = 0,\ y_{N+1} = 0,$$

then (if $\lambda \neq 0$) $y_k = a(\alpha^k - \beta^k)$ for $0 \leqslant k \leqslant N + 1$. Show that no non zero solution is possible if $\lambda = 0$.

(v) Show that, if $a \neq 0$ in (iv), then $\alpha^{N+1} = \beta^{N+1}$ and conclude, using (i), that $\alpha = \exp(\pi ir/(N + 1))$, $\beta = \exp(-\pi ir/(N + 1))$ for some integer r with $r \neq 0 \bmod N + 1$.

(vi) Conclude that the system of this question has eigenvalues $\lambda_r = -(2\sin(\pi r/(2(N + 1))))^2$ with associated eigensolutions $y(r)$ given by $y_k(r) = b\sin(\pi kr/(N + 1))$ $[1 \leqslant k \leqslant N]$. What values can b take to yield $\sum_{k=1}^{N} y_k(r)^2 = 1$?

39.4 (i) Show that, if f is three times continuously differentiable,

$$\frac{f(x + h) - 2f(x) + f(x - h)}{h^2} \to f''(x).$$

(ii) By rescaling the results of the previous two questions, show that the system of equations

$$\frac{y(\pi(r+1)/(N+1)) - 2y(\pi r/(N+1)) + y(\pi(r-1)/(N+1))}{(\pi/(N+1))^2} - \lambda y(\pi r/(N+1)) = 0\ [1 \leqslant r \leqslant N]$$

with $y(0) = y(\pi) = 0$ has a non zero solution given by $y(\pi r/(N+1)) = s_m(\pi r/(N+1))$ when $\lambda = \lambda_m$, where

$$s_m(t) = 2^{1/2} \sin(mt) \text{ for } 0 \leqslant t \leqslant \pi,\ \lambda_m = -\left(\frac{N+1}{\pi}\right)^2 \sin\left(\frac{\pi m}{N+1}\right)^2$$

$$[1 \leqslant m \leqslant N].$$

Show further that $(N+1)^{-1}\sum_{r=0}^{N+1} (s_m(\pi r/(N+1))^2 = 1$ and

$$(N+1)^{-1} \sum_{r=0}^{N+1} (s_m(\pi r/(N+1)))(s_n(\pi r/(N+1))) = 0$$

for $m \neq n$ $[1 \leqslant m,\ n \leqslant N]$. Show also that, if $f: [0, \pi] \to \mathbb{R}$ is given with $f(0) = f(\pi) = 0$, we have

$$f(r\pi/(N+1)) = \sum_{m=1}^{N} a_{mN}\, s_m(\pi r/(N+1)) \quad [0 \leqslant r \leqslant N+1],$$

where $a_{mN} = (N+1)^{-1}\sum_{r=0}^{N+1} f(\pi r/(N+1))\, s_m(\pi r/(N+1))$ and

$$\sum_{m=1}^{N} a_{mN}^2 = (N+1)^{-1} \sum_{r=0}^{N+1} f(\pi r/(N+1))^2.$$

(iii) Show that $\lambda_m \to -m^2$ and, if f is continuous,

$$a_{mN} \to (1/\pi) \int_0^\pi f(t) s_m(t)\, dt,\ \sum_{m=1}^{N} a_{mN}^2 \to (1/\pi) \int_0^\pi f(t)^2\, dt$$

as $N \to \infty$.

(iv) Let us now return to the differential equation

$$y''(t) - \lambda y(t) = 0, \quad y(0) = y(\pi) = 0.$$

Show, using the results of (iii), that

$$s''_m(t) + m^2 s_m(t) = 0, \quad s_m(0) = s_m(\pi) = 0,$$

$$\frac{1}{\pi}\int_0^\pi s_m(t)^2\, dt = 1, \quad \frac{1}{\pi}\int_0^\pi s_m(t) s_n(t)\, dt = 0 \quad [m \neq n],$$

and, if $f: [0, \pi] \to \mathbb{R}$ is continuous with $f(0) = f(\pi) = 0$, then

$$\sum_{m=0}^{\infty} \left(\frac{1}{\pi}\int_0^\pi f(t)\, s_m(t)\, dt\right)^2 \leqslant \frac{1}{\pi}\int_0^\pi f(t)^2\, dt. \qquad (*)$$

(v) (It would be very nice if we could find a simple direct proof, using only the ideas of this question, that (∗) is an equality, but I cannot see one.) Use the ideas or the results of Chapter 34 to show that, if $f: [0, \pi] \to \mathbb{R}$ is continuous and $f(0) = f(\pi) = 0$, then $\sum_{m=0}^{\infty}((1/\pi)\int_0^\pi f(t)\, s_m(t)\, dt)^2 = (1/\pi)\int_0^\pi f(t)^2\, dt$. Hence show that, with the notation of (iii),

$$\left(\frac{1}{\pi}\int_0^\pi \left|\sum_{m=1}^{N} a_{mN} s_m(t) - f(t)\right|^2 dt\right)^{1/2} \to 0 \text{ as } N \to \infty.$$

39.5 The attack on Sturm–Liouville differential equations outlined in the previous questions does not seem very promising. However, parts (iii) to (v) of Question 39.2 contain the germ of a very attractive idea – that of extracting eigenvalues and eigenfunctions as the solution of minimising problems. In what follows we consider the system

$$(pu')' - \lambda(r + q)\, u = 0 \quad \text{on } [a, b] \tag{1}$$

$$\alpha_a u(a) + \beta_a u'(a) = 0, \quad \alpha_b u(b) + \beta_b u'(b) = 0 \tag{2}$$

with the standard conditions laid out in Chapter 39.

(i) Show that, if $u_1, u_2, \ldots, u_n$ are orthonormal eigenfunctions associated with eigenvalues $\lambda_1, \lambda_2, \ldots, \lambda_n$ where $0 < \lambda_1 < \lambda_2 < \ldots < \lambda_n$, then, if $u = \sum_{j=1}^n a_j u_j$, we have

$$L(u, u) = \int_a^b p(u')^2 - qu^2\, dx = \sum_{j=1}^{n} \lambda_j |a_j|^2$$

and so that the unique (up to multiplication by -1) $u \in \{\sum_{j=1}^n a_j u_j \colon a_j \in \mathbb{C}\}$ which minimises $\int_a^b p(u')^2 - qu^2\, dx$ subject to $\int_a^b ru^2\, dx = 1$ is given by $u = u_1$.

(ii) Suppose u is a twice continuously differentiable function satisfying the boundary conditions (2) which minimizes $\int_a^b p(u')^2 - qu^2\, dx$ subject to $\int_a^b ru^2 dx = 1$ (i.e. such that, if v is a twice differentiable function with $\|v\|_2 = 1$, then $L(v, v) \geqslant L(u, u)$). Show by imitating part (iv) of Question 39.2 that u is an eigenfunction for the system.

(iii) More generally, show that, if $u_1, u_2, \ldots, u_n$ are orthonormal eigenfunctions of the system and u is a twice continuously differentiable function satisfying the boundary conditions (2) which minimizes (Lu, u) subject to $\|u\|_2 = 1$ and $(u, u_j) = 0$ for $1 \leqslant j \leqslant n$, then u is an eigenfunction.

(If we could establish that the required minimum exists then we could obtain all eigenfunctions by repeated applications of (iii). This is the approach adopted in Courant and Hilbert *Methods of Mathematical Physics*, Volume 1, but, although this gives a great deal of insight, the direct proof that a minimum exists is not all that easy. (It is not all that difficult either: see Chapter 8 of Gelfand and Fomin's *Calculus of Variations*.) The more indirect standard proof via the spectral theorem can also be considered as an energy argument.

Unfortunately most textbooks which give this powerful general theorem fail to show the link with any particular application. For a good treatment of the spectral theorem which links it with applications see N. Young *An Introduction to Hilbert Space*.)

39.6 (i) By using results on Fourier series such as Parseval's equality, show that the ideas of the previous question actually work for the system

$$u'' - \lambda u = 0 \quad \text{on} \quad [0, \pi], \quad u(0) = u(\pi) = 0.$$

Specifically show that we have eigenfunctions $s_1, s_2, \ldots$ and eigenvalues $\lambda_1, \lambda_2, \ldots$ such that (using the notation of Question 39.4)

(a) $\int_0^\pi s_n(x)^2\, dx = 1$, $\int_0^\pi s_n(x)\, s_m(x)\, dx = 0$ for $1 \leq m \leq n-1$,

(b) $(Ls_n, s_n) \leq (Lu, u)$ whenever u is a twice continuously differentiable function with $\int_0^\pi u(x)^2\, dx = 1$, $\int_0^\pi u(x)\, s_m(x)\, dx = 0$ for $1 \leq m \leq n-1$.

(ii) Observe that in (i)(b) we can replace the condition twice continuously differentiable by once continuously differentiable and prove the Wirtinger type inequality

$$\int_0^\pi y(t)^2\, dt \leq \int_0^\pi y'(t)^2\, dt$$

for all once continuously differentiable functions $y: [0, \pi] \to \mathbb{R}$ with $y(0) = y(\pi) = 0$. Show that we have equality if and only if $y(t) = C \sin t$ for some $C \in \mathbb{R}$.

(iii) Show that the ideas of Question 39.4 also work for the system

$$u'' - \lambda u = 0 \text{ on } [0, \pi/2], \quad u(0) = 0, \, u'(\pi/2) = 0.$$

Show that

$$\int_0^{\pi/2} y(t)^2\, dt \leq \int_0^{\pi/2} y'(t)^2\, dt$$

for all once continuously differentiable functions $y: [0, \pi/2] \to \mathbb{R}$ with $y(0) = 0$, $y'(\pi/2) = 0$. When does equality occur?

(iv) Is it true that, if $y: [0, \pi] \to \mathbb{R}$ is a once continuously differentiable function with $y'(0) = y'(\pi) = 0$, then

$$\int_0^\pi y(t)^2\, dt \leq \int_0^\pi y'(t)^2\, dt?$$

If not, what extra condition do you need to impose (compare Question 36.1)? Explain this in terms of lowest eigenvalues.

39.7 (i) Show that, if $a, b, g: \mathbb{R} \to \mathbb{R}$ are continuous functions such that $a(x) \neq 0$ for all x, then the differential equation

$$a(x)\frac{dy}{dx} + b(x)y = g(x) \qquad (*)$$

can be rewritten in the form

$$p(x)\frac{dy}{dx} + p'(x)y = f(x)$$

by a suitable choice of p and f.

(ii) Show that we can always solve $(*)$ for the inital condition $y(x_0) = y_0$ and write down the solution. Can we always solve $(*)$ for the initial condition $y'(x_0) = y_0$?

39.8 Suppose that a, b: $[0, 1] \to \mathbb{R}$ are continuous. Consider the differential equation

$$y'' + a(x)y' + b(x)y = 0 \qquad (*)$$

(i) Show that, if y_1, y_2 are solutions of $(*)$ on $[0, 1]$ and we set

$$W(x) = \det \begin{pmatrix} y_1(x) & y_2(x) \\ y_1'(x) & y_2'(x) \end{pmatrix},$$

then $W'(x) + a(x)W(x) = 0$. Deduce that either $W(x) = 0$ for all $x \in [0, 1]$ or $W(x) \neq 0$ for all $x \in [0, 1]$.

(ii) Suppose that $W(x) \neq 0$ for all $x \in [0, 1]$ and $y_1(0) = 0$, $y_2(1) = 0$. Show that, if f: $[0, 1] \to \mathbb{R}$ is continuous, the differential equation

$$y'' + a(x)y' + b(x)y = f(x)$$

with boundary conditions $y(0) = y(1) = 0$ has

$$y(x) = y_1(x)\int_x^1 \frac{y_2(t)f(t)}{W(t)}\,dt + y_2(x)\int_0^x \frac{y_1(t)f(t)}{W(t)}\,dt$$

as a solution.

(iii) In the particular case $a(x) = 0$, $b(x) = \lambda^2$, investigate for which real positive values of λ we can find y_1, y_2 of the form required by (ii).

(W is called the Wronskian of $(*)$. If we write

$$G(x, t) = y_1(t)y_2(x)W(t)^{-1} \quad \text{for } 0 \leqslant t \leqslant x,$$

$$G(x, t) = y_2(t)y_1(x)W(t)^{-1} \quad \text{for } x \leqslant t \leqslant 1,$$

then the final formula of (ii) takes the more symmetric form

$$y(x) = \int_0^1 G(x, t)f(t)\,dt.$$

G is called the Green's function. More general forms of the Green's function play an important role in advanced work.)

40

ORTHOGONAL POLYNOMIALS

40.1 Use Rolle's theorem to discuss the position and multiplicity of zeros for $d^r(x^2-1)^n/dx^r$ for $r=0,1,\ldots$ and hence show that the zeros of the Legendre polynomial P_n discussed in Lemma 40.7 are all real, simple and lie in $(-1,1)$.

40.2 Let r be as in Theorem 40.6.

(i) Show that there exists a unique sequence $\phi_0, \phi_1, \phi_2, \ldots$ of orthogonal polynomials such that ϕ_n is of degree exactly n with leading coefficient 1.

(ii) Show that, for each integer $j \geqslant 1$, we can find real numbers β_j, γ_j such that $\gamma_j > 0$ and

$$x\,\phi_j(x) = \phi_{j+1}(x) + \beta_j\phi_j(x) + \gamma_j\phi_{j-1}(x)$$

for all $x \in (a,b)$ and so (why?) for all $x \in \mathbb{R}$.

(iii) Deduce by induction that all the coefficients of ϕ_j are real and that between any two zeros of ϕ_j there lies a zero of ϕ_{j-1}. (The result of Theorem 40.8 will be useful.)

(iv) The recurrence relation given in (ii) can be useful in practice. Derive the recurrence relation for the Tchebychev polynomial and check it against the formula given at the end of the proof of Lemma 5.3.

(v) Show that, if P_m is the mth Legendre polynomial, then

$$nP_n(x) = (2n-1)x\,P_{n-1}(x) - (n-1)P_{n-2}(x) \quad [n \geqslant 2].$$

Compute $P_4(x)$ using this relation.

(vi) By using the formula of (v), or otherwise, show that

$$\int_{-1}^{1} xP_n(x)P_{n-1}(x)\,dx = 2n/(4n^2-1).$$

Compute $\int_{-1}^{1} x(P_n(x))^2\,dx$ and $\int_{-1}^{1} x^2(P_n(x))^2\,dx$.

(vii) Evaluate $\int_{-1}^{1} x^m P_n(x)\,dx$ for all integers $m, n \geqslant 0$.

40.3 Show that the following definitions of P_n are equivalent:

(i) $P_n(x) = \dfrac{1}{2^n n!} \dfrac{d^n}{dx^n} (x^2 - 1)^n$,

(ii) P_n is the unique polynomial solution of

$$(1 - x^2) P''_n(x) - 2x \cdot P'_n(x) + n(n+1) P_n(x) = 0 \text{ with } P_n(1) = 1,$$

(iii) $P_n(x)$ is the coefficient of z^n in the power series expansion of $(1 - 2xz + z^2)^{1/2}$, i.e.

$$(1 - 2xz + z^2)^{1/2} = \sum_{n=0}^{\infty} P_n(x)\, z^n,$$

(iv) $P_0(x) = 1$, $P_1(x) = x$ and

$$nP_n(x) = (2n-1)\, xP_{n-1}(x) - (n-1) P_{n-2}(x) \text{ for all } n \geqslant 2 \text{ and all } x.$$

(v) The P_n are polynomials of degree exactly n such that $\int_{-1}^{1} P_n(x)\, P_m(x)\, dx = 2\delta_{nm}/(2n+1)$, where $\delta_{nm} = 1$ if $n = m$, $\delta_{nm} = 0$ otherwise $[n, m \geqslant 0]$ and such that their leading coefficients are positive,

(vi) P_n is the polynomial given by

$$P_n(x) = \frac{1}{\pi} \int_0^{\pi} (x + (x^2 - 1)^{1/2} \cos \phi)^n \, d\phi \quad \text{for } |x| \geqslant 1,$$

(vii) $P_n(x) = \displaystyle\sum_{r=0}^{n} \frac{(-1)^r (n+r)!}{(n-1)!(r!)^2 2^{r+1}} ((1-x)^r + (-1)^n (1+x)^r)$

(viii) $P_0(x) = 1$ for all $x \in \mathbb{R}$, $P_n(1) = 1$ for all $n \geqslant 0$ and

$$xP'_n(x) - P'_{n-1}(x) = nP_n(x) \quad \text{for all } x \in \mathbb{R},\ n \geqslant 1,$$

(ix) $P_0(x) = 1$ for all $x \in \mathbb{R}$, $P_n(1) = 1$ for all $n \geqslant 0$ and

$$P'_n(x) - xP'_{n-1}(x) = nP_{n-1}(x) \quad \text{for all } x \in \mathbb{R},\ n \geqslant 1,$$

(x) $P_n(x) = \dfrac{1}{2^{n+1} \pi i} \displaystyle\int_C \frac{(w^2-1)^n}{(w-x)^{n+1}}\, dx$ where C is a circle in the complex plane centred at x.

(xi) P_n is the polynomial of degree n with leading coefficient positive such that $\int_{-1}^{1} P_n(t) f(t)\, dt = 0$ whenever f is a polynomial of degree less than n and

$$\int_{-1}^{1} P_n(t)^2\, dt = \frac{2}{2n+1}.$$

(xii) There is a unique polynomial Q_n of degree at most $n - 1$ which minimises

$$\int_{-1}^{1} |t^n + Q(t)|^2\, dt$$

over all polynomials of degree at most $n - 1$. If A_n is the minimum value of $\int_{-1}^{1} |t^n + Q(t)|^2\, dt$, then $P_n(t) = ((2n+1)A_n)/2)^{-1/2} (t^n + Q_n(t))$.

40.4 Let $w: [0, \infty) \to \mathbb{R}$ be a continuous function such that $\int_0^\infty x^n w(x)\, dx$ converges for all integers $n \geqslant 0$ and such that $w(x) > 0$ for all $x \in [0, \infty)$.

(i) Show that we can find polynomials $p_0, p_1, \ldots$ such that p_n has degree exactly n and

$$\int_0^\infty p_n(x)^2 w(x)\, dx = 1 \quad \text{for } n \geqslant 0,$$

$$\int_0^\infty p_n(x) p_m(x) w(x)\, dx = 0 \quad \text{for } n \neq m,\ n, m \geqslant 0.$$

(ii) Show that, if $f: [0, \infty) \to \mathbb{R}$ is a continuous function such that $\int_0^\infty f(x)^2 w(x)\, dx$ converges, then there is a unique polynomial f_N of degree N such that

$$\int_0^\infty (f(x) - f_N(x))^2 w(x)\, dx \leqslant \int_0^\infty (f(x) - g(x))^2 w(x)\, dx$$

for all polynomials g of degree N and that $f_N = \sum_{j=0}^N a_j p_j$ with $a_j = \int_0^\infty f(x) p_j(x) w(x) dx$.

(iii) Recall Example 6.2. Let $w(x) = \exp(-x^{-1/4}/2)$, $f(x) = \exp(-x^{1/4}/2) \sin x^{1/4}$. Show that w and f satisfy the hypotheses above and that $f(x) \to 0$ as $x \to \infty$, yet

$$\int_0^\infty (f(x) - f_N(x))^2 w(x)\, dx \not\to 0.$$

40.5 (The Laguerre polynomials) We do some preliminary work for the next question. Let $L_n(x) = e^x d^n(x^n e^{-x})/dx^n$.

(i) Show that L_n is a polynomial of degree exactly n with leading coefficient $(-1)^n$.

(ii) By repeated integration by parts, or otherwise, show that

$$\int_0^\infty e^{-x} L_n(x) L_m(x)\, dx = 0 \quad \text{if } m \neq n,$$

$$\int_0^\infty e^{-x} L_n(x)^2\, dx = (n!)^2.$$

(iii) Conclude that, in the notation of Question 40.4, if we take $w(x) = e^{-x}$, we can take $p_n(x) = (-1)^n (n!)^{-1} L_n(x)$. (The factor $(-1)^n$ is, of course, optional.)

40.6 In this question we continue the ideas and notation of Question 40.4 and show that, in spite of the counter example in part (iii), there do exist weight functions w for which $\int_0^\infty (f(x) - f_N(x))^2 w(x)\, dx \to 0$ for all continuous f with $\int_0^\infty f(x)^2 w(x)\, dx < \infty$.

(i) Let E be the set of continuous functions $f: [0, \infty) \to \mathbb{R}$ such that $\int_0^\infty f(x)^2 w(x)\, dx < \infty$ and $\int_0^\infty (f(x) - f_N(x))^2 w(x)\, dx \to 0$ as $N \to \infty$.

(a) Show that E is a vector space.

(b) Show that E consists of those continuous $f: [0, \infty) \to \mathbb{R}$ such that, writing $\|g\|_2 = (\int_0^\infty g(x)^2 w(x)\, dx)^{1/2}$, we have $\|f\|_2 < \infty$ and

$$\inf\{\|f - p\|_2 : p \text{ a polynomial}\} = 0.$$

(c) Show that E is closed in the sense that, if $f: [0, \infty) \to \mathbb{R}$ is continuous with $\|f\|_2 < \infty$ and we can find $f_n \in E$ with $\|f_n - f\|_2 \to 0$, then $f \in E$.

(d) Show that, if $f: [0, \infty) \to \mathbb{R}$ is continuous with $\|f\|_2 < \infty$, then $f \in E$ if and only if Bessel's inequality for f is an equality, i.e. if and only if

$$\|f\|_2^2 = \sum_{j=0}^{\infty} \left(\int_0^\infty f(x) p_j(x) w(x)\, dx \right)^2. \tag{*}$$

(ii) We now specialise for the remainder of the question to the case $w(x) = e^{-x}$. The p_j for this weight function were found in Question 40.5. Let $f(x) = e^{-rx}$ with $r \geqslant 0$ an integer. By repeated integration by parts, evaluate $\int_0^\infty f(x) p_j(x) w(x)\, dx$ and $\|f\|_2^2$. Show that (*) holds if and only if

$$\frac{1}{2r+1} = \sum_{j=0}^{\infty} \frac{r^{2j}}{(r+1)^{2j+2}}. \tag{**}$$

Check that (**) does indeed hold for all $r \geqslant 0$ and deduce from (i) (d) that, if $e_r(x) = e^{-rx}$, then $e_0, e_1, e_2, \ldots \in E$.

(iii) Let $f: [0, \infty) \to \mathbb{R}$ be a continuous function with $f(x) = 0$ for $x \geqslant a$ for some $a > 0$. By making a change of variable $x = -\log t$, show that

$$\int_0^\infty \left(f(x) - \sum_{j=0}^{M} c_j e_j(x) \right)^2 e^{-x}\, dx = \int_0^1 \left(\phi(t) - \sum_{j=0}^{M} c_j t^j \right)^2 dt$$

for some continuous $\phi: [0, 1] \to \mathbb{R}$. Use the Weierstrass approximation theorem together with (ii), (i) (a) and (i) (c) to show that $f \in E$.

(iv) Use (iii) and (i) (c) to conclude that, if $f: [0, \infty) \to \mathbb{R}$ is continuous and $\|f\|_2 < \infty$, then $\|f - f_N\|_2 \to 0$ as $N \to \infty$.

(Thus the Laguerre polynomials are mean square dense in the appropriate space. The elegant proof is due to V. A. Steklov and is taken from Natanson *Constructive Function Theory*. (This book contains much well-presented and interesting material on the approximation of functions which is not easily available elsewhere.) It should however be pointed out that the techniques of measure theory and functional analysis produce general theorems which enable us to organise apparently isolated results like this one.)

41

GAUSSIAN QUADRATURE

41.1 (i) The reader should find it a matter of pure routine to extend the theory of this chapter to prove the following result. Suppose that $r: (-1, 1) \to \mathbb{R}$ is a weight function in the sense of Chapter 40. Then we can find $-1 < x(1, n) < x(2, n) < \ldots < x(n, n) < 1$ and $A(1, n), A(2, n), \ldots, A(n, n) \in \mathbb{R}$ such that

$$\int_{-1}^{1} P(x)r(x)\,dx = \sum_{j=1}^{n} A(j, n)P(x(j, n))$$

for every polynomial of degree $2n - 1$ or less. Moreover the $x(j, n)$, $A(j, n)$ $[1 \leqslant j \leqslant n]$ are uniquely specified and

$$\sum_{j=1}^{n} A(j, n)P(x(j, n)) \to \int_{-1}^{1} f(x)r(x)\,dx$$

and $n \to \infty$ for every continuous $f: [-1, 1] \to \mathbb{R}$.

(ii) Hermite noticed a particularly elegant case of this result. Show by direct calculation that

$$\frac{1}{\pi}\int_{0}^{\pi} \cos mx dx = \frac{1}{\pi}\sum_{j=1}^{n} \cos m\frac{(2j-1)\pi}{2n}$$

for all $0 \leqslant m \leqslant 2n - 1$. By considering Tchebychev polynomials and making a suitable change of variable, deduce that, when $r(x) = (1 - x^2)^{-1/2}$, we get $x(j, n) = \cos((2j - 1 - n)\pi/2n)$ and $A(j, n) = 1/n$. We thus obtain the following suggested approximation known as Hermite's quadrature formula

$$\int_{-1}^{1} f(x)(1 - x^2)^{-1/2}\,dx \approx \frac{\pi}{n}\sum_{j=1}^{n} f(\cos(2j-1)\pi/2n).$$

(iii) By considering $f(x) = (1 - x^2)^{1/2}g(x)$, show that, if $g: [-1, 1] \to \mathbb{R}$ is continuous,

$$\frac{\pi}{n}\sum_{j=1}^{n}|\sin((2j-1)\pi/2n)|g(\cos((2j-1)\pi/2n)) \to \int_{-1}^{1} g(x)\,dx$$

as $n \to \infty$. (However, we should not expect to get something for nothing, particularly in numerical analysis. To adapt Press, Flannery, Teukolsky and Vetterling in their book *Numerical Recipes*: 'Where did the function $(1-x^2)^{-1/2}$ go? It is lurking there ready to give high-order accuracy to integrands close to the form polynomials times $(1-x^2)^{-1/2}$, and ready to *deny* high-order accuracy to integrands that are otherwise perfectly smooth and well behaved.' In the year 1987 *Numerical Recipes* accounted for half the sales of all mathematics titles of Cambridge University Press. The reader may find it instructive to compare *Numerical Recipes* with the other books in the numerical analysis section of her university library.)

41.2 (Pólya's criterion for numerical quadrature) Suppose $-1 \leqslant x(1, n) < x(2, n) < \ldots < x(n, n) \leqslant 1$ and $A(1, n), A(2, n), \ldots, A(n, n)$ are real. If $f: [-1, 1] \to \mathbb{R}$ is continuous, let us write

$$T_n f = \sum_{r=1}^{n} A(r, n) f(x(r, n))$$

and $If = \int_{-1}^{1} f(x)\,dx$.

(i) Show that, if

(A) $T_n P \to IP$ as $n \to \infty$, whenever P is a polynomial, and

(B) there exists a K independent of n with $\sum_{r=1}^{n} |A(r, n)| \leqslant K$ for all n, then

$$T_n f \to If \text{ as } n \to \infty \quad \text{for all continuous } f: [-1, 1] \to \mathbb{R}.$$

(ii) Show that, if (B) is false, we can find inductively a sequence $n(1) < n(2) < \ldots$ and continuous functions $f_j: [-1, 1] \to \mathbb{R}$ such that

(a) $\sup |f_j(x)| \leqslant 2^{-j}$,
(b) $|T_{n(k)} f_j| \leqslant 2^{-j-k}$ for all $1 \leqslant k \leqslant j-1$,
(c) $T_{n(j)} f_j \geqslant 2^j$.

Show that, if $\varepsilon_j \in \{-1, 1\}$, then $\sum_1^N \varepsilon_j f_j$ converges uniformly to a continuous function f. Show that

$$\left| T_{n(N)} f - \sum_{j=1}^{N} \varepsilon_j T_{n(N)} f_j \right| \leqslant \sum_{j=N+1}^{\infty} |T_{n(N)} f_j| \leqslant 2^{-N}$$

and conclude that, if $\varepsilon_1, \varepsilon_2, \ldots$ are chosen appropriately, then $|T_{n(N)} f| \to \infty$ as $N \to \infty$.

(iii) Conclude that (A) and (B) together form a necessary and sufficient condition for the convergence $T_n f \to If$ as $n \to \infty$ for all continuous $f: [-1, 1] \to \mathbb{R}$.

41.3 (i) Show that, if $x_1, x_2, \ldots, x_n$ and $A_1, A_2, \ldots, A_n$ are as in Lemma 41.1 then, if $f: [-1, 1] \to \mathbb{R}$ is given, there is a unique polynomial F_n of degree $n-1$ or less with $F_n(x_j) = f(x_j)$ $[1 \leq j \leq n]$ and further

$$\sum_{j=1}^{n} A_j f(x_j) = \int_{-1}^{1} F_n(x)\,dx.$$

(ii) Now consider the Newton–Cotes rules in which we take $x(j, n) = -1 + 2(j-1)/(n-1)$ and choose $A(j, n)$ in such a way that

$$\sum_{j=1}^{n} A(j, n)P(x(j, n)) = \int_{-1}^{1} P(x)\,dx$$

whenever P is a polynomial of degree $n-1$ or less. Observe that

$$A(j, n) = \int_{-1}^{1} \prod_{k \neq j} \frac{(x + 1 - 2(k-1)/(n-1))}{2(j-k)/(n-1)}\,dx.$$

(iii) In Question 75.21 we shall show, following a calculation of Pólya, that

$$A(2n+1, 4n+1) \to \infty \quad \text{as } n \to \infty.$$

(If you are ambitious there is no reason why you should not be able to prove this for yourself without looking at that question.) Assuming this result, prove

(A) there is a continuous function $f: [-1, 1] \to \mathbb{R}$ such that

$$\limsup_{n\to\infty} \left| \sum_{j=1}^{n} A(j, n) f(x(j, n)) \right| = \infty,$$

(B) if F_n is the unique polynomial of degree $n-1$ or less with $F_n(x(j, n)) = f(x(j, n))$, then

$$\limsup_{n\to\infty} \sup_{t\in[-1,1]} |F_n(t)| = \infty.$$

(The results of Pólya come from his paper 'Uber die Konvergenz von Quadraturverfahren' to be found in Volume 3 of his *Collected works*. This paper is a classic, but a classic of pure mathematics rather than of practical numerical analysis since, for example, a glance at the values $A(r, 21)$ quoted just before Lemma 41.5 would suffice to convince most numerical analysts that something very unpleasant happens when we attempt to use high degree Newton–Cotes formulae.)

43

TCHEBYCHEV AND UNIFORM APPROXIMATION I

43.1 I gave a 'very free sketch' in Figure 43.1 because Watt's parallelogram works too well to provide an interesting diagram. Find and sketch (you could use good computer graphics if available) the actual path of Q described in the third paragraph of Chapter 42.

43.2 Let $\pi(n)$ be the number of primes p with $1 \leqslant p \leqslant n$. Numerical experimentation lead several mathematicians including Gauss to guess that $n^{-1} \log n\, \pi(n) \to 1$ as $n \to \infty$ (i.e. for large n there are about $(\log n)^{-1} n$ primes between 1 and n). The first real progress in this matter was due to Tchebychev.

(i) Show that $\binom{2n}{n} = 2^n \prod_{r=0}^{n-1} (2 - r/(n-r))$ and deduce that $2^n < \binom{2n}{n} < 2^{2n}$.

(ii) Throughout this question p will be restricted to be prime. We write $t(p)$ for $[\log 2n/\log p]$ the integer part of $\log 2n/\log p$. Show by simple arguments that

(a) $\binom{2n}{n}$ divides $\prod_{p<2n} p^{t(p)}$,

(b) $\prod_{n<p\leqslant 2n} p$ divides $\binom{2n}{n}$.

(iii) Conclude that

(a) $2^n < \prod_{p<2n} p^{t(p)}$,

(b) $\prod_{n<p\leqslant 2n} p < 2^{2n}$.

(iv) Show that (iii) (a) implies $2^n < (2n)^{\pi(2n)}$. By taking logs conclude that $\pi(2n) > n \log 2/\log 2n$ and hence deduce that we can find an $A > 0$ such that

$\pi(n) \geq An(\log n)^{-1}$ for all $n \geq 2$. (You may need a special argument for the unimportant cases $2 \leq n \leq 4$.)

(v) Use (iii) (b) repeatedly to show that, if $m = 2^k$, then $\prod_{p \leq 2m} p < 2^{4m}$. Deduce that there exists a constant B_1 such that $\sum_{p \leq n} \log p < B_1 n$ for all $n \geq 2$.

(vi) Observing that $\sum_{n \geq p \geq \sqrt{n}} 1 \geq \pi(n) - \sqrt{n}$ and that $\log p \geq 2^{-1} \log n$ for $p \geq \sqrt{n}$, show that $\sum_{p \leq n} \log p \geq 2^{-1}(\pi(n) - \sqrt{n}) \log n$.

(vii) Combine (v) and (vi) to show that $\pi(n) \leq Bn(\log n)^{-1}$ for all $n \geq 2$ and some constant $B > 0$.

43.3 Let p_m be the mth prime.

(i) Use the previous question to show that there exist α, $\beta > 0$ such that $\alpha m \geq p_m/\log p_m \geq \beta m$.

(ii) Hence deduce successively, first that for m large enough $m^2 \geq p_m \geq m$, and then, that there exist constants a, b with $am \log m > p_m > bm \log m$ for all $m \geq 2$.

(iii) Show that $\sum_{p<N} p^{-\alpha}$ tends to a finite limit as $N \to \infty$ if and only if $\alpha > 1$.

43.4 So far as is known Tchebychev's ideas cannot be made to prove that $(n^{-1} \log n)\pi(n) \to 1$. In Questions 109.1 to 109.4 I shall outline a proof of this famous theorem of Hadamard and De la Vallée Poussin. We shall however use some of the ideas from here.

(i) Let $\theta(n) = \sum_{p \leq n} \log p$. Explain why $\theta(n) \leq \pi(n) \log n$ and by modifying the idea of Question 43.2 (vi) show that $\theta(n) \geq (1 - \varepsilon) \log n(\pi(n) - n^{1-\varepsilon})$. By using Question 43.2 (iv) deduce that given any $1 > \delta > 0$ there exists an $n_0(\delta)$ such that

$$\pi(n) \log n \geq \theta(n) \geq (1 - \delta)\pi(n) \log n$$

for all $n \geq n_0(\delta)$

(ii) Conclude that $n^{-1}(\log n)\pi(n) \to 1$ if and only if $n^{-1}\theta(n) \to 1$.

43.5 As was noted in Question 18.6 it has been an open question since the time of Euler whether $\zeta(2r + 1) = \sum_{n=1}^{\infty} n^{-(2r+1)}$ is transcendental or, indeed, whether $\zeta(2r + 1)$ is rational. However, at a conference held in 1978. . . 'The board of programme changes informed us that R. Apéry (Caen) would speak Thursday 14.00 "Sur l'irrationalité de $\zeta(3)$." Though there had been earlier rumours of his claiming a proof, scepticism was general. The lecture tended to strengthen this view to rank disbelief. Those who listened casually, or who were afflicted with being non-Francophone, appeared to hear only a sequence of unlikely assertions.' But after a long evening's discussion in which some of the more remarkable results were verified numerically in a number of specific cases . . .

'We came away convinced that Professeur Apéry had indeed found a quite miraculous and magnificent demonstration of the irrationality of $\zeta(3)$' (quoted from van der Porten's article in Volume 1 of the *Mathematical Intelligencier*, p. 195). Later Beukers was able to replace Apéry's long and mysterious proof by a short and mysterious proof (*Bulletin of the London Mathematical Society* 1979). We start with some miscellaneous results leaving the main proof to the next two questions.

(i) Suppose p and q are integers with no common factor and $q \geq 1$. Show that, if A and B are integers, then either $A + Bp/q = 0$ or $|A + Bp/q| \geq q^{-1}$. Deduce that, if x is real and we can find A_n, B_n integers with

$$0 < |A_n + B_n x| \to 0,$$

as $n \to \infty$ then x is irrational.

(ii) Let d_n be the lowest common multiple of 1, 2, ..., n. Show that $d_n = \prod_{p \leq n} p^{[\log n/\log p]} \leq n^{\pi(n)}$ and deduce from the prime number theorem stated in Question 43.4 (and proved in Questions 109.1 to 109.4) that $d_n \leq 3^n$ for all n sufficiently large. (This can also be proved by refining the constant in Question 43.2.)

(iii) Let

$$f(x, y, z) = x(1 - x)y(1 - y)z(z - 1)(1 - (1 - xy)z)^{-1} \text{ for } 0 < x, y, z < 1,$$

$f(x, y, z) = 0$ if any of x, y, z takes the value 0 or the value 1. Prove successively

(a) if $xy = \alpha^2$ for some fixed α with $0 \leq \alpha \leq 1$, then $x(1 - x)y(1 - y)$ attains its maximum for $0 \leq x, y \leq 1$ when $x = y$;

(b) when $f(x, y, z)$ attains its maximum for $0 \leq x, y, z \leq 1$, then $x = y$;

(c) (by considering $\log(z(1 - z)(1 - \alpha z)^{-1})$, or otherwise) that if $0 < \alpha < 1$ the function $z(1 - z)(1 - \alpha z)^{-1}$ attains its maximum for $0 \leq z \leq 1$ when $z = (1 - \sqrt{(1 - \alpha)})\,\alpha^{-1}$;

(d) $f(x, y, z) \leq (\sqrt{2} - 1)^4$ for all $0 \leq x, y, z \leq 1$.

(iv) If $p_n(x) = (1/n!)d^n x^n(1 - x)^n/dx^n$ show that p_n is a polynomial of degree n and $\int_0^1 p_n(x)p_m(x)\,dx = 0$ if $n \neq m$ (so the p_n are (scalar multiples of) the Legendre polynomials for [0, 1]). Show that the coefficients of p_n are integers.

43.6 Our proof centres round the evaluation of $\int_0^1\int_0^1 x^r y^s \log xy/(1 - xy)\,dxdy$ for r, s integers with $r, s \geq 0$. This is an improper Riemann integral and should be interpreted as

$$\lim_{\varepsilon \to 0+} \int_\varepsilon^{1-\varepsilon}\left(\int_\varepsilon^{1-\varepsilon} \frac{x^r y^s \log xy}{1 - xy}\,dx\right) dy.$$

(Thus this question assumes a slightly higher level of sophistication in handling repeated integrals than elsewhere in this book, but I think the extra effort required of the reader is worth it.)

(i) Evaluate $\int_0^1 x^r \log x \, dx$ and $\int_0^1 \int_0^1 x^r y^s \log xy \, dx \, dy$ for $r, s \geqslant 0$.

(ii) Show that $(1-xy)^{-1} \geqslant \sum_{n=0}^N x^n y^n$ for all $N \geqslant 1$ and $1 \geqslant x,\ y \geqslant 0$ $xy \neq 1$, that $\sum_{r=0}^N x^n y^n \to (1-xy)^{-1}$ uniformly for $1-\delta \geqslant x,\ y \geqslant 0$ as $N \to \infty$ for each $1 > \delta > 0$ and that $\sum_{r=0}^N x^n y^n$ is an increasing function of N for each $1 \geqslant x,\ y \geqslant 0$. Deduce that

$$\sum_{n=0}^{N} \int_0^1 \int_0^1 x^{n+r}\ y^{n+s} \log xy \ dx \ dy \to \int_0^1 \int_0^1 \frac{x^r y^s}{1-xy} \log xy \ dx \ dy$$

as $N \to \infty$.

(iii) Conclude that, if $r, s \geqslant 0$,

$$-\int_0^1 \int_0^1 \frac{x^r y^s}{1-xy} \log xy \, dx \, dy$$

$$= \sum_{n=0}^{\infty} \frac{1}{(n+r+1)(n+s+1)} \left(\frac{1}{n+r+1} + \frac{1}{n+s+1} \right).$$

(iv) Show from (iii) that

$$\int_0^1 \int_0^1 \frac{1}{1-xy} \log xy \ dx \ dy = -2\zeta(3),$$

$$\int_0^1 \int_0^1 \frac{x^r y^r}{1-xy} \log xy \ dx \ dy = -2\left\{ \zeta(3) - \frac{1}{1^3} - \ldots - \frac{1}{r^3} \right\} \quad \text{for } r \geqslant 1,$$

and that, if $r > s$,

$$\int_0^1 \int_0^1 \frac{x^r y^r}{1-xy} \log xy \ dx \ dy$$

$$= \frac{1}{r-s} \sum_{n=0}^{\infty} \left(\frac{1}{n+r+1} - \frac{1}{n+s+1} \right) \left(\frac{1}{n+r+1} + \frac{1}{n+s+1} \right)$$

$$= \frac{1}{r-s} \sum_{n=0}^{\infty} \left(\frac{1}{(n+r+1)^2} - \frac{1}{(n+s+1)^2} \right)$$

$$= \frac{-1}{r-s} \left\{ \frac{1}{(s+1)^2} + \ldots + \frac{1}{r^2} \right\}.$$

(v) Conclude that, if $r \geqslant 1$,

$$\int_0^1 \int_0^1 \frac{x^r y^r}{1-xy} \log xy \ dx \ dy = -2\zeta(3) + \frac{a_r}{d_r^3}$$

and, if $r > s \geqslant 0$,

$$\int_0^1 \int_0^1 \frac{x^r y^r}{1-xy} \log xy \, dx \, dy = \frac{a_{rs}}{d_r^3},$$

where a_r, a_{rs} are integers.

43.7 (i) Let p_n be as in Question 43.5 (iv). Let us write

$$I_n = \int_0^1 \int_0^1 \frac{-\log xy}{1-xy} p_n(x) p_n(y) \, dx \, dy.$$

Show, using Question 43.6 (iv) and (v), that $I_n = A_n d_n^{-3} + B_n \zeta(3)$, where A_n, B_n are integers.

(ii) Show that $\displaystyle\int_0^1 \frac{1}{1-(1-xy)z} dz = \frac{-\log xy}{1-xy}$ and so

$$I_n = \int_0^1 \int_0^1 \int_0^1 \frac{p_n(x) p_n(y)}{1-(1-xy)z} dx \, dy \, dz.$$

By integrating by parts n times with respect to x, show that

$$I_n = \int_0^1 \int_0^1 \int_0^1 \frac{(xyz)^n (1-x)^n \, p_n(y)}{(1-(1-xy)z)^{n+1}} dx \, dy \, dz.$$

Now make the substitution $w = (1-z)(1-(1-xy)z)^{-1}$ to obtain

$$I_n = \int_0^1 \int_0^1 \int_0^1 \frac{(1-x)^n (1-w)^n p_n(y)}{1-(1-xy)w} dx \, dy \, dw.$$

By integrating by parts n times with respect to w, conclude that

$$I_n = \int_0^1 \int_0^1 \int_0^1 \frac{(x(1-x)w(1-w)y(1-y))^n}{(1-(1-xy)w)^{n+1}} dx \, dy \, dw.$$

(iii) Deduce, using Question 43.5 (iii), that

$$0 < I_n \leqslant (\sqrt{2}-1)^{4n} \int_0^1 \int_0^1 \int_0^1 \frac{1}{1-(1-xy)w} dx \, dy \, dw,$$

and so, using the first result of part (ii) together with Question 43.6(iv), show that $0 < I_n \leqslant 2(\sqrt{2}-1)^{4n} \zeta(3)$.

(iv) Combine (i), (iii) and Question 43.5(ii) to show that there are integers A_n, C_n such that

$$0 < |A_n + C_n \zeta(3)| = 2\zeta(3) d_n^3 (\sqrt{2}-1)^{4n} < 2\zeta(3)\, 3^{3n} (\sqrt{2}-1)^{4n}$$
$$< (4/5)^n \zeta(3)$$

for all sufficiently large n.

(v) Conclude, using Question 43.5(i), that $\zeta(3)$ is irrational.

(Let me end by reminding the reader that we still do not know whether $\zeta(3)$ is transcendental nor whether $\zeta(2n+1)$ is irrational for any integer $n \geqslant 2$.)

43.8 (Bernstein's inequality) The object of this question is to show that, if P is a trigonometric polynomial of degree n or less, then $|P'(x)| \leqslant n \sup_{t\in\mathbb{T}} |P(t)|$ for all $x \in \mathbb{T}$.

(i) Show, by considering a simple trigonometric polynomial, that the inequality cannot be improved.

(ii) By choosing $\theta \in \mathbb{R}$ so that $e^{i\theta} P'(x)$ is real and positive and considering $Q(t) = e^{i\theta} P(t)$, or otherwise, show that the result for general P follows from that for real trigonometric polynomials P.

(iii) Explain why a non zero trigonometric polynomial of degree n has at most $2n$ zeros.

(iv) Now suppose P is a real trigonometric polynomial of degree n or less. Suppose that $|P(t)| \leqslant M$ for all $t \in \mathbb{T}$ but that $\sup_{x\in\mathbb{T}}|P'(x)| = nL$ with $L > M$. Explain why there exists an $x_0 \in \mathbb{T}$ with $|P'(x_0)| = nL$. Explain why we may suppose without loss of generality that $x_0 = 0$ and $P'(0) = nL$.

(v) Now set $S(t) = L \sin nt - P(t)$. By considering the values of $S((2r + 1)\pi/2n)$ show that S has at least $2n$ distinct zeros and deduce, using Rolle's theorem, that S' has at least $2n$ distinct zeros.

(vi) Show that 0 is a zero of S'. Using Rolle's theorem again, show that S'' has at least $2n$ distinct zeros in the set $\mathbb{T}\backslash\{0\}$. Conclude that 0 cannot be a zero of S'' unless $S''(t) = 0$ for all t. Derive a contradiction and conclude the truth of Bernstein's inequality.

(vii) Show, using the ideas of Chapter 5, or otherwise, that, if $P: [-1, 1] \to \mathbb{T}$ is a polynomial of degree at most n (i.e. $P(t) = \sum_{p=0}^{n} a_r t^r$), then

$$|P'(x)| \leqslant \frac{n}{\sqrt{(1 - x^2)}} \sup_{t\in[-1,1]} |P(t)| \quad \text{for all } |x| < 1.$$

(viii) Show, using the maximum modulus theorem, or otherwise, that, if $P: \{z: |z| \leqslant 1\} \to \mathbb{C}$ is a polynomial of degree at most n (i.e. $P(z) = \sum_{r=0}^{n} a_n z^n$), then

$$|P'(w)| \leqslant n \sup_{|z|\leqslant 1} |P(z)| \quad \text{for all } |w| \leqslant 1.$$

43.9 Let $X = \mathbb{T}$ or $X = [a, b]$ and write $C(X)$ for the space of real continuous functions on X. If $f \in C(X)$, let us write $\|f\|_\infty = \sup_{x\in\mathbb{T}} |f(x)|$. Since $C(X)$ is a vector space, we can consider subspaces V of $C(X)$. Let us say that f 'sticks out of' V if $\|f\|_\infty \leqslant \|f + g\|_\infty$ for all $g \in V$ (i.e. if 0 is a closest element in V to f). If $f \in C(X)$, let us write

$$E^+(f) = \{x \in X: f(x) = \|f\|_\infty\},$$

$$E^-(f) = \{x \in X: f(x) = -\|f\|_\infty\},$$

and

$$E(f) = E^+(f) \cup E^-(f).$$

(i) Suppose f is not identically zero and $g(x) > 0$ for all $x \in E^+(f)$ and $g(x) < 0$ for all $x \in E^+(f)$. Explain why we can find ε in $\varepsilon > 0$ such that $f(x)g(x) > 2\varepsilon$ for all $x \in E(f)$. Explain why we can find an open set $G \supseteq E(f)$ such that $f(x)g(x) > \varepsilon$ for all $x \in G$. Show also that we can find an $\eta > 0$ such that $|f(x)| \leqslant \|f\|_\infty - \eta$ for all $x \notin G$. By looking at $(f(x) - \delta g(x))^2$, or otherwise, deduce that there exists a $\delta > 0$ such that $\|f - \delta g\|_\infty < \|f\|_\infty$.

(ii) Show that $f \in C(X)$ sticks out of V if and only if there does not exist a $g \in V$ with $g(x) > 0$ for $x \in E^+(f)$, $g(x) < 0$ for $x \in E^-(f)$.

(iii) Let $F \in C(X)$. Show that $P \in V$ satisfies $\|P - F\|_\infty \leqslant \|Q - F\|_\infty$ for all $Q \in V$ if and only if there does not exist a $G \in V$ with $G(x) > 0$ for $x \in E^+(P - F)$, $G(x) < 0$ for $x \in E^-(P - F)$.

(iv) By taking $V = \mathcal{P}_{n-1}$ and using Lemma 43.3, obtain Theorem 43.5. (This argument is of course not really different from that of Chapter 43 but by casting it in a more general form some aspects of it become clearer.)

44

THE EXISTENCE OF THE BEST APPROXIMATION

44.1 Let us consider the problem of approximating a continuous function $f: [a, b] \to \mathbb{R}$ by rational functions. More precisely let $\mathcal{R}_n$ be the set of functions of the form P/Q with $P, Q \in \mathcal{P}_n$ and Q non zero on $[a, b]$.

(i) Let $[a, b] = [-1, 1]$, $R_m(x) = (1 + mx^2)^{-1}$. Observe that $R_m \in \mathcal{R}_2$ for all m and that $\sup_{x\in[-1,1]} |R_m(x)| \leq 1$ but show that we cannot find $R \in \mathcal{R}_2$ and $m(k) \to \infty$ such that $\sup_{x\in[-1,1]} |R_{m(k)}(x) - R(x)| \to 0$.

(ii) In spite of the problem raised by (i), we can use the methods of this chapter to show the existence of a best possible approximation. Explain why $\lambda = \inf \{\sup_{x\in[a,b]} |R(x) - f(x)|; R \in \mathcal{R}_n\}$ exists and choose $R_m \in \mathcal{R}_n$ such that $\sup_{x\in[a,b]} |R_m(x) - f(x)| \to \lambda$.

(iii) Show that we can find an A such that $\sup_{x\in[a,b]} |Q_m(x)| = 1$, for all $m \geq 1$ and that we can pick $P_m, Q_m \in \mathcal{P}_n$ such that $\sup_{x\in[a,b]} |Q_m(x)| = 1$, $\sup_{x\in[a,b]}|P_m(x)| \leq A$, $R_m = P_m/Q_m$.

(iv) Use the results of the chapter to show that there are $P, Q \in \mathcal{P}_n$ and $m(k) \to \infty$ such that

$$\sup_{x\in[a,b]} |P_{m(k)}(x) - P(x)| \to 0 \quad \text{and} \quad \sup_{x\in[a,b]} |Q_{m(k)}(x) - Q(x)| \to 0.$$

(v) Show that $\sup_{x\in[a,b]} |Q(x)| = 1$.

(vi) Show that, if $c \in [a, b]$, $r \geq 1$ and $(x - c)^r$ divides Q, then $(x - c)^r$ divides P.

(vii) Deduce that we can write $Q(x) = h(x)Q_0(x)$ $P(x) = h(x)P_0(x)$ where h, Q_0, $P_0 \in \mathcal{P}_n$ and Q_0 has no zeros on $[a, b]$. Set $R = Q_0/P_0$ and show that $R \in \mathcal{R}_n$, $\sup_{x\in[a,b]}|R(x) - f(x)| = \lambda$ so R is a best rational approximation in $\mathcal{R}_n$. (Under many circumstances rational approximation is a better practical method of approximation than polynomial approximation.)

45

TCHEBYCHEV AND UNIFORM APPROXIMATION II

45.1 Let $f: [-1, 1] \to \mathbb{R}$ be a smooth function. Suppose $x_1, x_2, \ldots, x_n \in [-1, 1]$ are distinct points and we know $y_1 = f(x_1)$, $y_2 = f(x_2)$, $\ldots$, $y_n = f(x_n)$. We are then asked for a good polynomial approximatiion P to f. How should we proceed? It is natural (though, as it turns out, foolish) to let P be that polynomial of degree at most $n-1$ with $P(x_j) = y_j$ $[1 \leqslant j \leqslant n]$.

(i) Explain why P is unique.

(ii) Suppose now that f is n times differentiable and we wish to bound the error $|f(x) - P(x)|$ for some fixed $x \in [-1, 1]$ with $x \neq x_1, x_2, \ldots, x_n$. Define $e: [-1, 1] \to \mathbb{R}$ by

$$e(t) = (f(t) - P(t)) - (f(x) - P(x)) \prod_{j=1}^{n} \frac{(t - x_j)}{(x - x_j)}.$$

Show that e has at least $n + 1$ zeros in $[-1, 1]$. By applying Rolle's theorem, deduce that e' has at least n zeros in $(-1, 1)$ and, proceeding by induction, that $e^{(n)}$ has at least one zero $\zeta = \zeta(x)$, say, in $(-1, 1)$.

(iii) Show that $f(x) - P(x) = f^{(n)}(\zeta)\prod_{j=1}^{n}(x - x_j)/n!$.

(iv) Conclude that

$$\sup_{|t| \leqslant 1} |f(t) - P(t)| \leqslant \sup_{|t| \leqslant 1} |f^{(n)}(t)| \sup_{|t| \leqslant 1} |Q(t)|/n!$$

where Q is the unique polynomial of degree n having leading coefficients 1 and roots $x_1, x_2, \ldots, x_n$.

(v) Suppose $n = 2N$ and $x_j = -1 + (j - 1/2)N^{-1}$. Show that $\log Q(-1) = \sum_{r=0}^{2N-1} \log((r + 1/2)/N)$ and, by using an integral to bound the sum below, or otherwise, deduce that $Q(-1) \geqslant A(2e^{-1})^n$ for some $A > 0$.

(vi) Explain why the choice of $x_1, x_2, \ldots, x_n$ which minimises $\sup_{|t| \leqslant 1} |Q(t)|$ is given by taking $x_1, x_2, \ldots, x_n$ to be the roots of the nth Tchebychev polynomial and why we then have

$$\sup_{|t| \leqslant 1}|f(t) - P(t)| \leqslant 2^{-n+1}(n!)^{-1} \sup_{|t| \leqslant 1}|f^{(n)}(t)|.$$

45.2 (i) Let $f\colon[-1,1]$ be an n times differentiable function. Suppose $m(1), m(2), \ldots, m(k)$ are integers with $m(j) \geqslant 1$ $[1 \leqslant j \leqslant k]$, $\sum_{j=1}^{k} m(j) = n$ and suppose that $x_1, x_2, \ldots, x_k \in [-1,1]$ are distinct. Show that there is a unique polynomial P of degree at most $n-1$ such that

$$P^{(r)}(x_j) = f^{(r)}(x_j) \quad \text{for } 0 \leqslant r \leqslant m(j)-1, 1 \leqslant j \leqslant k.$$

By imitating the previous question, show that

$$\sup_{|t|\leqslant 1} |f(t) - P(t)| \leqslant \sup_{|t|\leqslant 1} |f^{(n)}(t)| \sup_{|t|\leqslant 1} \left| \prod_{j=1}^{k} (t - x_j)^{m(j)} \right| \Big/ n!$$

(ii) Show that, if $f\colon(-1,1) \to \mathbb{R}$ is n times differentiable, then

$$f(t) = f(0) + f'(0)t + \ldots + \frac{f^{(n-1)}(0)}{(n-1)!} t^{n-1} + \frac{f^{(n)}(\zeta)}{n!} t^n$$

where $\zeta = \zeta(t)$ lies between 0 and t for all $|t| < 1$.

(iii) Suppose that $f\colon[-1,1] \to \mathbb{R}$ is $2n$ times differentiable and we require the polynomial P of degree $2n-1$ such that $P(x_j) = f(x_j)$, $P'(x_j) = f'(x_j)$ at n distinct points $x_1, x_2, \ldots, x_n \in [-1,1]$. Show that, by the appropriate choice of $x_1, x_2, \ldots, x_n$, we can ensure that

$$\sup_{|t|\leqslant 1} |f(t) - P(t)| \leqslant \sup_{|t|\leqslant 1} |f^{(2n)}(t)| 2^{-2n+2}/(2n)!.$$

(Results like those of Questions 45.1 (vi) and 45.2 (iii) look much better than they are. In theory $\sup_{|t|<1} |f^{(n)}(t)|$ can grow arbitrarily fast as $n \to \infty$ (see Question 11.4) and, in practice, we may have no reason to suppose that a physically observed function is many times differentiable. The next couple of questions reinforce this warning.)

(iv) We can also obtain an estimate for the error in Gaussian Quadrature (see Chapter 41) in this way. From now on let $x_1, x_2, \ldots, x_n$ and G_n be as in Theorem 41.6. Suppose f is $2n$ continuously differential. Let P be the unique polynomial of degree at most $2n-1$ such that $P(x_j) = f(x_j)$, $P'(x_j) = f'(x_j)$ $[1 \leqslant j \leqslant n]$. Show that

$$\left| G_n f - \int_{-1}^{1} f(x)dx \right| = \left| \int_{-1}^{1} (f(x) - P(x))\, dx \right|$$

$$\leqslant \sup_{|t|\leqslant 1} \frac{|f^{(2n)}(t)|}{(2n)!} \int_{-1}^{1} \left(\prod_{j=1}^{n} (x - x_j) \right)^2 dx = \frac{2}{2n+1} \frac{(2^n (n!)^2)^2}{((2n)!)^3} \sup_{|t|\leqslant 1} |f^{(2n)}(t)|.$$

(You may find some of the results of Lemma 40.7 useful.)

(v) By considering what happens if f is a polynomial of degree $2n$ show that the inequality of (iv) cannot be improved. Use Stirling's formula (Question 18.16 (i)) to estimate

$$\left|G_n f - \int_{-1}^{1} f(x)\,dx\right| \Big/ \sup_{|t|\leq 1} |f^{(2n)}(t)|$$

in the case when $f(t) = t^{2n}$.

45.3 (The questions that follow are based on Fejér's proof of theorem of Faber and N. Bernstein. Many related results will be found in the fascinating three-volume work by Natanson entitled *Constructive Function Theory* which I praised earlier.)

(i) Let $\theta_1, \theta_2, \ldots, \theta_n$ be distinct points with $0 \leq \theta_k \leq \pi$. Explain why we can find real trigonometric polynomials P_k with $P_k(\theta) = \sum_{j=0}^{n-1} a_{jk} \cos j\theta$ such that $P_k(\theta_r) = 0$ if $k \neq r$ and $P_k(\theta_k) = 1$. (Look at Chapter 5 if necessary.)

(ii) Let

$$A(\theta) = \frac{\cos\theta}{n-1} + \frac{\cos 2\theta}{n-2} + \ldots + \frac{\cos(n-1)\theta}{1},$$

$$B(\theta) = \frac{\cos(n+1)\theta}{1} + \frac{\cos(n+2)\theta}{2} + \ldots + \frac{\cos(2n-1)\theta}{n-1}.$$

Recall from Question 18.19 that $|A(\theta) - B(\theta)| \leq 4\sqrt{\pi}$ for all $\theta \in \mathbb{T}$ and that $A(0) \geq \log n$. Set

$$U(\theta) = A(2\theta) - \sum_{k=1}^{n} (B(\theta_k + \theta) + B(\theta_k - \theta))P_k(\theta)$$

and show that $U(\theta) = U(-\theta)$. Use orthogonality to show that $\int_0^\pi U(\theta)d\theta = \frac{1}{2}\int_{-\pi}^{\pi} U(\theta)d\theta = 0$ and deduce that there exists an α with $0 \leq \alpha \leq \pi$ and $U(\alpha) = 0$.

(iii) Now set

$$T(\theta) = (A(\theta + \alpha) + A(\theta - \alpha)) - \sum_{k=1}^{n} (B(\theta_k + \alpha) + B(\theta_k - \alpha))P_k(\theta).$$

Show that $T(\theta) = \sum_{j=0}^{n-1} a_j \cos j\theta$ for some real a_j, that $|T(\theta_k)| \leq 8\sqrt{\pi}$ for $1 \leq k \leq n$ and that $T(\alpha) \geq \log n$.

(iv) Conclude that, given $t_1, t_2, \ldots, t_n \in [a, b]$, we can find a real polynomial $S(t) = \sum_{j=0}^{n-1} b_j t^r$ with $|S(t_j)| \leq 8\sqrt{\pi}$ yet $S(c) \geq \log n$ for some $c \in [a, b]$.

45.4 The result of Question 45.3 (iv) stands in much the same relation to the theory of interpolation as Lemma 18.7 does to the theory of Fourier sums. We now try to obtain a result corresponding to Theorem 18.1. For this we need some notation. Let $[a, b]$ be a fixed interval and, for each $n \geq 1$, let $x_{1n}, x_{2n}, \ldots, x_{nn} \in (a, b)$ be distinct points. If f is a real continuous function on $[a, b]$, we write $P_n f$ for the unique polynomial of degree $n-1$ with $(P_n f)(x_{jn}) = f(x_{jn})$.

(i) By writing $(P_nf)(x) = \sum_{j=1}^{n} f(x_{jn})\prod_{i\neq j}(x - x_{in})/(x_{jn} - x_{in})$, or otherwise, show that there exists a $K(n)$ such that

$$|P_nf(x)| \leq K(n) \sup_{t\in[a,b]} |f(t)| \quad \text{for all } x \in [a, b].$$

Show also that, if f and g are continuous and λ, $\mu \in \mathbb{R}$, then $P_n(\lambda f + \mu g) = \lambda P_nf + \mu P_ng$.

(ii) Show from Question 45.3 that we can find a real continuous function f_n: $[a, b] \to \mathbb{R}$ and a $c'_n \in [a, b]$ such that $|f_n(t)| \leq 8\sqrt{\pi}$ for all $t \in [a, b]$ but $(P_nf_n)(c'_n) \geq \log n$. Explain further why we can find a $c_n \in [a, b]$ such that $|(P_nf_n)(c_n)| \geq |(P_nf_n)(t)|$ for all $t \in [a, b]$.

(iii) Write $\alpha(n) = |P_nf_n(c_n)|$. Show that we can find a sequence of integers $1 < n(1) < n(2) < \ldots$ such that

(a) $\alpha(n(r))^{1/2} \geq 2^r K(n(s))$ for all $r > s \geq 1$,
(b) $2^{-r}\alpha(n(r))^{1/2} \geq r + \sum_{s=1}^{r-1} 2^{-s}\alpha(n(s))^{1/2}$.

(iv) Show that $\sum_{r=1}^{\infty} 2^{-r}\alpha(n(r))^{-1/2} f_{nr}$ converges uniformly to a continuous function F. Using (a) show that

$$|(P_{n(r)} \sum_{s=1}^{r} 2^{-s}\alpha(n(s))^{-1/2} f_{n(s)})(x) - (P_{n(r)}F)(x)| \leq 1$$

for all $x \in [a, b]$. Using (b) show that

$$|(P_{n(r)} \sum_{s=1}^{r} 2^{-s}\alpha(n(s))^{-1/2} f_{n(s)})(c_{n(r)})| \geq r$$

for all $r \geq 1$. Deduce that $|P_{n(r)}F(c_{n(r)})| \to \infty$ as $r \to \infty$.

(v) State the result of (iv) in plain words.

45.5 However, Fejér observed that there is an interpolation procedure which always works.

(i) By recalling the definition $T_n(\cos\theta) = \cos n\theta$, or otherwise, show that, if ξ is a zero of T_n, the nth Tchebychev polynomial polynomial, then $T''_n(\xi)/T'_n(\xi) = \xi/(1 - \xi^2)$.

(ii) Suppose $x_{1n}, x_{2n}, \ldots, x_{nn}$ are distinct points of $[-1, 1]$. If f: $[-1, 1] \to \mathbb{R}$ is continuous, let us write Q_nf for the unique polynomial of degree $2n - 1$ with $(Q_nf)(x_n) = f(x_{jn})$, $(Q_nf)'(x_{jn}) = 0$. Show that

$$Q_nf(x) = \sum_{j=1}^{n} f(x_{jn})e_{jn}(x),$$

where

$$e_{jn}(x) = \left(1 - \frac{\omega''(x_{jn})}{\omega'(x_{jn})}(x - x_{jn})\right)\prod_{k\neq j}\left(\frac{x - x_{kn}}{x_{jn} - x_{kn}}\right)^2$$

and $\omega(x) = \prod_{j=1}^{n}(x - x_{jn})$. Explain why

(a) $\sum_{j=1}^{n} e_{jn}(x) = 1$ for all $x \in [-1, 1]$.

(iii) From now on let $x_{1n}, x_{2n}, \ldots, x_{nn}$ be the zeros of T_n. Show that

$$e_{jn}(x) = ((1 - x_{jn}x)/n^2)/(T_n(x)/(x - x_{jn}))^2.$$

Deduce that

(b) $e_{jn}(x) \geq 0$ for all $x \in [-1, 1]$ and $1 \leq j \leq n$.

Show that, if $|x - x_{jn}| > \delta$, then $|e_{jn}(x)| < 2(n^2\delta^2)^{-1}$. Deduce that

(c) $\sum_{|x - x_{jn}| > \delta} e_{jn}(x) \to 0$ uniformly for $x \in [-1, 1]$ as $n \to \infty$.

(iv) Combining (a), (b), (c) and the recollection of Chapter 2 (this is, after all, a result of Fejér), show that $Q_n f \to f$ uniformly on $[-1, 1]$ as $n \to \infty$.

(v) Observe that we have another proof of the Weierstrass polynomial approximation theorem.

(vi) Would the eye see $Q_n f$ as a good approximation to f, and why? (Think carefully.)

(In many ways this result points in the wrong direction. We remember from physics and statistics the general principle that a good model should contain few parameters. This suggests that, instead of seeking a polynomial P of degree $n - 1$ or more with $P(x_j) = y_i$ $[1 \leq j \leq n]$, we should seek a polynomial Q of degree m, with m much smaller than n, such that $Q(x_j)$ and y_i are close. Thus, for example, we might seek to minimize $\sum_{j=1}^{n} |Q(x_j) - y_i|^2$, in which case we have a least squares problem, or to minimize $\max_{1 \leq j \leq n} |Q(x_j) - y_i|$, in which case we have a linear programming problem, which before 1945 would have been computationally infeasible, but in the new age of the electronic computer may well be a matter of simple routine.)

45.6 (A simple set of wavelets) In this question we shall write

$$\langle f, g \rangle = \int_{-\infty}^{\infty} f(t)g(t)^* dt,$$

provided the integral we well defined. Let us define function ψ, ϕ:$\mathbb{R} \to \mathbb{R}$ by the conditions

$$\begin{aligned} \psi(x) &= 1 && \text{for } 0 \leq x < 1/2, \\ \psi(x) &= -1 && \text{for } 1/2 < x \leq 1, \\ \psi(x) &= 0 && \text{otherwise} \\ \phi(x) &= 1 && \text{for } 0 \leq x < 1, \\ \phi(x) &= 0 && \text{otherwise.} \end{aligned}$$

Define $\psi_{n,m}(x) = 2^{n/2}\psi(2^n x - m)$ and $\psi_m(x) = \phi(x - m)$ for $x \in \mathbb{R}$ and n, m integers with $n \geq 0$.

(i) Show that $\langle \psi_{n,m}, \psi_{p,q} \rangle = 1$ if $n = p$ and $m = q$, $\langle \psi_{n,m}, \psi_{p,q} \rangle = 0$ otherwise; $\langle \phi_m, \phi_q \rangle = 1$ if $m = q$, $\langle \phi_m, \phi_q \rangle = 0$ otherwise; $\langle \phi_m, \psi_{p,q} \rangle = 0$ always $[n, m, p, q \in \mathbb{Z}, n, p \geq 0]$.

(ii) Let $f:\mathbb{R} \to \mathbb{R}$ be continuous. Explain why there are no convergence problems if we set

$$f_N(t) = \sum_{m=-\infty}^{\infty} \langle f, \phi_m \rangle \, \phi_m(t) + \sum_{n=0}^{N} \sum_{m=-\infty}^{\infty} \langle f, \psi_{n,m} \rangle \, \psi_{n,m}(t).$$

Show, by induction or otherwise, that f_{N-1} is that function which takes the value $\lambda_{m,N}$ on the interval $[m2^{-N}, (m+1)2^{-N})$, where

$$\int_{m2^{-N}}^{(m+1)2^{-N}} (f(x) - \lambda_{m,N})dx = 0$$

i.e. $\lambda_{m,N} = 2^N \int_{m2^{-N}}^{(m+1)2^{-N}} f(x)dx \; [m \in \mathbb{Z}]$.

(iii) Show that $f_N(t) \to f(t)$ uniformly as $N \to \infty$ on each interval $[-R, R]$. If f is uniformly continuous on $\mathbb{R}$, show that $f_N(t) \to f(t)$ uniformly on $\mathbb{R}$ as $N \to \infty$. Would these results be altered in any way if instead of considering functions $f:\mathbb{R} \to \mathbb{R}$, we considered functions $f:\mathbb{R} \to \mathbb{C}$?
(Thus we have a 'wavelet' family of orthonormal functions with excellent convergence properties. The family also has good localisation. If we wish to know about f only on $[a, b]$, then we need only consider

$$\sum_{m \in \Lambda(0)} \langle f, \phi_m \rangle \, \phi_m(t) + \sum_{n=0}^{N} \sum_{m \in \Lambda(n)} \langle f, \psi_{n,m} \rangle \, \psi_{n,m}(t)$$

where $\Lambda(n) = \{m:[a, b] \cap [m2^{-n}, (m+1)2^{-n}) \neq \emptyset\}$. Can we find smoother wavelets ϕ and ψ which give rise to similar families? I refer the reader to Meyer's book *Wavelets and Operators*.)

45.7 (i) By using the fact that the non trivial polynomial $\sum_{m=0}^{2n} a_m z^m$ has at most $2n$ roots (multiple roots counted multiply), or otherwise, show that the equation $A_0 + \sum_{r=1}^{n} A_r \cos r\theta + \sum_{r=1}^{n} B_r \sin r\theta = 0$ either has every θ as a solution or at most $2n$ solutions (multiple solutions counted multiply).

(ii) Let us write $\mathcal{A}_n$ for the set of real trigonometric polynomials of degree n, i.e. let $\mathcal{A}_n$ consist of those $f:\mathbb{T} \to \mathbb{R}$ such that $f(\theta) = A_0 + \sum_{r=1}^{n} A_r \cos r\theta + \sum_{r=1}^{n} B_r \sin r\theta$ for some $A_0, A_1, \ldots, A_n, B_1, B_2, \ldots, B_n \in \mathbb{R}$. By looking at P', or otherwise, show that, if $|P(\theta_i)| = \sup_{\phi \in \mathbb{T}} |P(\phi)|$ for some θ_i with $0 \leq \theta_1 < \theta_2 < \ldots < \theta_{2n} < 2\pi$, then either P is constant or the $P(\theta_i)$ alternate in sign.

(iii) By imitating the arguments of Chapter 45, or otherwise, show that, if $P \in \mathcal{A}_n$ is non constant and $|P(\theta)| = \sup_{\phi \in \mathbb{T}} |P(\phi)|$ for $2n$ distinct values of θ, then $P(\theta) = A\cos(n\theta + \alpha)$ for some $A, \alpha \in \mathbb{R}$ and all θ.

45.8 (Carathéodory's lemma) The sequence of questions which follow are based on work of Rivlin and Shapiro. In them I try to give an example of how the geometrical reasoning associated with functional analysis can give elegant concrete results. The reader will need to do Questions 32.1 to 32.3 and Question 43.9 before starting this sequence. If the reader is ignorant of elementary functional analysis, she may find the arguments long and opaque, but, even so, I believe that she will find it worthwhile to persevere.

(i) Show that, if $\mathscr{A}$ is a non empty collection of convex sets in $\mathbb{R}^m$, then $\bigcap_{A\in\mathscr{A}} A$ is convex. Deduce, by choosing a suitable $\mathscr{A}$, that, if B is a set in $\mathbb{R}^m$, there exists a unique set $h(B)$ such that

(1) $h(B) \supseteq B$,
(2) $h(B)$ is convex,
(3) if C is convex and $C \supseteq B$, then $C \supseteq h(B)$.

We call $h(B)$ the convex hull of B.

(ii) if B is a set in $\mathbb{R}^m$, show that $h(B)$ consists precisely of the points of the form $\sum_{j=1}^N \lambda_j b_j$ with $\sum_{j=1}^N \lambda_j = 1$; $\lambda_j \geqslant 0$, $b_j \in B[1 \leqslant j \leqslant N]$ and N a positive integer.

(iii) Now let us refine (ii). Suppose $0 \in h(B)$. Then

$$0 = \sum_{j=1}^N \lambda_j b_j \quad \text{with } \lambda_j,\, b_j \text{ as in (ii).}$$

Suppose that $N \geqslant n+2$. Explain why $b_2, b_2, \ldots, b_N$ must be linearly dependent and so we can find $\mu_2, \mu_3, \ldots, \mu_N \in \mathbb{R}$ not all zeros such that

$$0 = \sum_{j=2}^N \mu_j b_j.$$

Setting $\mu_1 = 0$ observe that for all $\alpha \in \mathbb{R}$

$$0 = \sum_{j=1}^N (\lambda_j + \alpha\mu_j) b_j.$$

By choosing α suitably, obtain

$$0 = \sum_{j=1}^N \theta_j b_j$$

with $\theta_j \geqslant 0$ $[1 \leqslant j \leqslant N]$, not all θ_j zero but at least one θ_j zero. Conclude that we can find $\lambda'_j \geqslant 0$, $b'_j \in B$ $[1 \leqslant j \leqslant N-1]$ with $\sum_{j=1}^{N-1} \lambda'_j = 1$ and $0 = \sum_{j=1}^{N-1} \lambda'_j b'_j$.

(iv) Show that, if $0 \in h(B)$, then we can find $\lambda_j \geqslant 0$, $b_j \in B$ $[1 \leqslant j \leqslant m+1]$ such that $\sum_{j=1}^{m+1} \lambda_j = 1$ and $0 = \sum_{j=1}^{m+1} \lambda_j b_j$. More generally, prove Carathéodory's lemma which states that, if $b \in h(B)$, then we can find $\lambda_j \geqslant 0$, $b_j \in B$ $[1 \leqslant j \leqslant m+1]$ such that $\sum_{j=1}^{m+1} \lambda_j = 1$ and $b = \sum_{j=1}^{m+1} \lambda_j b_j$.

(v) Illustrate Carathéodory's lemma in the case $m = 2$ by drawing diagrams. Show that we cannot replace $m+1$ by m.

45.9 (i) Let A be a closed bounded subset of $\mathbb{R}^M$ and $F:A \to \mathbb{R}^N$ a continuous map. By using the methods of Appendix B, or otherwise, show that $F(A)$ is a closed bounded subset of $\mathbb{R}^N$.

(ii) Let B be a closed bounded subset of $\mathbb{R}^m$. Set $M = (m+1)^2$ and let A be the subset of $\mathbb{R}^M$ consisting of points

$$\mathbf{a} = (\lambda_0, \lambda_1, \ldots, \lambda_m, \mathbf{b}_0, \mathbf{b}_1, \ldots, \mathbf{b}_m)$$

with $\lambda_j \geqslant 0$ $[0 \leqslant j \leqslant m]$, $\sum_{j=0}^{m} \lambda_j = 1$, $\mathbf{b}_j \in B$ $[0 \leqslant j \leqslant m]$. Show that A is closed and bounded.

Let $F:A \to \mathbb{R}^m$ be defined by

$$\mathbf{F}(\lambda_0, \lambda_1, \ldots, \lambda_m, \mathbf{b}_0, \mathbf{b}_1, \ldots, \mathbf{b}_m) = \sum_{j=0}^{m} \lambda_j \mathbf{b}_j$$

(with λ_j, $\mathbf{b}_j$ as in the previous paragraph). Use Carathéodory's lemma to show that $h(B) = F(A)$ and conclude that $h(B)$ is closed and bounded. Thus the hull of a closed bounded set in $\mathbb{R}^m$ is closed and bounded.

(iii) Which of the following are true, and which are false, for a subset B of $\mathbb{R}^m$ with $m \geqslant 2$? Give reasons.

(a) If B is open, $h(B)$ is open.
(b) If B is closed, $h(B)$ is closed.
(c) If B is bounded, $h(B)$ is bounded.
(d) If B is not open, $h(B)$ is not open.
(e) If B is not closed, $h(B)$ is not closed.

(You may find if helpful to look at $B = \{(0,1)\} \cup \{(x,0) : x \in \mathbb{R}\}$ in $\mathbb{R}^2$.) What happens in each case when $m = 1$ and why?

45.10 This question uses the notation and results of parts (i) and (ii) of Question 43.9. Thus, for example, V will be an m-dimensional subspace of $C(X)$.

(i) Suppose that $f \in C(X)$ and there exist $x_1, x_2, \ldots, x_k \in E(f)$ together with $\lambda_1, \lambda_2, \ldots, \lambda_k \geqslant 0$ satisfying $\sum_{j=1}^{k} \lambda_j = 1$ and

$$\sum_{j=1}^{k} \lambda_j f(x_j) g(x_j) = 0 \quad \text{for all } g \in V.$$

Show that

$$\|f\|_\infty^2 = \sum_{j=1}^{k} \lambda_j f(x_j)(f(x_j) + g(x_j)) \leqslant \|f\|_\infty \|f+g\|_\infty$$

for all $g \in V$ and conclude that f sticks out of V.

(ii) Suppose, conversely, that f sticks out of V. Let $\phi_1, \phi_2, \ldots, \phi_m$ form a basis for V and consider the map $F: E(f) \to \mathbb{R}^m$ given by

$$F(x) = (f(x)\phi_1(x), f(x)\phi_2(x), \ldots, f(x)\phi_m(x)).$$

By using part (i) and part (ii) Question 45.9, show that $h(F(E(f)))$ is closed and convex. Write $\Gamma = h(F(E(f)))$.

(iii) We wish to know whether $\mathbf{0} = (0, 0, \ldots, 0)$ lies in Γ or not. By using the finite-dimensional Hahn–Banach theorem given in Question 32.3 (iv), show that, if $\mathbf{0} \notin \Gamma$, then we can find $a_1, a_2, \ldots, a_m \in \mathbb{R}$ such that $\sum_{r=1}^{m} a_r y_r > 0$ for all $\mathbf{y} = (y_1, y_2, \ldots, y_m) \in \Gamma$. Observe that this implies $\sum_{r=1}^{m} a_r f(x)\phi_r(x) > 0$ for all $x \in E(f)$ and, by considering $g = \sum_{r=1}^{m} a_r \phi_r$ in the context of Question 43.9 (ii), obtain a contradiction.

(iv) We now know that $\mathbf{0} \in \Gamma$. By using Carathéodory's lemma (Question 45.8), conclude that we can find $\lambda_1, \lambda_2, \ldots, \lambda_{m+1} \geqslant 0$ and $x_1, x_2, \ldots, x_{m+1} \in E(f)$ such that $\sum_{j=1}^{m+1} \lambda_j = 1$ and $\sum_{j=1}^{m+1} \lambda_j f(x_j)\phi_r(x_j) = 0$ for $1 \leqslant r \leqslant m$. Show that we now have $\sum_{j=1}^{n+1} \lambda_j f(x_j) = 0$ for all $g \in V$.

(v) Combine (i) and (iv) to show that f sticks out of V if and only if there exists an $M \leqslant m+1$ and $\lambda_1, \lambda_2, \ldots, \lambda_m > 0$ together with $x_1, x_2, \ldots, x_M \in E(f)$ such that $\sum_{j=1}^{M} \lambda_j f(x_j) g(x_j) = 0$ for all $g \in V$.

45.11 (The interpolation formula of Rivlin and Shapiro) Let $X = [a, b]$ or $X = \mathbb{T}$, let U be an n-dimensional subspace of $C(X)$ and let $T: U \in \mathbb{R}$ be a non zero linear map.

(i) Use the finite dimensionality of U to show that T is continuous and that there exists an $f_0 \in U$ such that $\|f_0\| \leqslant 1$ and $|Tf_0| \geqslant |Tg|$ for all $g \in U$ with $\|g\|_\infty \leqslant 1$. Show that, in fact, $\|f_0\|_\infty = 1$.

(ii) Let $V = \{g \in U: Tg = 0\}$. Show that V is an $n-1$ dimensional subspace of $C(X)$ and that f_0 sticks out of V. Hence, by Question 45.10, we can find $N \leqslant n$ and $\lambda_1, \lambda_2, \ldots, \lambda_N > 0$ together with $x_1, x_2, \ldots, x_N \in E(f_0)$ such that $\sum_{j=1}^{N} \lambda_j f_0(x_j) g(x_j) = 0$ for all $g \in U$. Show that, if $f \in U$, then $(Tf_0)f - (Tf)f_0 \in V$ and deduce that

$$Tf = (Tf_0)\left(\sum_{j=1}^{N} \lambda_j f_0(x_j) f(x_j)\right) \Bigg/ \left(\sum_{j=1}^{N} \lambda_j f_0(x_j)^2\right).$$

(iii) Deduce that, if $T: U \to \mathbb{R}$ is a non zero linear map, then we can find $N \leqslant n$, $x_1, x_2, \ldots, x_N \in X$ distinct, $c_1, c_2, \ldots, c_N \neq 0$ and $f_1 \in U$ with $Tf_1 \neq 0$ such that

(a) $Tf = \sum_{j=1}^{N} c_j f(x_j)$ for all $f \in U$,
(b) $Tf_1 = \sum_{j=1}^{N} |c_j| \|f_1\|_\infty$.

(iv) Under the circumstances described in (iii), show that

$$|Tf| \leqslant \sum_{j=1}^{N} c_j \|f\|_\infty \quad \text{for all } f \in U$$

and that, if $f_2 \in U$ is such that $Tf_2 \neq 0$ and $Tf_2 = \sum_{j=1}^{N} |c_j| \, \|f_2\|_\infty$, then $x_1, x_2, \ldots, x_N \in E(f_2)$ and $\operatorname{sgn} f_2(x_j) = \operatorname{sgn} c_j$ $[1 \leq j \leq N]$. (Recall that $\operatorname{sgn} c = 1$, if $c > 0$, $\operatorname{sgn} c = -1$, if $c < 0$.)

(v) Using the notation and results of Chapters 43 and 45, show that in the particular case when $X = [-1, 1]$, $U = \mathcal{P}_n$ and $N = n + 1$ then either f_2 is constant or f_2 is a non zero multiple of T_n.

(vi) Use Question 45.7 (iii) to state and prove a similar result when $U = \mathcal{A}_n$ and $N = 2n$.

45.12 We have now completed our investment in abstract analysis and can draw some concrete dividends.

(i) If t is a fixed real number with $|t| \geq 1$, define $S: \mathcal{P}_n \to \mathbb{R}$ by $SP = P(t)$. (Formally, if $P \in \mathcal{P}_n$, then $P(x) = \sum_{r=0}^{n} a_r x^r$ for all $x \in [-1, 1]$ and some $a_0, a_1, \ldots, a_n \in \mathbb{R}$. Set $SP = \sum_{r=0}^{n} a_r t^r$.) Show that S is linear and so, from Question 45.11, there exists an $N \leq n + 1$, $x_1, x_2, \ldots, x_N \in [-1, 1]$ distinct, $c_1, c_2, \ldots c_N \neq 0$ and $Q \in \mathcal{P}_n$ with $SQ \neq 0$ such that

(a) $SP = \sum_{j=1}^{N} c_j P(x_j)$ for all $P \in \mathcal{P}_n$,
(b) $SQ = \sum_{j=1}^{N} |c_j| \sup_{|x| \leq 1} |Q(x)|$.

Show that, if $N \leq n$, we can find a $Q_1 \in \mathcal{P}_n$ such that $Q_1(x_j) = 0$ $[1 \leq j \leq N]$ and $Q_1(t) \neq 0$. Deduce that $N = n + 1$ and hence, using Question 45.11, that either Q is constant or Q is a non zero multiple of T_n.

Show that Q is a non zero multiple of T_n and hence obtain the following inequality of Tchebychev

$$|P(t)| \leq |T_n(t)| \sup_{|x| \leq 1} |P(x)| \quad \text{for all } P \in \mathcal{P}_n.$$

(ii) Show that the result just given may fail if $|t| < 1$. Explain where the proof given in (i) breaks down.

(iii) Show that, if $P \in \mathcal{P}_n$ and $|t| \geq 1$, then

$$|P^{(j)}(t)| \leq |T_n^{(j)}(t)| \sup_{|x| \leq 1} |P(x)|$$

for $j = 0, 1, \ldots, n$. Show, moreover, that equality occurs if and only if P is a multiple of T_n.

(iv) By taking $j = n$ show that, if $P \in \mathcal{P}_n$, then, writing $a_n(P)$ for the coefficient of x^n in P, we have

$$|a_n(P)| \leq 2^{n-1} \sup_{|x| \leq 1} |P(x)|$$

with equality if and only if P is a multiple of T_n. Deduce Theorem 45.3.

(v) (This is just a remark.) Show that, if $|t| \geq 1$, $T_n(t) = \cosh nw$ with $w = \cosh^{-1} t$.

(vi) Let $\psi \in \mathbb{T}$. Prove Bernstein's inequality

$$|T'(\psi)| \leqslant n \max_{\theta \in \mathbb{T}} |T(\theta)| \quad \text{for all } T \in \mathcal{A}_n$$

with equality if and only if $T(\theta) = K \sin n(\theta - \psi)$ for some $K \in \mathbb{R}$.

(vii) By looking at the $x_1, x_2, \ldots, x_N$ which arise in the proof of (vi), show that we can find $c_j(\psi) \in \mathbb{R}$ $[1 \leqslant j \leqslant 2n]$ such that

$$T'(\psi) = \sum_{r=1}^{2n} c_r(\psi) T(\psi + (2r-1)\pi/2n) \quad \text{for all } T \in \mathcal{A}_n,$$

and $\sum_{r=1}^{2n} |c_r(\psi)| = n$.
(This is a result of Riesz who also determined the $c_r(\psi)$ explicitly.) By exploiting symmetries, show that

$$\sum_{k=1}^{2n} |T'(\psi + (2k-1)\pi/2n)| \leqslant 2n^2 \sum_{r=1}^{2n} |T(\psi + (2r-1)\pi/2n)|$$

and deduce the integral form of Bernstein's inequality

$$\int_0^{2\pi} |T'(\psi)| d\psi \leqslant n \int_0^{2\pi} |T(\psi)| d\psi \quad \text{for all } T \in \mathcal{A}_n.$$

(viii) By considering $S:\mathcal{P}_n \to \mathbb{R}$ given by $SP = \int_{-1}^{1} P(t)dt$, show that we can find $N \leqslant n+1$, $x_1, x_2, \ldots, x_N \in [-1, 1]$ and $c_1, c_2, \ldots, c_n > 0$ such that

$$\int_{-1}^{1} P(t)dt = \sum_{j=1}^{N} c_j P(x_j) \quad \text{for all } P \in \mathcal{P}_n.$$

(This result is weaker than those of Chapter 43, but by no means self evident. Rivlin has written a book entitled *The Chebychev Polynomials* in which he takes these matter much further as well as presenting many enjoyable results concerning the polynomials of his title. The book *Smoothing and Approximation of Functions* and *Topics in Approximation Theory* by Shapiro are also worth looking at, but assume a knowlege of Lebesgue measure and a smattering of functional analysis.)

46

INTRODUCTION TO FOURIER TRANSFORMS

46.1 (i) Suppose that g is a twice continuously differentiable function such that $g(x) = 0$ for $|x| \geqslant 2$. Show that there exists a constant A such that $|\hat{g}(\zeta)| \leqslant A\zeta^{-2}$ for $\zeta \neq 0$. Deduce in particular that $\int_{-\infty}^{\infty} |\hat{g}(\zeta)|\, d\zeta$ converges.

(ii) With g as in (i), write $g_N(t) = (\exp iNt) g(t)$. Show that, if R is fixed, $\int_{-R}^{R} |\hat{g}_N(\zeta)|\, d\zeta \to 0$ as $N \to \infty$ and, if N is fixed $\int_{|x| \geqslant R} |\hat{g}_N(\zeta)|\, d\zeta \to 0$ as $R \to \infty$.

(iii) Let f_n be a twice continuously differentiable function such that $1 \geqslant f_n(x) \geqslant 0$ for all x, $f_n(x) = 1$ for $|x| \leqslant 1$ and $f_n(x) = 0$ for $|x| \geqslant 1 + n^{-1}$. Show, using Example 46.4, for otherwise, that $\int_{-\infty}^{\infty} |\hat{f}_n(\zeta)|\, d\zeta \to \infty$ as $n \to \infty$.

(iv) Show that we can find integers $n(1) < n(2) < \dots$ and real numbers $0 = R(1) < R(2) < \dots$ such that, taking f_n as in (iii) and writing $g_{n(j)}(t) = \exp(in(j)t) f_{n(j)}(t)$, we have $\int_{|t| \leqslant R(j)} |\hat{g}_{n(j)}(t)|\, dt < 1$, $\int_{-\infty}^{\infty} |\hat{g}_{n(j)}(t)|\, dt \geqslant 2^{2j}$ and $\int_{|t| \leqslant R(j+1)} |\hat{g}_{n(j)}(t)|\, dt < 1$. Show that $\sum_{j=1}^{\infty} 2^{-j} g_{n(j)}$ converges uniformly to a positive continuous function G, say, with $G(t) = 0$ for $|t| \geqslant 2$ and $\int_{-\infty}^{\infty} |\hat{G}(\zeta)|\, d\zeta$ divergent. Is it possible that G is twice continuously differentiable?

(v) With the notation of (iv), set $h_{n(j)} = g(t - 2j)$. Show that $\sum_{j=1}^{\infty} 2^{-j} h_{n(j)}$ converges to a positive twice continuously differentiable function H with $H(t) \to 0$ as $|t| \to \infty$. Show that $\int_{-\infty}^{\infty} H(t)\, dt$ converges and $\int_{-\infty}^{\infty} |H(t) - \sum_{j=1}^{m} 2^{-j} h_{n(j)}(t)|\, dt \to 0$. Deduce that $\hat{H}(\zeta) = \sum_{j=1}^{\infty} 2^{-j} \hat{h}_{n(j)}(\zeta)$ and hence, or otherwise, show that $\int_{-\infty}^{\infty} |\hat{H}(\zeta)|\, d\zeta$ is divergent.

46.2 Follow the ideas of Chapter 18 to produce an $f \in L^1 \cap C$ such that $f(x) = 0$ for $|x| > 1$ and

$$\limsup_{R \to \infty} \int_{-R}^{R} \hat{f}(\zeta) \exp(i\zeta t)\, d\zeta = \infty$$

for a given t.

47

CHANGE IN THE ORDER OF INTEGRATION I

47.1 **By interchanging the order of integration in**

$$\int_0^x \left(\int_a^b e^{-tx}\, dt \right) dx \quad [b > 0,\, x > 0]$$

show that

$$\int_0^x \frac{e^{-ax} - e^{-bx}}{x}\, dx = \int_a^b \frac{1 - e^{-tx}}{t}\, dt.$$

Show that

$$\int_a^b \frac{e^{-tx}\,dt}{t} \leqslant e^{-ax} \log(b/a) \text{ and hence find } \int_0^\infty \frac{(e^{-ax} - e^{-bx})\,dx}{x}.$$

48

CHANGE IN THE ORDER OF INTEGRATION II

48.1 A less obviously contrived example than that in Example 48.1 runs as follows. Set $a_{nm} = (m^2 - n^2)^{-1}$ for $m \neq n$, $a_{nn} = 0$. By using the relation

$$a_{nm} = (2m)^{-1}((m-n)^{-1} + (m+n)^{-1}),$$

or otherwise, show that $\sum_{n=1}^{\infty} a_{nm} = -3/(4m^2)$. Using Lemma 33.6 (iv) compute $\sum_{m=1}^{\infty}(\sum_{n=1}^{\infty} a_{nm})$ and $\sum_{n=1}^{\infty}(\sum_{m=1}^{\infty} a_{nm})$.

48.2 An immediate use for Lemma 48.2 occurs if we wish to develop complex variable theory using power series as our main tool.

(i) Explain why, if $c_{nm} \in \mathbb{C}$, $z, h \in \mathbb{C}$ and $\sum_{n=0}^{\infty}\sum_{m=0}^{\infty} |c_{nm}||z|^n|h|^m$ exists, then $\sum_{n=0}^{\infty}\sum_{m=0}^{\infty} c_{nm} z^n h^m$ and $\sum_{m=0}^{\infty}\sum_{n=0}^{\infty} c_{nm} z^n h^m$ exist and are equal.

(ii) if $a_n \in \mathbb{C}$ and $\sum_{n=0}^{\infty} a_n z^n$ has radius of convergence R, show that, if $|z_1| + |h| < R$, then

$$\sum_{m=0}^{\infty} a_m(z_1 + h)^m = \sum_{n=0}^{\infty} b_n h^n$$

where $b_n \in \mathbb{C}$, $\sum_{n=0}^{\infty} b_n z^n$ has radius of convergence at least $R - |z_1|$, and $b_0 = \sum_{n=0}^{\infty} a_n z_1^n$, $b_1 = \sum_{n=1}^{\infty} n a_n z_1^{n-1}$.

(iii) If $a_n = 1$ for all n, compute $b_0, b_1, b_2, \ldots$. By considering the cases $z_1 = 1/2$ and $z_1 = -1/2$, show that (in the notation of (ii)) the radius of convergence can be exactly $R - |z_1|$ or strictly greater than $R - |z_1|$.

(iv) Returning to the case of general a_n show that $|b_n| \leqslant M(R - |z_1|)^{-n}$ for some $M > 0$ (otherwise, what could we say about the associated radius of convergence?) and deduce that

$$\left|\left(\sum_{n=0}^{\infty} b_n h^n - b_0\right)h^{-1} - b_1\right| \to 0 \quad \text{as } |h| \to 0.$$

Conclude that $\sum_{n=0}^{\infty} a_n z^n$ is differentiable term by term within its radius of convergence. (Compare Question 75.1.)

48.3 (The volume of the n-dimensional sphere)

(i) Let $\chi_{[0,a]}(x) = 1$ for $0 \leqslant x \leqslant a$, $\chi_{[0,a]}(x) = 0$ otherwise. Explain why, if V_n is the volume of the n-dimensional sphere,

$$\int_{-\infty}^{\infty}\int_{-\infty}^{\infty}\cdots\int_{-\infty}^{\infty}\chi_{[0,a]}(\sqrt{}(x_1^2 + x_2^2 + \ldots + x_n^2))dx_1dx_2\ldots dx_n =$$

$$nV_n\int_0^{\infty} r^{n-1}\chi_{[0,a]}(r)dr$$

i.e., in a more compact notation,

$$\int\chi_{[0,a]}(|\mathbf{x}|)dV(\mathbf{x}) = nV_n\int_0^{\infty} r^{n-1}\chi_{[0,a]}(r)dr.$$

(ii) Show that if $a(1), a(2), \ldots, a(m) \geqslant 0$ and $\lambda_1, \lambda_2, \ldots, \lambda_m \in \mathbb{R}$, then

$$\int\sum_{j=1}^{m}\lambda_j\chi_{[0,a(j)]}(|\mathbf{x}|)dV(\mathbf{x}) = nV_n\int_0^{\infty} r^{n-1}\sum_{j=1}^{m}\lambda_j\chi_{[0,a(j)]}(r)dr.$$

Hence show that, if f is a continuous function $f:[0,\infty)\to\mathbb{R}$ with $f(r) = 0$ for $r > b$, say, then

$$\int f(|\mathbf{x}|)dV(\mathbf{x}) = nV_n\int_0^{\infty} r^{n-1}f(r)dr.$$

Deduce that, if $g:[0,\infty)\to\mathbb{R}$ is a continuous function with $r^{n+2}g(r)\to 0$ as $r\to\infty$, then

$$\int g(|\mathbf{x}|)dV(\mathbf{x}) = nV_n\int_0^{\infty} r^{n-1}g(r)dr.$$

(iii) Let $g(r) = \exp(-r^2/2)$. Show that

$$\int g(|\mathbf{x}|)dV(\mathbf{x}) = \left(\int_{-\infty}^{\infty}\exp(-x^2/2)dx\right)^n = (2\pi)^{n/2}$$

and hence use (ii) to obtain

$$V_{2n} = \pi^n/n!, \quad V_{2n-1} = n!2^{2n}\pi^{n-1}/(2n)!.$$

48.4 (i) Show that $\int_{-\infty}^{\infty}\exp(-(ax^2 + 2hxy + by^2))dxdy = \pi/(ab - h^2)^{1/2}$, provided that a and $ab - h^2$ are positive. What happens if $a > 0$, $ab - h^2 \leqslant 0$? What happens if $a \leqslant 0$?

(ii) Show that the product of $\int_{-\infty}^{\infty}\exp(-x^4)dx$ and $\int_{-\infty}^{\infty}x^2\exp(-x^4)dx$ is $\frac{1}{4}\pi\sqrt{}2$.

48.5$^{(-)}$ For which real x and y does

$$1 + (1 + y)x + \left(1 + y + \frac{y^2}{2!}\right)x^2 + \left(1 + y + \frac{y^2}{2!} + \frac{y^3}{3!}\right)x^3 + \ldots$$

converge, and what is its sum?

48.6 (i) Let $a_{rs} \in \mathbb{C}$. Show that, if $\sum_{r=-\infty}^{\infty}|a_{rs}|$ converges for each s and $\sum_{s=-\infty}^{\infty}\sum_{r=-\infty}^{\infty}|a_{rs}|$ converges, then $\sum_{r=-\infty}^{\infty}a_{rs}$ converges for each s, $\sum_{r=-\infty}^{\infty}a_{r,t-r}$ converges for each t, $\sum_{s=-\infty}^{\infty}\sum_{r=-\infty}^{\infty}a_{rs}$ converges $\sum_{t=-\infty}^{\infty}\sum_{r=-\infty}^{\infty}a_{r,t-r}$ converges and

$$\sum_{s=-\infty}^{\infty}\sum_{r=-\infty}^{\infty} a_{rs} = \sum_{t=-\infty}^{\infty}\sum_{r=-\infty}^{\infty} a_{r,t-r}.$$

(ii) Deduce that, if $\sum_{r=0}^{\infty}a_r$, $\sum_{s=0}^{\infty}b_s$ are absolutely convergent and $c_t = \sum_{r=0}^{t}a_r b_{t-r}$ $[t \geqslant 0]$, then $\sum_{t=0}^{\infty}c_t$ is absolutely convergent and

$$\sum_{r=0}^{\infty} a_r \sum_{s=0}^{\infty} b_s = \sum_{t=0}^{\infty} c_t.$$

(iii) Show, using the power series expansions for cos and cosh, that

$$\cos x \cosh x = \sum_{n=0}^{\infty} (-1)^n \frac{4^n}{(4n)!} x^{4n}.$$

(iv) Obtain the result of (iii) by using Taylor's theorem with remainder. (See e.g. Question 4.3.)

49

FEJÉR'S THEOREM FOR FOURIER TRANSFORMS

49.1 Evaluate $\int_{-\infty}^{\infty}(\sin t/t)^2\,dt$ by complex variable methods. (You may find it useful to recall that $(\sin t)^2 = (1 - \cos 2t)/2$.) Hence obtain Lemma 49.2 (iii).

49.2 (i) Let us write

$$\Delta_R(x) = 1 - |x|/R \text{ for } |x| \leqslant R, \quad \Delta_R(x) = 0 \text{ otherwise},$$

$$\chi_{[a,b]}(x) = 1 \text{ if } x \in [a, b], \quad \chi_{[a,b]}(x) = 0 \text{ otherwise}.$$

Show that, if R is fixed,

$$\left|\Delta_R(x) - N^{-1}\sum_{j=1}^{N}\chi_{[-jR/N,jR/n]}(x)\right| \to 0$$

uniformly in x as $N \to \infty$. Deduce that, if f is continuous,

$$\int_{-\infty}^{\infty}\Delta_R(x)f(x)dx - N^{-1}\sum_{j=1}^{n}\int_{-jR/N}^{jR/N}f(x)dx \to 0 \quad \text{as } N \to \infty.$$

Conclude that

$$\int_{-\infty}^{\infty}\Delta_R(x)f(x)\,dx = \frac{1}{R}\int_0^R\left(\int_{-y}^{y}f(x)\,dx\right)dy$$

(ii) Hence show that, if $\int_{-R}^{R}f(x)dx \to L$ as $R \to \infty$, then

$$\int_{-\infty}^{\infty}\Delta_R(x)f(x)\,dx \to L \text{ as } R \to \infty.$$

(iii) Decide each of the following questions by giving a proof or a counter example.

(a) Is it true that, if f is continuous and bounded, then, if $\int_{-\infty}^{\infty}\Delta_R(x)f(x)\,dx \to L$, it follows that $\int_{-R}^{R}f(x)\,dx \to L$ as $R \to \infty$?

(b) Is it true that, if f is continuous and $f(x) \to 0$ as $x \to \infty$, then, if $\int_{-\infty}^{\infty} \Delta_R(x) f(x) \to L$, it follows that $\int_{-R}^{R} f(x)\,dx \to L$ as $R \to \infty$?

(c) Is it true that, if f is continuous and we can find an interval $[a, b]$ such that $f(x) = 0$ for $x \notin [a, b]$, then, if $\int_{-\infty}^{\infty} \Delta_R(x) f(x)\,dx \to L$, it follows that $\int_{-R}^{R} f(x)\,dx \to L$ as $R \to \infty$?

50

SUMS OF INDEPENDENT RANDOM VARIABLES

50.1$^{(-)}$ Let X be a random variable with the Cauchy distribution (see Lemma 50.2 (ii)). Show that, if $X = X(1), = X(2)$, then $X(1)$ and $X(2)$ are not independent, yet $\hat{f}_{X(1)+X(2)}(\zeta) = \hat{f}_{X(1)}(\zeta)\hat{f}_{X(2)}(\zeta)$.

50.2 Let B be the collection of random variables X taking values in $\mathbb{R}$ and having bounded continuous probability density f_X.

(i) Show that, if $X \in B$, then $-X \in B$ and $\hat{f}_{-X}(\zeta) = (\hat{f}_X(\zeta))^*$.

(ii) Deduce that, if $X, Y \in B$ are independent identically distributed random variables, then $X - Y \in B$ and $\hat{f}_{X-Y}(\zeta) \geqslant 0$.

(iii) Let Z be a random variable taking values in $\mathbb{R}$ and having probability density $f_Z(t) = (\sin t)/2$ for $|t| < \pi/2$, $f_Z(t) = 0$ for $|t| \geqslant \pi/2$. Show that we cannot find independent identically distributed random variables $X, Y \in B$ such that $Z = X - Y$.

50.3 (i) Let $p \geqslant 1$. Suppose that Y is a real valued random variable such that $\mathbb{E}|Y|^p$ exists. Show that

$$Pr(|Y| > a) \leqslant \mathbb{E}|Y|^p/a^p \quad [a > 0].$$

(ii) State (giving examples or proofs as appropriate) whether the following statements are true or false.

(a) If $X_1, X_2, \ldots$ are identically distributed (but not necessarily independent) real valued random variables with $|X_j| \leqslant 1$, $\mathbb{E}X_j = 0$, then

$$Pr\left(\left|\sum_{j=1}^{n} X_j/n\right| > a\right) \to 0 \quad \text{as } n \to \infty \text{ for all } a > 0.$$

(b) If $X_1, X_2, \ldots$ are identically distributed real valued random variables with $|X_j| \leqslant 1, \mathbb{E}X_j = 0$ and $\mathbb{E}(X_jX_{j+2}) = 1$ for all $j \geqslant 1$, then

$$Pr\left(\left|\sum_{j=1}^{n} X_j/n\right| > 1/2\right) \nrightarrow 0 \quad \text{as } n \to \infty.$$

(iii) Suppose n letters are placed at random in n envelopes (one letter to one envelope). Define $X_k = 1$ if the kth letter goes in the kth envelope, $X_k = 0$ otherwise. Find $\mathbb{E}X_k$ and $\mathbb{E}X_iX_j$. Are $X_1, X_2, \ldots, X_n$ independent? Explain the meaning of the random variable $Y_n = X_1 + X_2 + \ldots X_n$ and show that Y_n has mean and variance 1. Hence prove that, if $\phi(n) \to \infty$ as $n \to \infty$,

$$Pr(|Y_n - 1| \geqslant \phi(n)) \to 0 \quad \text{as } n \to \infty.$$

(iv) Suppose $X_1, X_2, \ldots$ are independent random variables with $Pr(X_k = k^\delta) = Pr(X_k = -k^\delta) = 1/2$, where $0 \leqslant \delta < 1/2$. Show that, if $\varepsilon > 0$, then

$$Pr\left(\left|\sum_{k=1}^{n} X_k\right| > \varepsilon n\right) \to 0 \quad \text{as } n \to \infty.$$

50.4 (i) Suppose $g:\mathbb{R} \to \mathbb{R}$ satisfies $g(x+y) = g(x) + g(y)$ for all $x, y \in \mathbb{R}$. Show that, if p and q are integers with $q \neq 0$, we have $g(p/q) = g(1)p/q$. Deduce that if, in addition, g is continuous we must have $g(x) = xg(1)$ for all $x \in \mathbb{R}$.

(ii) Suppose $f:\mathbb{R} \in \mathbb{R}$ is a continuous function with $f(x+y) = f(x)f(y)$ for all $x, y \in \mathbb{R}$. Show that either f is identically zero or $f(x) > 0$ for all $x \in \mathbb{R}$. Deduce by using (i), or otherwise, that either $f(x) = 0$ for all $x \in \mathbb{R}$ or we can find an $\alpha \in \mathbb{R}$ such that $f(x) = e^{\alpha x}$ for all $x \in \mathbb{R}$.

(iii) Let X_1, X_2 be independent identically distributed random variables with associated continuous probability densities. Show that, if $X_\theta = X_1 \cos\theta + X_2 \sin\theta$ has the same distribution as X_1 for all θ, then the common distribution must be normal with zero mean (i.e. X_1 has probability density $(2\pi\sigma^2)^{-1/2} \exp(-x^2/(2\sigma^2))$. Can we say any thing further about the variance σ^2?

50.5 (There are so many natural proofs of the result $\int_{-\infty}^{\infty} \exp(-x^2/2)dx = (2\pi)^{1/2}$ that the proof contained in the question may appear superfluous. None the less, in view of the nature of the complex variable proof of Lemma 50.2 (ii), it is reasonable to ask for a contour integration proof of this identity. The following proof was given in 1920 by Mordell.)

By integrating $\exp(\pi i z^2) \operatorname{cosec} \pi z$ round the parallelogram whose four corners are $\pm 1/2 \pm R \exp(i\pi/4)$ with R real and allowing $R \to \infty$, show that

$$\int_{-\infty}^{\infty} \exp(-\pi t^2)dt = 1$$

and so $\int_{-\infty}^{\infty} \exp(-x^2/2)dx = (2\pi)^{1/2}$.

51
CONVOLUTION

51.1 Let $f(x) = e^{-x}$ for $x \geqslant 0$, $f(x) = 0$ for $x < 0$. Write $f_1 = f$ and $f_n = f_{n-1} * f$ $[n \geqslant 2]$.

(i) Find f_n for all $n \geqslant 1$.

(ii) Verify by direct computation that $\hat{f}_n(\lambda) = \hat{f}(\lambda)^n$ for all $\lambda \in \mathbb{R}$, $n \geqslant 1$.

52

CONVOLUTION ON $\mathbb{T}$

52.1 (i) Let $\varepsilon(n) > 0$ and $\varepsilon(n) \to 0$ as $n \to \infty$. Show that we can find $n(j)$ with $1 \leq n(1) < n(2) < \ldots$ and $\varepsilon(n(j)) < 2^{-j}$. Show that

$$f(t) = \sum_{j=1}^{\infty} \varepsilon(n(j)) \exp(in(j))t$$

is a well-defined continuous function with $\varepsilon(n)^{-1}\hat{f}(n) \not\to 0$.

(Thus the Riemann–Lebesgue lemma cannot be improved to give some fixed rate of decrease.)

(ii) Are the following statements true or false? Give a proof or a counter example.

(a) Given $\varepsilon(n) > 0$ with $\varepsilon(n) \to 0$ as $n \to \infty$, we can find a continuous function $f: \mathbb{T} \to \mathbb{R}$ with $\varepsilon(n)^{-1}\hat{f}(n) \to 0$ as $n \to \infty$.
(b) Given $\varepsilon(n) > 0$, we can find a continuous function $f: \mathbb{T} \to \mathbb{C}$ with $\varepsilon(n)^{-1}\hat{f}(n) \not\to 0$ as $n \to \infty$.
(c) Given $\varepsilon(n) > 0$ with $\varepsilon(n) \to 0$ as $n \to \infty$, we can find a continuously differentiable function $f: \mathbb{T} \to \mathbb{C}$ with $\varepsilon(n)^{-1}\hat{f}(n) \not\to 0$ as $n \to \infty$.

52.2$^{(-)}$ Obtain the Riemann–Lebesgue lemma for $\mathbb{T}$ (Theorem 52.4) from Bessel's inequality (Theorem 33.1).

52.3 (Convergence of Fourier sums via the Riemann–Lebesgue lemma) The following pretty proof is given in Helson's *Harmonic Analysis* and is due to Paul Chernoff. Suppose $f \in C(\mathbb{T})$ and f is differentiable at t_0. We shall show that

$$\sum_{r=-m}^{n} \hat{f}(r) \exp(irt_0) \to f(t_0)$$

as $m, n \to \infty$ (the reader may take $m = n$ but nothing is gained thereby).

(i) Observe, without loss of generality, that we may take $t_0 = 0$ and $f(0) = 0$.

(ii) Show that, if we set $g(t) = f(2t)/\sin t$ for $t \neq 0$, $g(0) = 2f'(0)$, then $g: \mathbb{T} \to \mathbb{C}$ is a well-defined continuous function.

(iii) Observe that $2if(2t) = (e^{it} - e^{-it})g(t)$ and deduce that $2i\hat{f}(r) = \hat{g}(2r-1) - \hat{g}(2r+1)$. Hence, using the Riemann–Lebesgue lemma, show that

$$\sum_{r=-m}^{n} \hat{f}(r) = \frac{\hat{g}(-2m-1) - \hat{g}(2n+1)}{2i} \to 0 \quad \text{as } m, n \to \infty,$$

as required.

(Notice that the hypotheses are of a slightly different kind to those for the convergence theorems in the text.)

52.4 (The Riemann–Lebesgue lemma for Riemann integrable functions)

(i) By arguing from the definitions, or otherwise, show that, if $f: \mathbb{T} \to \mathbb{R}$ is Riemann integrable then, given any $\varepsilon > 0$, we can find $I(1), I(2), \ldots, I(k)$ intervals in $\mathbb{T}$ and $\lambda_1, \lambda_2, \ldots, \lambda_k \in \mathbb{R}$ such that

$$\frac{1}{2\pi} \int_{\mathbb{T}} \left| f(t) - \sum_{j=1}^{k} \lambda_j \chi_{I(j)}(t) \right| dt < \varepsilon.$$

(Here $\chi_I(t) = 1$ if $t \in I$, $\chi_I(t) = 0$ otherwise.)

(ii) By modifying the second proof of Theorem 52.4, or otherwise, show that, if $f: \mathbb{T} \to \mathbb{R}$ is Riemann integrable, then $\hat{f}(n) \to 0$ as $|n| \to \infty$. Extend this result to Riemann integrable functions $f: \mathbb{T} \to \mathbb{C}$.

52.5 (A weak form of the Riemann localisation principle)

(i) (Revision of Chapter 18) Show that, if $f: \mathbb{T} \to \mathbb{C}$ is Riemann integrable, then $S_n(f, t) = f * D_n(t)$, where D_n, the Dirichlet kernel, is given by

$$D_n(t) = \sum_{r=-n}^{n} e^{irt} = \sin((n + 1/2)t)/\sin(t/2) \quad [t \neq 0].$$

(ii) Use the Riemann–Lebesgue lemma to show that, if $\varepsilon > 0$,

$$\frac{1}{2\pi} \int_{|t| \geq \varepsilon} f(t) D_n(-t)\, dt \to 0 \quad \text{as } n \to \infty.$$

(iii) Deduce that, if $f, g: \mathbb{T} \to \mathbb{C}$ are Riemann integrable and $f(t) = g(t)$ for $|t| < \varepsilon$, then $|S_n(f, 0) - S_n(g, 0)| \to 0$ as $n \to \infty$. Conclude that, if $f, g: \mathbb{T} \to \mathbb{C}$ are Riemann integrable and $f = g$ in a neighbourhood of x, then $S_n(f, x)$ and $S_n(g, x)$ either both converge or both diverge as $n \to \infty$.

52.6 (Rate of divergence of Fourier sums)

(i) Suppose $f: \mathbb{T} \to \mathbb{C}$ is Riemann integrable and that f is continuous at the origin with $f(0) = 0$. Thus, given any $\eta > 0$, we can find $1 > \varepsilon > 0$ such that $|f(t)| \leqslant \eta$ for $|t| < \varepsilon$. Use the argument of Question 18.3 to show that

$$|S_n(f, 0)| \leqslant A\,\eta \log n + \frac{1}{2\pi}\left|\int_{|t|\geqslant\varepsilon} f(t)\, D_n(-t)dt\right| \quad \text{for } n \geqslant 2$$

where A is some constant independent of η and n.

($A = 10$ will certainly do the trick, but the precise value of A you obtain is irrelevant.)

(ii) Using part (ii) of Question 52.5, deduce that

$$|S_n(f, 0)| \leqslant (A + 1)\eta \log n \quad \text{for } n \text{ sufficiently large.}$$

Deduce that

$$(\log n)^{-1}\, S_n(f, 0) \to 0 \quad \text{as } n \to \infty.$$

(iii) By using the ideas of Chapter 16, deduce that, if $f: \mathbb{T} \to \mathbb{C}$ is Riemann integrable, then, for each $t \in \mathbb{T}$,

$$(\log n)^{-1}\, S_n(f, t) \to 0 \quad \text{as } n \to \infty.$$

(iv) Why is the result of (iii) compatiable with the example given in Question 18.4 of a Riemann integrable function F and a sequence $M(j) \to \infty$ with

$$|S_{M(j)}\,(F, 2^{-j})| \geqslant \pi^2 \log M(j)?$$

(v) By modifying the arguments of parts (i) to (iii) above, show that, if $f: \mathbb{T} \to \mathbb{C}$ is continuous, then

$$(\log n)^{-1}\, S_n(f, t) \to 0 \quad \text{uniformly as } n \to \infty.$$

52.7 (Mean value theorems for integrals)

(i) (First mean value theorem) Let $f,\ g: [a, b] \to \mathbb{R}$ be continuous and suppose $g(x) \geqslant 0$ for $x \in [a, b]$. Show that

$$\inf_{x\in[a,b]} f(x) \int_a^b g(x)dx \leqslant \int_a^b f(x)\, g(x)\, dx \leqslant \sup_{x\in[a,b]} f(x) \int_a^b g(x)\, dx$$

and, by carefully applying the intermediate value theorem, deduce that there exists a $c \in [a, b]$ with

$$\int_a^b f(x)\, g(x)\, dx = f(c)\int_a^b g(x)\, dx.$$

(ii) Suppose $f: [a, b] \to \mathbb{R}$ is continuous and $g: [a, b] \to \mathbb{R}$ is increasing with continuous derivative. If we write $F(x) = \int_a^x f(t)\, dt$, apply (i) to show that there exists a $c \in [a, b]$ with

$$\int_a^b F(x) g'(x)\, dx = F(c)\,[\, g(b) - g(a)].$$

Hence use integration by parts to show that

$$\int_a^b f(x)\, g(x)\, dx = g(a) \int_a^c f(x)\, dx + g(b) \int_c^b f(x)\, dx.$$

(iii) (Second mean value theorem) Suppose now $f: [a, b] \to \mathbb{R}$ is continuous and $g: [a, b] \to \mathbb{R}$ is increasing and continuous. Show that we can find $g_n: [a, b] \to \mathbb{R}$ increasing with continuous derivative such that $g_n \to g$ uniformly as $n \to \infty$. Using the fact that any sequence $c(n)$ in the closed bounded interval $[a, b]$ must have a convergent subsequence $c(n(j)) \to c$, say, show that there exists a $c \in [a, b]$ with

$$\int_a^b f(x)\, g(x)\, dx = g(a) \int_a^c f(x)\, dx + g(b) \int_c^b f(x)\, dx.$$

(See also Question 18.5(vi).)

(iv) Show that in (i) we can replace the condition g continuous by g Riemann integrable. Is the result still true if we have g continuous but f merely Riemann integrable?

(v) Show that in (iii) we can replace the condition f continuous by f Riemann integrable. (The mean value theorems do not lie anywhere near as deep as the mean value theorem of the differential calculus, but, as the next questions show, they are occasionally useful.)

(vi)$^{(+)}$ Show that (iii) can be improved further so that we only demand g increasing and f integrable.

52.8 (An estimate needed for the next question)

(i) Sketch D_n for yourself.

(ii) Let us write $u_r = (1/2\pi) \int_{2r\pi/(2n+1)}^{2(r+1)\pi/2n+1} D_n(x)\, dx$. Show that $u_0 \geqslant -u_1 \geqslant u_2 \geqslant -u_3 \geqslant \ldots \geqslant |u_{n-1}| \geqslant 0$ and hence (recall the alternating series test) show that $|\sum_{r=j}^k u_r| \leqslant u_0$ for all $0 \leqslant j \leqslant k \leqslant n-1$. Deduce that

$$\left| \frac{1}{2\pi} \int_a^b D_n(x)\, dx \right| \leqslant 3u_0 \quad \text{for all } 0 \leqslant a \leqslant b \leqslant \pi$$

and so

$$\left| \frac{1}{2\pi} \int_a^b D_n(x)\, dx \right| \leqslant 6u_0 \quad \text{for all } -\pi \leqslant a < b \leqslant \pi.$$

Conclude that

$$\left|\frac{1}{2\pi}\int_a^b D_n(x)\,dx\right| \leqslant 6 \quad \text{for all } -\pi \leqslant a < b \leqslant \pi.$$

(iii)$^{(+)}$ What is the smallest value of A such that

$$\left|\frac{1}{2\pi}\int_a^b D_n(x)\,dx\right| \leqslant A \quad \text{for all } -\pi \leqslant a < b \leqslant \pi$$

and all $n \geqslant 1$?

52.9 (Pointwise convergence for functions of bounded variation) In this question we give an alternative proof of the main result of Question 16.3.

(i) Let $0 < \eta < \pi$. Consider the following four statements.

(a) If $g: [-\pi, \pi) \to \mathbb{R}$ is increasing and $g(t) = g(\eta)$ for $\eta < t \leqslant \pi$, $g(t) = g(-\eta)$ for $-\pi \leqslant t < -\eta$, then, if g is continuous at 0 with $g(0) = 0$, it follows that $S_n(g, 0) \to 0$ as $n \to \infty$.

(b) If $g: [-\pi, \pi) \to \mathbb{R}$ is increasing and $g(t) = g(\eta)$ for $\eta < t \leqslant \pi$, $g(t) = g(-\eta)$ for $-\pi \leqslant t < -\eta$, then

$$S_n(g, 0) \to (\lim_{t\to 0+} g(t) + \lim_{t\to 0-} g(t))/2.$$

(c) If $g: [-\pi, \pi) \to \mathbb{R}$ is of bounded variation and $g(t) = g(\eta)$ for $\eta < t \leqslant \pi$, $g(t) = g(-\eta)$ for $-\pi \leqslant t < -\eta$, then

$$S_n(g, 0) \to (\lim_{t\to 0+} g(t) + \lim_{t\to 0-} g(t))/2.$$

(d) If $f: \mathbb{T} \to \mathbb{C}$ is such that f is of bounded variation on $[x - \eta, x + \eta]$, then $S_n(f, 0) \to (\lim_{t\to x+} f(t) + \lim_{t\to x-} f(t))/2$.

Explain why (a) implies (b) and (b) implies (c). (If you have done Question 16.3, just sketch this quickly. If you have not done Question 16.3 use parts (i), (ii) and (iv) of that question as a guide. Note that we are not now interested in the behaviour of $\hat{f}(n)$.) Use the Riemann localisation principle (Question 52.5) to show that (c) implies (d).

(ii) Suppose now g satisfies the hypotheses of (a). Then, given $\varepsilon > 0$, we can find a δ with $0 < \delta < \eta$ such that $|g(x)| < \varepsilon$ for $|x| < \delta$. We observe that

$$S_n(g, 0) = \frac{1}{2\pi}\int_{|t|\leqslant\delta} g(t)D_n(-t)\,dt + \frac{1}{2\pi}\int_{|t|>\delta} g(t)D_n(-t)\,dt.$$

Use the second mean value theorem (Question 52.7) to show that

$$\left|\frac{1}{2\pi}\int_{|t|\leqslant\delta} g(t)D_n(-t)\,dt\right| \leqslant \frac{\varepsilon}{\pi}\sup_{-\pi\leqslant a<b\leqslant\pi}\left|\int_a^b D_n(-t)\,dt\right|$$

Hence use question 52.8 and question 52.5 (ii) to show that for n sufficiently large

$$|S_n(g,0)| \leqslant 13\,\varepsilon.$$

Deduce that $S_n(g,0) \to 0$ as $n \to \infty$ and hence that (d) is true.

(This proof is not that far from Dirichlet's original proof. To the reader who has just been chivied through it, the proof may appear tedious and complex, but if she stands back she will see that the approach is a natural one in which obstacles are tackled one at a time as they arise.)

52.10 By imitating the ideas of Chapter 11, construct a continuous function $h: \mathbb{T} \to \mathbb{R}$ such that $S_n(h,t)$ converges uniformly to $h(t)$ as $n \to \infty$, but h is not of bounded variation on any interval (i.e. given $a < b$ and M we can fnd $a = x_0 < x_1 < \ldots < x_m = b$ with $\sum_{j=1}^{m} |h(x_j) - h(x_{j-1})| > M$) and $n|\hat{h}(n)|$ is not bounded as $n \to \infty$.

52.11 (i) Let $f, g: \mathbb{T} \to \mathbb{C}$ be continuous. Let us write $g_m(t) = g(mt)$ for $m \geqslant 1$ an integer. By using the ideas of Question 2.4, or otherwise, show that

$$f * g_m \to \frac{1}{2\pi}\int_{\mathbb{T}} f(t)\,dt\,\frac{1}{2\pi}\int_{\mathbb{T}} g(t)\,dt$$

uniformly as $m \to \infty$.

(ii) Suppose now $f, g: \mathbb{R} \to \mathbb{C}$ are continuous and there exists a T such that $f(t) = g(t) = 0$ for all $|t| > T$. Which, if any, of the following statements are always true?

(a) $f * g_m \to 0$ uniformly as $m \to \infty$.
(b) $f * g_m(t) \to 0$ as $m \to \infty$ for each $t \in \mathbb{R}$.
(c) $f_m * g_m \to \int_{-\infty}^{\infty} f(x)\,dx \int_{-\infty}^{\infty} g(x)\,dx$ uniformly as $m \to \infty$.
(d) $f_m * g_m(t) \to \int_{-\infty}^{\infty} f(x)\,dx \int_{-\infty}^{\infty} g(x)\,dx$ as $m \to \infty$ for each $t \in \mathbb{R}$.

Prove your statements.

52.12 (i) For which real α is it true that

$$\int_{-\infty}^{\infty} f(t) \sin \lambda t^{\alpha}\,dt \to 0 \quad \text{as } \lambda \to \infty$$

for all $f \in L^1 \cap C$? Prove your answer.

(ii) For which continuous functions $g: \mathbb{R} \to \mathbb{R}$ which are periodic with period 2π is it true that

$$\int_{-\infty}^{\infty} f(t) g(\lambda t)\,dt \to 0 \quad \text{as } \lambda \to \infty$$

for all $f \in L^1 \cap C$? Prove your answer.

53
DIFFERENTIATION UNDER THE INTEGRAL

53.1$^{(-)}$ By applying Theorem 53.3 to $g_n(x) = x^{2n} \cos \pi x/2$, or otherwise, show that

$$(2n)!\left(\frac{1}{(2n-2)!}\frac{2^2}{\pi^2} - \frac{1}{(2n-4)!}\frac{2^4}{\pi^4} + \ldots + (-1)^{n-1}\frac{2^{2n}}{\pi^{2n}}\right) \to 1 \quad \text{as } n \to \infty.$$

53.2 In this question and the one following, we give a different approach to Theorem 53.4. This is based on a uniform version of the local Taylor's theorem given in Question 4.1 (iv). Specifically, suppose $f: [a, b] \to \mathbb{R}$ has a continuous derivative. We claim that, given $\varepsilon > 0$, we can find a $\delta > 0$ such that

$$|f(x+h) - f(x) - f'(x)h| < \varepsilon|h| \quad \text{for all } x, x+h \in [a, b], |h| < \delta.$$

(i) Show that, if the result is false, we can find an $\varepsilon > 0$ and $x_n, y_n \in [a, b]$ such that $x_n \neq y_n$, $x_n - y_n \to 0$, but

$$|f(y_n) - f(x_n) - f'(x_n)(y_n - x_n)| \geq \varepsilon\,|y_n - x_n|.$$

(ii) Explain why, by extracting a subsequence, we may suppose $x_n \to z \in [a, b]$.

(iii) Show that, provided n is large enough and $x_n, y_n \neq z$,

$$\max\left(\frac{|f(y_n) - f(z) - f'(z)(y_n - z)|}{|y_n - z|}, \frac{|f(x_n) - f(z) - f'(z)(x_n - z)|}{|x_n - z|}\right) \geq \frac{\varepsilon}{2}.$$

(iv) Conclude that

$$|h|^{-1}|f(z+h) - f(z) - f'(z)h| \not\to 0 \quad \text{as } h \to 0$$

and hence obtain a contradiction.

(v) Suppose $f: [a, b] \to \mathbb{R}$ has an nth continuous derivative. State and prove the appropriate generalisation of the result of this question.

(vi) What standard result of elementary analysis do we get if $n = 0$ in (v)?

53.3 (i) Suppose $g: \mathbb{R}^2 \to \mathbb{R}$ is a continuous function with continuous partial derivative g_2. Show, by modifying the ideas of the previous question, that, if $[a, b]$, $[c, d]$ are fixed intervals, then, given any $\varepsilon > 0$, we can find a $\delta > 0$ such that

$$|g(x, t + h) - g(x, t) - g_2(x, t)h| < \varepsilon\, |h|$$

whenever $x \in [a, b]$, $t \in [c, d]$, $t + h \in [c, d]$ and $|h| < \delta$.

(ii) Hence show that, if $[a, b]$ is a fixed interval, $\int_a^b g(x, t)dx$ is differentiable and

$$\frac{d}{dt}\int_a^b g(x, t)dx = \int_a^b g_2(x, t)dx.$$

(iii) Suppose $f: \mathbb{R}^2 \to \mathbb{R}$ is continuous. By differentiating

$$\int_a^b \left(\int_c^t f(x, s)ds\right) dx - \int_c^t \left(\int_a^b f(x, s)dx\right) ds,$$

show that

$$\int_a^b \left(\int_c^d f(x, s)ds\right) dx = \int_c^d \left(\int_a^b f(x, s)dx\right) ds$$

for all $a \leqslant b$, $c \leqslant d$. (This is a way of obtaining Lemma 47.2 but not a recommended one.)

53.4 In his book *Surely You're Joking, Mr Feynman!* the Nobel physicist Feynman tells the following anecdote:

One day [my high school teacher Mr Bader] told me to stay after class. 'Feynman,' he said, 'you talk too much and you make too much noise. I know why. You're bored. So I'm going to give you a book. You go up there in the back, in the corner, and study this book, and when you know everything that's in this book, you can talk again.'

So every physics class, I paid no attention to what was going on with Pascal's Law, or whatever they were doing. I was up in the back with this book: *Advanced Calculus*, by Woods. Bader knew I had studied *Calculus for the Practical Man* a little bit, so he gave me the real works – it was for a junior or senior course in college. It had Fourier series, Bessel functions, determinants, elliptic functions – all kinds of wonderful stuff that I didn't know anything about.

That book also showed how to differentiate parameters under the integral sign – it's a certain operation. It turns out that's not taught very much in the universities; they don't emphasize it. But I caught on how to use that method, and I used that one damn tool again and again. So because I was self-taught using that book, I had this peculiar method of doing integrals.

The result was, when guys at MIT or Princeton had trouble doing a certain integral, it was because they couldn't do it with the standard methods they had learned in school. If it was contour integration, they would have found it; if it was a simple series expansion, they would have found it. Then I come along and try differentiating under the integral sign, and often it worked. So I got a great reputation for doing integrals, only because my box of tools was different from everybody else's, and they had tried all their tools on it before giving the problem to me.

(i) Show that, if $a > 0$, then $\int_0^\infty dx/(x^2 + a^2) = \pi/2a$. By repeated differentiation, show that

$$\int_0^\infty \frac{dx}{(x^2 + a^2)^{n+1}} = \frac{\pi}{2} \frac{1.3 \ldots (2n-1)}{2.4 \ldots 2n} \frac{1}{a^{2n+1}},$$

whence $\int_0^\infty (1 + x^2)^{n+1}\, dx = (\pi/2)(2n)!/(2^n n!)^2$.

What can you say about $\int_0^\infty dx/(x^2 + a^2)^{n+1}$ if $a < 0$ or if $a = 0$?

(ii) A slightly more sophisticated use of this technique is illustrated in this alternative derivation of the Fourier transform of E where $E(x) = \exp(-x^2/2)$. In order to provide a fair comparison with the complex variable methods of Lemma 50.2, it is important that the reader sets out all the steps in full. Show that $\hat{E}: \mathbb{R} \to \mathbb{R}$ is differentiable, find its derivative and by integrating by parts, or otherwise, obtain the equation $\hat{E}'(\lambda) = -\lambda \hat{E}(\lambda)$. Hence show that

$$\frac{d}{d\lambda}(\exp(\lambda^2/2)\hat{E}(\lambda)) = 0$$

and (quoting the value of $\hat{E}(0)$ found, for example, in Lemma 48.10) deduce that $\hat{E}(\lambda) = (2\pi)^{1/2} E(\lambda)$.

(iii) If $\psi(a) = \int_{-\infty}^{\infty} \exp(-x^2 - a^2/x^2)dx$, show that ψ is differentiable on $(0, \infty)$ and $\psi'(a) = -2\psi(a)$ for $a > 0$. Show also that $\psi(a) \to \psi(0)$ as $a \to 0+$. Hence find $\psi(a)$ for $a \geqslant 0$ and thus $\psi(a)$ for all a. Is ψ differentiable at 0? What is the value of $\int_{-\infty}^{\infty} \exp(-x^2 - x^{-2})dx$?

(iv) Show that

$$\int_0^\infty e^{-x^2/2} \sin xy\, dx = e^{-y^2/2} \int_0^y e^{-t^2/2}\, dt$$

for $y \neq 0$.

53.5 (i) Show by comparison, or otherwise, that

$$\int_0^\infty \left| e^{-ax} \frac{\sin x}{x} \right| dx \quad \text{and} \quad \int_0^\infty |e^{-ax} \sin x|\, dx$$

converge for all $a > 0$. Show further by using the alternating nature of the function that, if $x_2 > x_1 > (N+1)\pi$,

$$\left| \int_{x_1}^{x_2} e^{-ax} \frac{\sin x}{x}\, dx \right| \leqslant \int_{N\pi}^{(N+1)\pi} e^{-ax} \left| \frac{\sin x}{x} \right| dx \leqslant \frac{1}{N}$$

for all integers $N \geqslant 1$ and all $a \geqslant 0$.

(ii) Deduce, or prove otherwise, that

(a) $\int_0^\infty (\sin x)/x \, dx$ converges,
(b) $\int_0^\infty (e^{-ax} \sin x)/x \, dx \to \int_0^\infty \sin x/x \, dx$ as $a \to 0+$,
(c) $\int_0^\infty (e^{-ax} \sin x)/x \, dx \to 0$ as $a \to \infty$.

(iii) Writing $F(a) = \int_0^\infty (e^{-ax} \sin x)/x \, dx$ $[a \geqslant 0]$, show that $F'(a)$ exists for $a > 0$ and that

$$F'(a) = \frac{-1}{1 + a^2}.$$

By solving this equation and using parts (c) and (b) of (ii), show that

$$\int_0^\infty \frac{\sin x}{x} \, dx = \frac{\pi}{2}.$$

53.6 (Another infinitely differentiable function without a Taylor expansion valid in a neighbourhood. One such example was given in Example 4.2.)

(i) Suppose $f: \mathbb{R} \to \mathbb{R}$ is continuous with $f(x) = 0$ for all $x \notin [0, 1]$. By using uniform continuity, or otherwise, show that, given $\varepsilon > 0$, we can find a piecewise linear function with $g(x) = 0$ for all $x \notin [0, 1]$ and $|g(x) - f(x)| < \varepsilon$ for all $x \in \mathbb{R}$. Hence, or otherwise, find a once continuously differentiable function h with $h(x) = 0$ for $x \notin [0, 1]$ and $|h(x) - f(x)| < 2\varepsilon$ for all $x \in \mathbb{R}$.

(ii) Now choose $f_0: \mathbb{R} \to \mathbb{R}$ a once continuously differentiable function with $f_0(1/2) = 2$ and $f_0(x) = 0$ for all $x \notin [0, 1]$. Show inductively that we can find once continuously differentiable functions $f_n: \mathbb{R} \to \mathbb{R}$ such that $f_n(x) = 0$ for all $x \notin [0, 1]$ and

$$|f_n(x) - f'_{n-1}(x)| < 2^{-n} \quad \text{for all } x \in \mathbb{R}.$$

(iii) Let F_n be the result of integrating f_n n times (more specifically let F_n be the $(n + 1)$ times continuously differentiable function defined on $[-1, 1]$ with $F_n(0) = F'_n(0) = F''_n(0) = \ldots = F_n^{(n-1)}(0) = 0$, $F_n^{(n)}(x) = f_n(x)$ for all x). Show that $F_n^{(n)}(x) = 0$ for all $x \leqslant 0$ and

$$|F_{n+1}^{(r)}(x) - F_n^{(r)}(x)| < 2^{-n} \quad \text{for all } 0 \leqslant r \leqslant n + 1 \text{ and all } x.$$

(You may wish to prove this first for $r = n + 1$, then $r = n$, and so on.)

(iv) Use the inequalities just obtained to show that $F_n^{(r)}$ converges uniformly as $n \to \infty$ to some continuous function G_r. Show that $G = G_0$ is infinitely differentiable with rth derivative G_r and that $G(x) = 0$ for $x \leqslant 0$. Show also that $|G(x) - f_0(x)| \leqslant 1$ and so $G(1/2) \geqslant 1$.

(v) Conclude that there exists a real number a with $0 \leqslant a < 1/2$ such that $G(x) = 0$ for $x \leqslant a$, but there exist $x_n \to a+$ such that $G(x_n) \neq 0$. Show that $G^{(n)}(a) = 0$ for all n and deduce that G has no Taylor expansion valid near a.

(vi) Set $H(x) = \int_0^x (G(t))^2\,dt$. Show that $H: [-1, 1] \to \mathbb{R}$ is infinitely differentiable, $H(x) = 0$ for $x \leqslant a$, $H(x) > 0$ for $x > a$. Finally, define $K: \mathbb{R} \to \mathbb{R}$ by $K(x) = H(x + a)H(4^{-1} + a - x)$ for $|x| \leqslant 1/2$, $K(x) = 0$ otherwise. Show that K is infinitely differentiable, $K(x) > 0$ for $0 < x < 1/4$ and $K(x) = 0$ otherwise. Conclude that K has no Taylor expansion valid near 0.

54

LORD KELVIN

54.1 (Euclid's proof that there exist an infinity of primes)

(i) Suppose $p_1, p_2, \ldots, p_n$ are primes. Show that, if $N = p_1p_2 \ldots p_n + 1$, then N is not divisible by any of $p_1, p_2, \ldots, p_n$. Since N is either a prime or divisible by a prime smaller than itself, conclude that there must be a prime which is none of $p_1, p_2, \ldots, p_n$. Thus there must be an infinity of primes.

(ii) Is it true that, if $p_1, p_2, \ldots, p_n$ are distinct primes, then $N = p_1p_2 \ldots p_n + 1$ is necessarily a prime?

(iii) Is it true that $n! + 1$ is necessarily a prime?

(iv) Is it true that, if $p_1, p_2, \ldots, p_n$ are the first n primes, then $N = p_1p_2 \ldots p_n + 1$ is necessarily a prime? (If you use a hand calculator and a table of primes up to 100 a counter example will appear just as your finger begins to get sore.)

(v) By considering expressions of the form $4p_1p_2 \ldots p_n + 3$ show that there are an infinity of primes of the form $4k - 1$. Can you find other a and d such that there are an infinity of primes of the form $a + kd$? (You should be able to find at least two more essentially different sequences. Do not give up the search until you are convinced that it becomes rather hard rather quickly. You will then appreciate chapters 106 to 109 even more.)

54.2 (Pólya's proof that there exist an infinity of primes) New proofs of the existence of an infinity of primes are always welcome. Here is one due to Pólya. Let

$$F_n = 2^{2^n} + 1.$$

The F_n are called Fermat numbers.

(i) If $k \geqslant 1$ show that F_n divides $F_{n+k} - 2$.

(ii) Deduce that if $m \neq n$ then F_n and F_m can have no prime factor in common.

(iii) Conclude that there are an infinity of primes. (Explain your reasoning with care.)

(iv) If p_n is the nth prime, show that $p_n \leqslant 2^{2^n}$.

55

THE HEAT EQUATION

55.1 (Dirichlet's problem for the half plane) In this question we attempt to find a solution for Laplace's equation in the half plane. Specifically, we seek $\phi: \mathbb{R} \times [0, \infty) \to \mathbb{R}$ such that ϕ is continuous on $\mathbb{R} \times [0, \infty)$ and has continuous second derivatives on $\mathbb{R} \times (0, \infty)$ with

$$\text{(I)} \quad \frac{\partial^2 \phi}{\partial x^2} + \frac{\partial^2 \phi}{\partial y^2} = 0 \quad \text{on } \mathbb{R} \times (0, \infty),$$

$$\text{(II)} \quad \phi(x, 0) = u(x),$$

for some well-behaved function $u: \mathbb{R} \to \mathbb{R}$. As in the first part of Chapter 55, our aim is to guess a solution, so rigour is to be ignored.

(i) By imitating the argument following Definition 55.2, show that

$$\mathcal{F}_1\phi(\zeta, y) = A(\zeta)e^{+\zeta y} + B(\zeta)e^{-\zeta y},$$

where $A(\zeta)$, $B(\zeta)$ are functions of ζ to be determined.

(ii) Making the reasonable assumption that $\mathcal{F}_1\phi(\zeta, y)$ remains bounded as $y \to \infty$, show that, in fact,

$$\mathcal{F}_1\phi(\zeta, y) = C(\zeta)e^{-|\zeta|y},$$

where $C(\zeta)$ is a function of ζ to be determined.

(iii) Now continue the argument, in the style proposed, to obtain

$$\phi(x, y) = \frac{y}{\pi} \int_{-\infty}^{\infty} \frac{u(s)}{(x - s)^2 + y^2} \, ds$$

(In the next question we verify our tentative solution for a wide class of u.)

55.2 (For this question it would be helpful to have glanced at Chapters 27 to 31.)

(i) Let $u: \mathbb{R} \to \mathbb{R}$ be a continuous function such that $u(x) \to u_0$, say, as $|x| \to \infty$. Write

$$\phi(x, y) = \frac{y}{\pi}\int_{-\infty}^{\infty} \frac{u(s)}{(x-s)^2 + y^2}\,ds \quad \text{for } x \in \mathbb{R},\ y > 0.$$

By differentiating under the integral sign, show that

$$\text{(I)}\quad \frac{\partial^2 \phi}{\partial x^2} + \frac{\partial^2 \phi}{\partial y^2} = 0 \quad \text{on } \mathbb{R} \times (0, \infty).$$

(ii) By imitating proofs like that of Lemma 27.3 and Lemma 27.4 (ii), show that, if we set

$$\text{(II)}\quad \phi(x, 0) = u(x) \quad \text{for } x \in \mathbb{R},$$

then $\phi\colon \mathbb{R} \times [0, \infty) \to \mathbb{R}$ is continuous.

(iii) Show also that

$$\text{(III)}\quad \phi(x, y) \to u_0 \text{ uniformly as } x^2 + y^2 \to \infty \text{ with } y \geqslant 0.$$

(iv) Suppose $\psi\colon \mathbb{R} \times [0, \infty) \to \mathbb{R}$ also satisfies condition (I), (II), (III) (i.e. $\nabla^2\psi = 0$ on $\mathbb{R} \times [0, \infty)$, ψ continuous on $\mathbb{R} \times [0, \infty)$ with $\psi(x, 0) = u(x)$, $\psi(x, y) \to u_0$ uniformly as $x^2 + y^2 \to \infty$). By applying the maximum principle (Theorem 31.2) to $\phi - \psi$ on large semicircles, centre the origin, show that $\phi = \psi$.

(v) Show that, if we drop the condition $u(x) \to u_0$ as $|x| \to \infty$ in (i) and replace it by the condition $|u(x)| \leqslant M$ for all x, the conclusions of (i) and (ii) still hold.

(vi) Show that, even if $u(x) \to 0$ as $|x| \to \infty$, part (iv) is false if we drop the hypothesis (III) on ψ. (Look at Example 31.4 if you cannot see how to do this.)

(vii) Show that parts (i), (ii), (iii), (iv) can be obtained from the results of Chapter 27 and 28 by considering a conformal map of the disc into the half plane (cf. Theorem 31.33). You should make explicit the connection between the function $Q(x,y) = y/\pi(x^2 + y^2)$ and the Poisson kernel $P_r(\theta) = (1 - r^2)/(1 - 2r\cos\theta + r^2)$.

55.3 Consider the problem of finding a continuous function $\phi : \mathbb{R} \times [0,\infty)$ such that ϕ is continuously four times differentiable on $\mathbb{R} \times (0,\infty)$ and

$$\text{(I)}\quad \frac{\partial^4 \phi}{\partial x^4} - K^2\frac{\partial^2 \phi}{\partial y^2} = 0 \text{ on } \mathbb{R} \times (0,\infty),$$

$$\text{(II)}\quad \phi(x,0) = u(x) \text{ for } x \in \mathbb{R},$$

$$\text{(III)}\quad \phi(x,y) \text{ is bounded on } \mathbb{R} \times [0,\infty),$$

where $u : \mathbb{R} \to \mathbb{R}$ is continuous and bounded.

(i) Use the methods of the first part of Chapter 55 and Question 55.1 to guess a solution.

(ii) Verify that you have a solution. (You should point out an immediate connection with Theorem 55.4. You are not asked to discuss uniqueness.)

55.4 Suppose $K>0$. Let us write $E_K(x,t)=2^{-1}(\pi Kt)^{-1/2}\exp(-x^2/4Kt)$ for $x\in\mathbb{R}$, $t>0$ and $E_K(x,t)=0$ for $x\in\mathbb{R}$, $t\leqslant 0$. We have seen in Theorem 55.4 that it is useful to look at

$$\phi(x,t)=(G*E_K(\ ,t))(x)\quad [x\in\mathbb{R},\ t>0].$$

Symmetry suggests that we look at

$$\psi(x,t)=(g*E_K(x,\ ))(t)\quad [x\in\mathbb{R}, t>0].$$

As it stands, the expression may not be defined, so we restrict g so that $g(t)=0$ for $t<0$. (This procedure is standard; see e.g. the discussion of the Laplace transform in Chapters 72 to 74.) We shall further assume that g is continuous and bounded on $[0,\infty)$ and define ψ by

$$\psi(x,t)=\int_{-\infty}^{\infty} g(s)E_K(x,t-s)\,dx$$

$$=\frac{1}{\sqrt{(2\pi)}}\int_0^t g(s)\frac{\exp(-x^2/4K(t-s))}{(2K(t-s))^{1/2}}ds\quad [x\in\mathbb{R},\ t>0].$$

Establish results (i)–(v). (It is easy, but not important, to supply rigour.)

(i) $\dfrac{\partial\psi}{\partial t}(x,t)=K\dfrac{\partial^2\psi}{\partial x^2}(x,t)$ for $t>0$, $x\neq 0$.

(ii) $\displaystyle\int_{-\infty}^{\infty}\psi(x,t)\,dx=\int_0^t g(s)\,ds$ for $t>0$.

(iii) If $\varepsilon>0$, $t>0$, then

$$\frac{1}{\delta t}\int_{|x|>\varepsilon}|\psi(x,t+\delta t)-\psi(x,t)|\,dx\to 0 \text{ as } \delta t\to 0.$$

(iv) If $\varepsilon>0$, $t>0$ then

$$\frac{1}{\delta t}\int_{|x|>\varepsilon}\psi(x,t+\delta t)-\psi(x,t)\,dx\to g(t)\quad\text{as } \delta t\to 0.$$

(v) $\psi(x,t)\to 0$ uniformly in x as $t\to 0+$.

(vi) This part of the question is adapted from an exam in biological mathematics and I cannot vouch for the accuracy of the biological information. A male insect sits in a very long uniform tube. At time $t=0$ a female is placed in the tube at a distance l from the male and continuously emits small quantities of a pheromone at concentration $g(t)$ at time t. The male responds (vigorously) to the pheromone when its concentration reaches C_1. Assuming that the diffusion of this substance is governed by the diffusion equation

$$\frac{\partial\psi}{\partial t}=K\frac{\partial^2\psi}{\partial x^2}$$

(where x is the distance from the female), explain why

$$\psi(x,t) = \frac{1}{\sqrt{(2\pi)}} \int_0^t g(s) \frac{\exp(-x/^2(4K(t-s)))}{(2K(t-s))^{1/2}}\, ds$$

is the required solution. You should explain the relevance of the results (i) to (v) above.

Suppose that $g(s) = 0$ for $s < 0, g(s) = C$ for $s \geqslant 0$. Show that

$$\psi(x,t) = \frac{ACx}{K} \int_\eta^\infty \frac{e^{-s^2/2}}{s^2}\, ds,$$

where $\eta = x/(2Kt)^{1/2}$ and A is a constant to be determined. Show that the male is galvanised into activity after a time $t_1 = \alpha K^{-1} l^2$, where $\alpha = \alpha(C/KC_1)$ depends only on the value of C/KC_1. (We shall return to these insects in Question 79.9 (ii).)

55.5 Consider the following model for a domestic hot water tank proposed in the 1985 Tripos exam. The tank is a vertical cylinder of length l with perfectly insulating boundaries and convection is ignored. The temperature θ is then a function of time t and the vertical coordinate z.

The water in the tank is initially at temperature θ_H. The bottom half tankful is drawn off and replaced by cold water at temperature θ_C. Explain why the study of $\theta(t, z)$ now reduced to solving

$$K\frac{\partial^2\theta}{\partial z^2} = \frac{\partial\theta}{\partial t} \text{ subject to the conditions}$$

(a) $\theta(t,z) \to \theta_C$ for $0 < z < l/2$, $\theta(t,z) \to \theta_H$ for $l/2 < z < l$ as $t \to 0+$,

(b) $\dfrac{\partial\theta}{\partial z}(t,0) = \dfrac{\partial\theta}{\partial z}(t,l) = 0.$

By seeking trial solutions $\theta(t,z) = h(z)f(t)$ to the equation $K\partial^2\theta/\partial z^2 = \partial\theta/\partial t$ subject to condition (b) only, and combining them to produce a solution satisfying (a), obtain the plausible solution

$$\theta(t,z) = \frac{\theta_C + \theta_H}{2} + \frac{2}{\pi}(\theta_C - \theta_H) \sum_{n=1}^\infty \frac{1}{n} \sin\left(\frac{n\pi}{2}\right) \cos\left(\frac{n\pi z}{l}\right) \exp\left(-K\left(\frac{n\pi}{l}\right)^2 t\right).$$

Verify directly that it is a solution.

What happens as $t \to \infty$? If the boundary conditions (b) are replaced by

(b)′ $\theta(t,0) = \theta_1, \theta(t,l) = \theta_2,$

state what the behaviour of $\theta(t,z)$ will be as $t \to \infty$ and give brief reasons for your answers.

57

THE AGE OF THE EARTH II

57.1 Consider what happens when a drop of dyed water falls into a glass of clear water. As it enters the water the sides are retarded and move slower than its centre. A vortex ring forms. Which way does the fluid rotate in the ring?

Smoke rings form in a similar manner. Suppose two equal, slowly moving, but rapidly rotating smoke rings blown in opposite directions collide head on. By considering which velocities cancel and which add, predict what will happen. When a smoke ring is blown towards a wall it expands as it approaches the wall. Explain this.

(This example is taken from Walker's *The Flying Circus of Physics*. I did get a friend to try the experiment and it worked.)

57.2 In Chapter 57, I asserted that the heat equation for the cooling sphere is not hard to solve. In this question and the next we solve them.

(i) Suppose $\Psi(\mathbf{x},t) = \phi(r, t)$ with $r = |\mathbf{x}|$. Verify that the heat equation $\frac{\partial \Psi}{\partial t}(\mathbf{x}, t) = K\nabla^2\Psi(\mathbf{x}, t)$ becomes $\frac{K}{r^2}\frac{\partial}{\partial r}\left(r^2\frac{\partial \phi}{\partial r}\right) = \frac{\partial \phi}{\partial t}$ and that, if we set $\eta(r, t) = r\phi(r, t)$, this reduces in turn to

$$K\frac{\partial^2 \eta}{\partial r^2} = \frac{\partial \eta}{\partial t}.$$

(ii) Explain why, if we wish to solve the problem with $\Psi(\mathbf{x}, 0) = \theta_0$ for $|\mathbf{x}| < R$, $\Psi(\mathbf{x}, t) = 0$ for all $|\mathbf{x}| = R$, $t > 0$, then the appropriate conditions are

(a) $\eta(r, 0) = \theta_0$ for $0 \leqslant r < R$,
(b) $\eta(0, t) = \eta(R, t) = 0$ for $t > 0$.

(iii) It turns out that there is a clever way of finding η in a form which is more relevant to the considerations of Chapter 57. We defer this to the next question and instead attack the problem in the obvious way. Use separation of variables to find solutions of

$$K\frac{\partial^2\eta}{\partial r^2} = \frac{\partial\eta}{\partial t},\quad \eta(0, t) = \eta(R, t) = 0$$

of the form $\eta_n(r, t) = \sin(n\pi r/R)\exp(-Kn^2\pi^2 t/R^2)$ with $n \in \mathbb{Z}$, $n \geqslant 1$. (If you do not know about separation of variables just verify the result.) Next find $a_n \in \mathbb{R}$ such that, if $\eta(r, t) = \sum_{n=1}^{\infty} a_n\eta_n(r, t)$, then η satisfies condition (b) of (ii). (Do not bother with rigour at this stage. If you are stuck, consider the Fourier series of g, where $g(x) = x$ for $-\pi < x < \pi$.) Conclude that

$$\phi(r, t) = \frac{2R\theta_0}{\pi r}\sum_{n=1}^{\infty}\frac{(-1)^{n+1}}{n}\sin\frac{n\pi r}{R}\exp\left(\frac{-Kn^2\pi^2 t}{R^2}\right).$$

(iv) If you so wish, verify rigorously that we have indeed found a solution. However, this is not our central concern.

(v) Show that the average temperature of the sphere at time t is

$$\frac{6\theta_0}{\pi^2}\sum_{n=1}^{\infty}\frac{1}{n^2}\exp\left(\frac{-Kn^2\pi^2 t}{R^2}\right).$$

(This has some relevance to the behaviour of thermometers.)

57.3 Here is another way of finding the Ψ of Question 57.2.

(i) Let us set $\Psi_0 = \Psi - \theta_0$ and, correspondingly, $\eta_0 = \eta - r\theta_0$. Check that we now have to solve

$$K\frac{\partial^2\eta_0}{\partial r^2} = \frac{\partial\eta_0}{\partial t} \text{ for } 0 \leqslant r \leqslant R,\ t > 0 \text{ subject to the conditions}$$

(a) $\eta_0(r, 0) = 0$ for $0 \leqslant r < R$.

(b) $\eta_0(0, t) = 0, \eta_0(R, t) = -\theta_0 R$ for $t > 0$.

(ii) Sketch the function

$$k(x) = \sum_{n=-\infty}^{\infty}(H(x - R - 4nR) - H(x - 3R - 4nR)),$$

where $H(x) = 1$ for $x > 0$, $H(x) = 0$ otherwise. Give reasons for supposing that, if we solve

$$K\frac{\partial^2\theta}{\partial x^2} = \frac{\partial\theta}{\partial t} \text{ for } x \in \mathbb{R},\ t \geqslant 0 \text{ subject to the condition}$$

(α) $\theta(x, 0) = -R\theta_0 k(x)$ for all $x \in \mathbb{R}$,

then we should expect $\theta(r, t) = \eta(r, t)$ for $0 \leqslant r \leqslant R$, $t > 0$. (Think of Figures 57.1 and 57.2.)

(iii) By using Lemma 55.6 (ii), or otherwise, obtain the plausible solution for ψ

$$\psi(\mathbf{r}, t) = \theta_0 - \frac{R\theta_0}{r} \sum_{n=-\infty}^{\infty} \left(E\left(\frac{r - R - 4nR}{(2Kt)^{1/2}}\right) - E\left(\frac{r - 3R - 4nR}{(2Kt)^{1/2}}\right)\right),$$

where $E(s) = (1/(2\pi)^{1/2})\int_s^\infty \exp(-u^2/2)\,du$. By using symmetry, obtain

$$\psi(\mathbf{r}, t) = \theta_0 - \frac{2R\theta_0}{r} \sum_{n=0}^{\infty} \left(E\left(\frac{r - R - 4nR}{(2Kt)^{1/2}}\right) - E\left(\frac{r - 3R - 4nR}{(2Kt)^{1/2}}\right)\right)$$

(iv) Question 57.2 (iii) and 57.3 (iii) give two expressions for ψ. Which is most suitable when t is large and which when t is small? Show that, when t is small and r is close to R, we recover the solution of Chapter 57.

(v) If you are interested (but this is not the point of the question) you may wish to

(a) verify rigorously that the expression for ψ given in (iii) is indeed a solution to our initial question,

(b) verify directly (i.e without using theorems on the uniqueness of solutions) by, for example, using power series expansions, that the solutions ψ of Question 57.2 (iii) and 57.3 (iii) are in fact identical.

57.4 (i) (Dido's problem) Dido, the legendary founder of Carthage, arriving as a fugitive on the coast of Africa purchased a piece of land on the seashore 'not larger than an oxhide can surround'. She cut the oxhide into fine narrow strips and joined them to form a very long string. 'Dido's problem' asks for the largest area bounded by a straight line (the seashore) and a curve of given length.

(ii) (The mirror) It is a good slogan (though the truth is more subtle) that 'Light always takes the shortest path'. Suppose we work in the plane $\mathbb{R}^2$. What is the shortest path between two points A and B which includes some point on a given line?

Suppose we work in $\mathbb{R}^3$. What is the shortest path between two points A and B which includes some point on a given plane?

(If you want an elementary yet beautiful and profound treatment of minimising problems (and much else beside) consult Pólya's *Mathematics and Plausible Reasoning* from which the above problems are taken.)

59

WEIERSTRASS'S PROOF OF WEIERSTRASS'S THEOREM

59.1 (Alternative proof of Weierstrass's theorem) Once we have the idea of Chapter 59, we can modify it to make the proof easier. Let us write

$$P_n(x) = c_n(L^2 - x^2)^n,$$

where c_n is chosen so that $\int_{-4}^{4} P_n(x)dx = 1$.

(i) Suppose $f: \mathbb{R} \to \mathbb{C}$ is continuous and $f(x) = 0$ for $|x| \geqslant 2$. Show that

$$f * P_n(x) = \int_{-\infty}^{\infty} f(x-t)P_n(t)dt$$

defines a polynomial and that $f * P_n \to f$ uniformly on $[-1,1]$.

(ii) Show further that, if f is m times continuously differentiable, then $(f * P_n)^{(m)}$ exists and tends to $f^{(m)}$ uniformly on $[-1,1]$.

(iii) Deduce carefully (note that the function F in Step 1 of the proof of Theorem 59.2 is continuous but not necessarily differentiable) that, if f: $[a,b] \to \mathbb{C}$ is m times continuously differentiable (with the appropriate conventions about left and right derivatives at end points), then there exist a sequence Q_j of polynomials with $Q_j^{(r)} \to f^{(r)}$ uniformly on $[a,b]$ for each r.

(iv) Deduce that, if f: $[a,b] \to \mathbb{C}$ is infinitely differentiable, we can find a sequence Q_j of polynomials with $Q_j^{(r)} \to f^{(r)}$ uniformly on $[a,b]$ for each r.

(v) It is true in general that, under the hypotheses of (iv), we can find a sequence of polynomials Q_j with

$$\sup_{r \geqslant 0} \sup_{x \in [a,b]} |Q_j^{(r)}(x) - f^{(r)}(x)| \to 0 \text{ as } j \to \infty?$$

60

THE INVERSION FORMULA

60.1 (Alternative proof of the Inversion theorem)

(i) If $f,\ g,\ \hat{f},\ \hat{g} \in L^1 \cap C$, show, using Theorem 47.5 applied to $F(x, y) = f(x)g(y)\exp(-ixy)$, that

$$\int_{-\infty}^{\infty} f(t)\hat{g}(t)\,dt = \int_{-\infty}^{\infty} \hat{f}(t)g(t)\,dt.$$

(ii) Write $E(t) = \exp(-t^2/2)$ for $t \in \mathbb{R}$ and $E_R(t) = RE(Rt)$ for $R > 0$, $t \in \mathbb{R}$. Using part (i) and Lemma 50.2, show that

$$(2\pi)^{1/2}\int_{-\infty}^{\infty} f(t)E_R(t)\,dt = \int_{-\infty}^{\infty} \hat{f}(t)E(R^{-1}t)\,dt.$$

(iii) Observe that $|E(R^{-1}t)| \leqslant 1$ and that $E(R^{-1}t) \to 1$ uniformly on closed bounded intervals as $R \to \infty$. Hence show that

$$\int_{-\infty}^{\infty} \hat{f}(t)E(R^{-1}t)\,dt \to \int_{-\infty}^{\infty} \hat{f}(t)\,dt$$

as $R \to \infty$. Conclude, using Theorem 5.1 and part (ii), that

$$f(0) = \frac{1}{2\pi}\int_{-\infty}^{\infty} \hat{f}(t)\,dt.$$

(iv) By applying (iii) to the function f defined by $f_x(t) = f(t + x)$, conclude that

$$f(t) = \frac{1}{2\pi}\int_{-\infty}^{\infty} \hat{f}(t)\exp(ixt)\,dt$$

whenever $f,\ \hat{f} \in L^1 \cap C$.

60.2 (The Fourier inversion theorem via the Riemann–Lebesgue lemma)

(i) Suppose $f \in C \cap L^1$, f is differentiable at 0 and $f(0) = 0$. By using an appropriate modification of Theorem 48.8, show that

$$\int_{-R}^{S} \hat{f}(\lambda)\, d\lambda = \int_{-\infty}^{\infty} \left(\int_{-R}^{S} f(x) \exp(-i\lambda x)\, d\lambda \right) dx \quad [R, S > 0].$$

Observe that, if $x \neq 0$,

$$\int_{-R}^{S} f(x) \exp(-i\lambda x)\, d\lambda = ig(x)(\exp(-ixS) - \exp(ixR))$$

where $g(x) = f(x)/x$ for $x \neq 0$. By imitating the arguments of Question 52.3, conclude that

$$\int_{-R}^{S} \hat{f}(\lambda)\, d\lambda \to 0 \text{ as } R, S \to \infty.$$

(ii) Suppose now that we only know that $f \in C \cap L^1$ and f is differentiable at 0. At the corresponding point in Question 52.3 we assumed, without loss of generality, that $f(0) = 0$. Why can we not do the same now?

(iii) Supopose f satisfies the hypotheses of (ii). Let $E(x) = \exp(-x^2/2)$. Recall that we know, by direct calculation (Lemma 50.2), and without using any version of the Fourier inversion theorem, that $\hat{E}(\lambda) = (2\pi)^{1/2} E(\lambda)$ and so $\int_{-\infty}^{\infty} \hat{E}(\lambda)\, d\lambda = \hat{\hat{E}}(0) = 2\pi E(0) = 2\pi$. By considering $f - f(0)E$ and applying (i), show that $\int_{-R}^{S} \hat{f}(\lambda)\, d(\lambda) \to 2\pi f(0)$.

(iv) Suppose, finally, that $f \in L^1 \cap C$ and f is differentiable at t. By considering f_1 defined by $f_1(x) = f(x - t)$, show that

$$\int_{-R}^{S} \hat{f}(\lambda) \exp(i\lambda t) dt \to 2\pi f(t) \quad \text{as } R, S \to \infty.$$

60.3 We continue with some of the themes and notations of the previous question.

(i) If $f : \mathbb{R} \to \mathbb{C}$ is periodic of period 2π, show, using Theorem 2.5, that we can find $N \geq 1$ and $a_n \in \mathbb{C}$ such that $|f(t) - \sum_{n=-N}^{N} a_r \exp(int)| < \varepsilon$.

(ii) Using the fact that $E(t) = E(-t)$, show that

$$\int_{-R}^{S} E_R(t) \exp(int)\, dt = \sqrt{2\pi} \exp(-n^2/(2R^2))$$

and deduce that, if f is as in (i),

$$\int_{-R}^{S} E_R(t) f(t)\, dt = \sqrt{2\pi} \sum_{n=-\infty}^{\infty} \hat{f}(n) \exp(-n^2/(2R)^2),$$

where $\hat{f}(n) = (1/2\pi) \int_{-\pi}^{\pi} f(t) \exp(int)\, dt$.

(iii) Hence conclude that, if $t \in \mathbb{R}$,

$$\sum_{n=-\infty}^{\infty} \exp(-n^2/(2R^2)) \hat{f}(n) \exp(int) \to f(t) \text{ as } R \to \infty$$

and so $\sum_{n=-\infty}^{\infty} \exp(-\delta n^2) \hat{f}(n) \exp(int) \to f(t)$ as $\delta \to 0+$.
(Thus we have yet another way of summing Fourier series.)

60.4 (The Poisson formula) This is one of the most remarkable and useful equalities in Fourier analysis.

(i) Suppose $f:\mathbb{R}\to\mathbb{R}$ is a continuous function such that $\sum_{m=-M}^{M}|\hat{f}(m)|$ converges as $M\to\infty$ and $\sum_{n=-N}^{N}|f(x+2\pi n)|$ converges uniformly on $[-\pi,\pi]$ as $N\to\infty$. Show that $\sum_{n=-N}^{N}f(x+2\pi n)$ converges uniformly to a continuous function $g(x)$ on $[-\pi,\pi]$ as $N\to\infty$. Show further that $g(\pi)=g(-\pi)$.

(ii) Show that

$$\frac{1}{2\pi}\int_{-\pi}^{\pi}g(x)\exp(-imx)\,dx=\frac{1}{2\pi}\sum_{n=-\infty}^{\infty}\int_{-\pi}^{\pi}f(2\pi n+x)\exp(-imx)\,dx=\frac{\hat{f}(m)}{2\pi}$$

and deduce, using Theorem 9.1, that $(2\pi)^{-1}\sum_{m=-M}^{M}\hat{f}(m)\exp(imx)\to g(x)$ uniformly on $[-\pi,\pi]$ as $M\to\infty$. In particular, by taking $x=0$, obtain Poisson's formula

$$\sum_{m=-\infty}^{\infty}\hat{f}(m)=2\pi\sum_{n=-\infty}^{\infty}f(2\pi n).$$

(iii) Show that the conditions of (i) certainly hold if f is twice continuously differentiable with $f', f''\in L^1\cap C$, $f'(x)\to 0$ as $|x|\to\infty$ and $x^{-2}f(x)\to 0$ as $x\to\infty$. Prove further that, under these conditions,

$$\sum_{m=-\infty}^{\infty}\hat{f}(Km)=2\pi K^{-1}\sum_{n=-\infty}^{\infty}f(2\pi K^{-1}n)\quad\text{for } K>0.$$

(iv) Using the notation of Appendix F, find the constants C_1, C_2 in the formula $\sum_{m=-\infty}^{\infty}(\mathcal{F}f)(m)=C_1\sum_{n=-\infty}^{\infty}f(C_2n)$ for well-behaved f. What values of A and B yield the symmetric form $\sum_{m=-\infty}^{\infty}(\mathcal{F}f)(Cm)=\sum_{n=-\infty}^{\infty}f(Cn)$?

(v) Suppose f satisfies the hypotheses of part (iii). Show carefully that

$$K\sum_{m=-\infty}^{\infty}\hat{f}(Km)\to\int_{-\infty}^{\infty}\hat{f}(x)\,dx$$

and $\sum_{n=-\infty}^{\infty}f(2\pi K^{-1}n)\to f(0)$ as $K\to 0+$. Hence conclude that $\hat{\hat{f}}(0)=2\pi f(0)$. By applying this result to f_t defined by $f_t(x)=f(x-t)$, obtain the full inversion formula $\hat{\hat{f}}=2\pi\check{f}$ for f satisfying the hypotheses of part (iii). (Remark. In Question 60.6(vi) we see a connection between the Poisson formula and Poisson summation but, generally speaking, the two ideas are unrelated.)

60.5 (Euler's formula (see Questions 8.7 and 18.8) via the Poisson summation formula)

(i) Let $t>0$ and let $f:\mathbb{R}\to\mathbb{R}$ be given by $f(x)=\exp(-t|x|/2\pi)$. Compute $\hat{f}$ and show that f and $\hat{f}$ satisfy the hypotheses of Question 60.4 (i). Now use the Poisson formula to show that

$$2(1-e^{-t})^{-1}-1=\sum_{n=-\infty}^{\infty} 2t(t^2+4\pi^2n^2)^{-1}. \qquad (*)$$

(ii) By expanding $(t^2+4\pi n^2)^{-1}$ in powers of t and then interchanging sums (you must justify both operations), show that, if $|t|<1$, we can rewrite (*) in the form

$$2(1-e^{-t})^{-1}-1=2t^{-1}+\sum_{m=0}^{\infty} c_m t^m,$$

with $c_{2m}=0$ and $c_{2m+1}=a_{2m+1}\sum_{n=1}^{\infty} n^{-2m}$ where a_{2m+1} is to be found explicitly $[m\geqslant 0]$.

(iii) Hence obtain Euler's formula

$$\sum_{n=1}^{\infty} n^{-2m}=(-1)^{m-1}2^{2m-1}b_{2m-1}\pi^{2m}/(2m-1)! \quad [m\geqslant 1],$$

where the b_j are defined by the formula

$$(e^y-1)^{-1}=y^{-1}-2^{-1}+\sum_{n=1}^{\infty} b_n y^n/n! \quad \text{for } y\neq 0.$$

60.6 (i) Calculate the Fourier transform of $e^{-|x|}$ and use the inversion theorem to find the Fourier transform of $1/(1+x^2)$. (We thus have a derivation of Lemma 50.2 which avoids complex variable techniques.)

(ii) Use a scaling argument to find the Fourier transform of $\alpha/(\alpha^2+x^2)$ for $\alpha>0$. What happens if $\alpha\leqslant 0$?

(iii) By differentiating with respect to α, find the Fourier transform of $(x^2-\alpha^2)/(x^2+\alpha^2)^2$ for $\alpha\neq 0$.

(iv) (The next two parts of this question are exercises in manipulation drawn from first year Tripos papers.) Find the Fourier transform of the function given by

$$g(x)=\cos x \text{ for } |x|\leqslant \pi/2, \quad g(x)=0 \text{ otherwise}.$$

Hence, or otherwise, find an expression as a convolution for the Fourier transform form of the continuous function h given by

$$h(x)=(x^4-1)^{-1}\cos(\pi x/2) \quad \text{for } |x|\neq 1.$$

Show that $\hat{h}$ is even and find $\hat{h}(\lambda)$ when $|\lambda|>\pi/2$.

(v) Find an $f\in L^1\cap C$ such that

$$\hat{f}(\lambda)=\lambda^2 e^{-|\lambda|}$$

and hence show that

$$\int_0^{\infty}\frac{x(1-3x^2)}{(1+x^2)^3}\sin\lambda x\,dx=\pi\frac{\lambda}{4}(|\lambda|-2)e^{-|\lambda|}.$$

(vi) Apply the Poisson formula (Question 60.4 (ii)) to show that

$$\frac{1}{\pi}\sum_{n=-\infty}^{\infty}\frac{y}{y^2+(x-2n\pi)^2}=\frac{1}{2\pi}\sum_{n=-\infty}^{\infty}e^{-y|n|}e^{inx}$$

for all $y>0$, $x\in\mathbb{R}$. Setting $r=e^{-y}$ obtain

$$\frac{1}{\pi}\sum_{n=-\infty}^{\infty}\frac{y}{y^2+(x-2n\pi)^2}=\frac{1}{2\pi}\frac{1-r^2}{1-2r\cos x+r^2},$$

thus connecting the functions Q and P_r of Question 55.2. (The observation is taken from J. S. Walker's *Fourier Analysis*, a book full of good things.)

60.7 (i) Extend the result of Question 60.1 (i) to give

$$\int_{-\infty}^{\infty}f(t)g(t)^*\,dt=\frac{1}{2\pi}\int_{-\infty}^{\infty}\hat{f}(\lambda)\hat{g}(\lambda)^*\,d\lambda$$

whenever f, g, $\hat{f}$, $\hat{g}\in L^1\cap C$. (This is a version of Parseval's theorem (Theorem 34.4).)

(ii) Calculate the Fourier transform of the function $f:\mathbb{R}\to\mathbb{R}$ given by

$$f(x)=\max(0,1-|x|)$$

and use the formula of part (i) to obtain

$$\int_{-\infty}^{\infty}\frac{(1-\cos\lambda)^4}{\lambda^4}\,d\lambda=\frac{\pi}{3}.$$

60.8$^{(-)}$ (i) Show that, if $f\in L^1\cap C$, then, if $\int_{-\infty}^{\infty}|t|^n|\hat{f}(t)|\,dt<\infty$, it follows that f is n times continuously differentiable.

(ii) Let $X_1, X_2, \ldots$ be independent random variables with the uniform distribution on $[0,1]$. Show that, if $n\geq 2$, then $n^{-1}\sum_{i=1}^{n}X_i$ has an $(n-2)$ times continuously differentiable probability density.

60.9 (i) Show that, if $\phi:\mathbb{R}\to\mathbb{R}$ is once continuously differentiable with $\phi\in L^1\cap C$ and $\phi(x)\to 0$ as $|x|\to\infty$, then $\phi'\in L^1\cap C$ and $(\phi')\hat{\ }(\zeta)=i\zeta\hat{\phi}(\zeta)$.

(ii) Suppose that $\phi:\mathbb{R}\to\mathbb{R}$ is twice continuously differentiable with $\phi\in L^1\cap C$ and $\phi(x)$, $\phi'(x)\to 0$ as $|x|\to\infty$. Suppose further that $f\in L^1\cap C$. Show, by taking Fourier transforms, that, if $k>0$ and

$$\phi''(x)-k^2\phi(x)=f(x)$$

then

$$\phi(x)=\frac{-1}{2\pi}\int_{-\infty}^{\infty}\frac{\hat{f}(\zeta)}{\zeta^2+k^2}e^{i\zeta x}\,d\zeta.$$

(iii) Conversely, suppose that we know that $f, \hat{f} \in L^1 \cap C$ and $k > 0$. Show, by differentiating under the integral sign, that, if we set

$$\phi(x) = -\frac{1}{2\pi}\int_{-\infty}^{\infty} \frac{\hat{f}(\zeta)}{\zeta^2 + k^2} e^{i\zeta x}\, d\zeta,$$

then ϕ is twice continuously differentiable with

$$\phi''(x) - k^2\phi(x) = f(x).$$

Using the Riemann–Lebesgue lemma (Theorem 52.11), show that $\phi(x)$, $\phi'(x) \to 0$ as $|x| \to \infty$.

(iv) By considering the general solution of $\phi''(x) - k^2\phi(x) = 0$, show that, if $k > 0$, the equation $\phi''(x) - k^2\phi(x) = f(x)$ can have at most one solution with $\phi(x) \to 0$ as $|x| \to \infty$.

(v) Suppose $k = 0$. Under what conditions does $\phi''(x) = f(x)$ have a solution with $\phi(x) \to 0$ as $|x| \to \infty$? Is it unique? Discuss informally whether there is a connection with the solutions obtained in (ii) and (iii)

60.10 (i) Show that, if $G: \mathbb{R}^2 \to \mathbb{R}$ is differentiable and $g(x) = G(x, x)$, then $g: \mathbb{R} \to \mathbb{R}$ is differentiable and $g'(x) = G_1(x, x) + G_2(x, x)$. Suppose now that $F: \mathbb{R}^2 \to \mathbb{R}$ is continuous, that F_2 is continuous and that

$$\int_{-\infty}^{\infty} |F(x, t)|\, dx, \int_{-\infty}^{\infty} |F_2(x, t)|\, dx < \infty.$$

Show that, if $f(t) = \int_{-\infty}^{t} F(x, t)\, dt$, then $f: \mathbb{R} \to \mathbb{R}$ is differentiable and find its derivative.

(ii) Suppose ϕ is given by the formula of Question 60.9 (iii). Show that $\phi = f * \psi_k$, where $\psi_k(x) = -(1/2k)e^{-k|x|}$.

(iii) Verify more generally that, if $f: \mathbb{R} \to \mathbb{R}$ is a bounded continuous function, then

$$\phi(x) = -\frac{1}{2k}\int_{-\infty}^{\infty} e^{-k|x-y|} f(y)\, dy$$

defines a twice continuously differentiable bounded function which satisfies $\phi''(x) - k^2\phi(x) = f(x)$. Is it the only bounded solution?

(If you have met Green's functions before, you will recognise $G(x, y) = (-1/2k)e^{-k|x-y|}$ as a Green's function.)

60.11 (i) Let $f: \mathbb{R} \to \mathbb{R}$ be a non constant twice continuously differentiable function with $f(x) = 0$ for $|x| > R$. Write $F = \hat{f}$ and $e_a(x) = \exp(iax)$ for $x \in \mathbb{R}$. Show that, if $|a| > 2R$, then $F, e_aF \in L^1 \cap C$ and F and e_aF are not identically zero yet $F*(e_aF) = 0$.

(ii) Show that, if $f \in L^1 \cap C$ and $f(-t) = f(t)^*$, then $\hat{f}(\lambda)$ is real for all $\lambda \in \mathbb{R}$. Show that we can find F, G: $\mathbb{R} \to \mathbb{R}$ with F, $G \in L^1 \cap C$ such that F and G are not identically zero yet $F * G = 0$.

(iii) Let f: $\mathbb{R} \to \mathbb{R}$ be a non constant, twice continuously differentiable function with $f(x) = 0$ for $|x| > \mathbb{R}$. Show that $(\mathcal{F}f)(z) = \int_{-\infty}^{\infty} f(x)e^{-ixz}\,dx$ is well defined for all $z \in \mathbb{C}$ and further that $\mathcal{F}f$ is an analytic function. (If you need a hint, look at the proof of Theorem 75.2) Deduce that, if f is not identically zero, $\mathcal{F}f$ is a non constant analytic function and so (see Lemma 75.5, if necessary) has no non isolated zeros. (In other words, if $\mathcal{F}f(w) = 0$, then we can find a $\delta_w > 0$ such that $(\mathcal{F}f)(z) \neq 0$ for $0 < |z - w| < \delta_w$.) Conclude that, if f is as above, $g \in L^1 \cap C$ and g is not identically zero, then $f * g$ is not identically zero.

60.12 (The black box) Engineers are frequently faced with a 'black box' into which they feed an input signal $f(t)$ to obtain an output signal $g(t)$. More formally, if f: $\mathbb{R} \to \mathbb{R}$ is a well-behaved function, the output g: $\mathbb{R} \to \mathbb{R}$ is a well-behaved function and we write $g = Tf$. What properties should T have? The most obvious is temporal homogeneity. If f: $\mathbb{R} \to \mathbb{R}$ is a function, write $S_a f(t) = f(t - a)$. We require

$$(A)\ \ TS_a f = S_a Tf,$$

i.e. if we delay the input by a, we obtain the same output but delayed by a. The second property we might demand is linearity

$$(B)\ \ T(\lambda_1 f_1 + \lambda_2 f_2) = \lambda_1 Tf_1 + \lambda_2 Tf_2.$$

Not all black boxes are linear, but those that are, are much easier to deal with than those that are not. It is also natural to demand that, in some sense, small input should produce small output. We formalise this demand by asking that

(C) There exists a κ such that, if $|f(t)| \leqslant 1$ for all $t \in \mathbb{R}$, then $|(Tf)(t)| \leqslant \kappa$ for all $t \in \mathbb{R}$.

(i) Show, using (B) and (C), that, if $|f(t)| \leqslant \varepsilon$ for all $t \in \mathbb{R}$, then $|(Tf)(t)| \leqslant \kappa\varepsilon$ for all $t \in \mathbb{R}$.

(ii) Let us write $\Delta(t) = 1 - |t|$ for $|t| \leqslant 1$, $\Delta(t) = 0$ otherwise, and let us set $\Delta_n(t) = \Delta(nt)$. Suppose now, until further notice, that F: $\mathbb{R} \to \mathbb{R}$ is a continuous function with $F(t) = 0$ for $t \notin [0, 1]$. Show that

$$\sum_{r=1}^{n} F(r/n)\Delta_n(t - r/n) \to F(t) \text{ uniformly on } \mathbb{R} \text{ as } n \to \infty.$$

(If you need a hint, see Question 5.4.)

(iii) Deduce that, for all $t \in \mathbb{R}$,

$$\sum_{r=1}^{n} F(r/n)k_n(t - r/n) \to (TF)(t)$$

where $k_n = T\Delta_n$.

(iv) With many systems the output corresponding to an input f which lasts only a short period of time (specifically $f(t) = 0$ for $|t| > \delta$ where δ is small) depends mainly on the 'amount of impulse' $\int_{-\delta}^{\delta} f(t)\,dt$ and very little on the precise form of f. This leads us to impose our final condition on our black box.

(D) $T(m\Delta_m)(t) \to K(t)$ uniformly on $\mathbb{R}$ as $m \to \infty$, where $K: \mathbb{R} \to \mathbb{R}$ is a bounded continuous function.

Show that, for all $t \in \mathbb{R}$,

$$\sum_{r=1}^{n} F(r/n)k_n(t - r/n) = \sum_{r=1}^{n} n^{-1} F(r/n)(nk_n(t - r/n))$$
$$\to \int_0^1 F(x)K(t - x)\,dx$$

as $n \to \infty$. Conclude that

$$TF(t) = \int_0^1 F(x)K(t - x)\,dx \quad \text{for all } t \in \mathbb{R}.$$

(v) Show more generally that, if $F: \mathbb{R} \to \mathbb{R}$ is a continuous function with $F(t) = 0$ for $t \notin [a, b]$, then

$$TF(t) = \int_a^b F(x)K(t - x)\,dx \quad \text{for all } t \in \mathbb{R}.$$

By choosing appropriate F and using condition (C), show that $\int_{-\infty}^{\infty} |K(t)|\,dt$ converges and indeed $\int_{-\infty}^{\infty} |K(t)|\,dt \leqslant \kappa$. Finally, by using (i), or otherwise, show that, if $F(t) \to 0$ as $|t| \to \infty$,

$$TF(t) = \int_{-\infty}^{\infty} F(x)K(t - x)\,dx.$$

(vi) Now let us set

$$L^1 \cap C_0 = \{f \in L^1 \cap C: f(x) \to 0 \text{ as } |x| \to \infty\}.$$

Show that, if we reverse our line of argument by taking an arbitrary $k \in L^1 \cap C_0$ and defining $Tf = f * k$ whenever $f \in L^1 \cap C_0$, then

(α) $Tf \in L^1 \cap C_0$ whenever $f \in L^1 \cap C_0$,

(A) $TS_a f = S_a Tf$ for all $a \in \mathbb{R}$, $f \in L^1 \cap C_0$,

(B) $T(\lambda_1 f_1 + \lambda_2 f_2) = \lambda_1 Tf_1 + \lambda_2 Tf_2$ for all $\lambda_1, \lambda_2 \in \mathbb{C}$, $f_1, f_2 \in L^1 \cap C_0$,

(*C*) $\sup_{t\in\mathbb{R}}|Tf(t)| \leqslant \int_{-\infty}^{\infty} |k(x)|\,dx \sup_{t\in\mathbb{R}}|f(t)|$ for all $f \in L^1 \cap C_0$,
(*D*) $(T(n\Delta_n))(t) \to k(t)$ uniformly on $\mathbb{R}$ as $n \to \infty$.

(Henceforward we will refer to the situation described in (vi) as a black box with *response function* k. As an example, consider a listener at a point A in a concert hall listening to a noise made at point B. The concert hall acts as a very complicated black box. The response function k_{AB} is sometimes found by firing a starting pistol held at B and recording the result at A.)

60.13 (We give some simple illustrations of the ideas of the previous question.)

(i) Consider a black box with response function k given by

$$k(t) = a_1 e^{-\gamma_2 t} \sin \omega_1 t + a_2 e^{-\gamma_2 t} \sin \omega_2 t \text{ for } t \geqslant 0$$

$$k(t) = 0 \text{ otherwise,}$$

where $\gamma_1, \gamma_2, \omega_1, \omega_2 > 0$. Find a relation between α_1 and α_2 which guarantees that, whatever the input f, the output Tf satisfies $\int_{-\infty}^{\infty} (Tf)(t)\,dt = 0$.

(ii) Consider a black box with response function k given by $k(t) = (2\pi)^{-1/2} e^{-t^2/2}$. If its output Tf is given by $Tf(t) = te^{-t^2/4}$, what was its input f? Is it possible to have an output Tf given by $Tf(t) = te^{-t^2}$?

(iii) Consider a black box with non-zero response function k. Is it always true that different inputs produce different outputs?

(iv) Suppose now we are told that the response function k is such that $\hat{k}(\lambda) \neq 0$ for all $\lambda \in \mathbb{R}$. Is it true mathematically that, given the output Tf, we can find f? Give a proof or a counter example. Discuss briefly any problems that might arise in trying to apply this idea in the real world.

(v) Suppose we merely know that the zeros of $\hat{k}$ are isolated (i.e., if $\hat{k}(\mu) = 0$, we can find a $\delta_\mu > 0$ such that $\hat{k}(\lambda) \neq 0$ for $0 < |\lambda - \mu| < \delta_\mu$). Would your answer to (iv) be affected and, if so, how? How would your answer to (iv) be affected, if instead, we know that $k(t) = 0$ for $|t| > R$, say, but k is not identically zero?

60.14 Black boxes can be connected up in various ways.

(i) Suppose the same input f is fed into two black boxes with response functions k_1 and k_2 and the resulting outputs $T_1 f$, $T_2 f$ added to produce $Sf = T_1 f + T_2 f$. Show that the new system is a black box with response function $k_1 + k_2$.

(ii) Suppose the input f is fed into a black box with response function k_1 and the resulting output $T_1 f$ is fed into a black box with response function k_2 to produce an output $Sf = T_2(T_1 f)$. Show that the new system is a black box with response function $k_2 * k_1$.

(iii) Consider the system in (ii). Suppose $\hat{k}_1(\lambda) \neq 0$ for $\lambda \in \mathbb{R}$. Is it possible to choose k_2 such that $Sf = f$ for all f?

(iv) Consider the system in (ii). Suppose $\hat{k}_1(\lambda) \neq 0$ for $\lambda \in \mathbb{R}$ and let $R > 0$ be fixed. Is it possible to choose k_2 such that $Sf = f$ for all f with $\hat{f}(\lambda) = 0$ for $|\lambda| \geqslant R$?

(v) Black boxes are not always connected as simply as in parts (i) and (ii). We are all familiar with the phenomenon of feedback in which part of the output is fed back as input. Sometimes the results are bad, as when an amplifying system starts to whistle, but often feedback is desirable. In a control system it is important to check that the actual output resembles the desired output and for this we must have feedback. Let us look at a simple feedback system.

Consider two black boxes, call them box 1 and box 2, with response functions k_1 and k_2 linked in such a way that, if the input to the entire system if f, then

(a) the output of box i is f_i $[i = 1, 2]$,
(b) the input to box 1 is $\alpha f_2 + f$ where α is some real number,
(c) the input to box 2 is f_1,
(d) the output Tf of the entire system is f_2.

Assuming that the entire system is a black box, show that its response function k is given by

$$\hat{k}(\lambda) = \frac{\hat{k}_2(\lambda)}{1 - \alpha \hat{k}_1(\lambda) \hat{k}_2(\lambda)}.$$

(vi) Suppose k_1 and k_2 are such that $\alpha \hat{k}_1(\lambda) \hat{k}_2(\lambda) = 1$ for $\lambda \in [a, b]$ where $a < b$. State briefly what you think would go wrong if we wired up such a system of black boxes.
(We shall return to the topic in Question 77.4 but, without going any further, it is clear that designers and users of black boxes will wish to pay attention not merely to the response function k, but to its Fourier transform. Ideally they will wish both k and $\hat{k}$ to be nice. However k and $\hat{k}$ cannot be specified independently and we shall see in the next few questions that, in certain respects, the better we get one to behave, the more difficult it is to get the other to behave.)

(vii) Before leaving the topic of black boxes, we remark that there is one further constraint that the response function k often has to satisfy. If t actually represents *time*, we would expect the value of the output $Tf(t_0)$ at time t_0 to depend only on the values of the input $f(t)$ for $t \leqslant t_0$. Show that

(e) $Tf(0) = 0$ for all f with $f(t) = 0$ for $t \leqslant 0$

if and only if $k(t) = 0$ for $t \leqslant 0$. Condition (e) is called a causality condition, since it is an application of the principle that effect cannot precede cause. (For further developments I refer again to Question 77.4.)

60.15 (The Heisenberg inequality) For many users of Fourier transforms one fact dominates all others. If f is thin, then $\hat{f}$ is fat.

(i) Let $g \in L^1 \cap C$ and write $g_a(x) = g(ax)$ for $x \in \mathbb{R}$ $[a > 0]$. Compute $\hat{g}_a$.

(ii) Let $g \in L^1 \cap C$, $g(x) \geqslant 0$ for all x and $\int_{-\infty}^{\infty} g(x)\,dx = 1$. Show that, if $g(x) = 0$ for $|x| \geqslant \delta$, then $|\hat{g}(\lambda)| \geqslant 1/2$ for $|\lambda| \leqslant 10^{-2}\delta^{-1}$. (The constant 10^{-2} is pretty obviously not best possible.)

(iii) A piano note lasts only for a short period of time. Can it be a pure note (i.e. a single frequency)?

The second sentence of this question can be given a variety of quantitative expressions. One of the most beautiful is the Heisenberg inequality.

(iv) Assuming that f is sufficiently well behaved, justify each step in the following chain of equalities and inequalities. (You will need Question 60.7(i).)

$$\frac{1}{2\pi}\int_{-\infty}^{\infty} x^2|f(x)|^2\,dx \int_{-\infty}^{\infty} \lambda^2|\hat{f}(\lambda)|^2\,d\lambda = \frac{1}{2\pi}\int_{-\infty}^{\infty} |xf(x)|^2\,dx \int_{-\infty}^{\infty} |\lambda\hat{f}(\lambda)|^2\,d\lambda$$

$$= \int_{-\infty}^{\infty} |xf(x)|^2\,dx \int_{-\infty}^{\infty} |f'(x)|^2\,dx \geqslant \left(\int_{-\infty}^{\infty} |xf'(x)f(x)|\,dx\right)^2$$

$$\geqslant \left[\int_{-\infty}^{\infty} \frac{x}{2}\,(f'(x)f^*(x) + f(x)f^{*\prime}(x))\,dx\right]^2 = \frac{1}{4}\left(\int_{-\infty}^{\infty} x\left(\frac{d}{dx}|f(x)|^2\right)dx\right)^2$$

$$= \frac{1}{4}\left(\int_{-\infty}^{\infty} |f(x)|^2\,dx\right)^2 = \frac{1}{8\pi}\int_{-\infty}^{\infty} |f(x)|^2\,dx \int_{-\infty}^{\infty} |\hat{f}(\lambda)|^2\,d\lambda$$

(v) Explain briefly why the formula

$$\frac{\int_{-\infty}^{\infty} x^2|f(x)|^2\,dx}{\int_{-\infty}^{\infty} |f(x)|^2\,dx}\;\frac{\int_{-\infty}^{\infty} \lambda^2|\hat{f}(\lambda)|^2\,d\lambda}{\int_{-\infty}^{\infty} |\hat{f}(\lambda)|^2\,d\lambda} \geqslant \frac{1}{4},$$

just obtained is a quantitative expression of the statement that, if f is concentrated near the origin, $\hat{f}$ cannot be.

(vi) Show, by examining the proof in (iv), that, if equality is attained in the Heisenberg inequality of (v) and f is well behaved, then

$$f'(x) = kxf(x) \quad \text{for all } x$$

and some constant k. Conclude that the only well-behaved functions for which equality is attained are those of the form

$$f(x) = A\exp(-|k|x^2/2)$$

with $A, k \in \mathbb{C}$, $k \neq 0$. Verify that equality is attained by such functions.

(The reader might indulge the fleeting hope that, since we have only proved the Heisenberg inequality for well-behaved functions, it might be possible to evade it by using ill-behaved functions. But it is not hard to approximate ill-behaved functions by well-behaved functions in such a way as to show that the inequality holds for the more general class. For details and a discussion of the relation with quantum mechanics, see the excellent book of Dym and McKean, *Fourier Series and Integrals*.)

60.16 (The Shannon sampling theorem)

(i) In Question 60.4(ii) we saw that, if $f: \mathbb{R} \in \mathbb{C}$ is well behaved, then

$$\sum_{n=-\infty}^{\infty} \hat{f}(m) \exp(imx) = 2\pi \sum_{n=-\infty}^{\infty} f(x + 2\pi n).$$

Deduce that, if f is well behaved and $f(x) = 0$ for $|x| \geqslant \pi$, then

$$2\pi f(x) = \sum_{m=-\infty}^{\infty} \hat{f}(m) \exp(imx) \quad \text{for } |x| \leqslant \pi,$$

$$f(x) = 0 \text{ for } |x| \geqslant \pi,$$

and conclude that $\hat{f}: \mathbb{R} \to \mathbb{R}$ is uniquely determined by its values at integer points (i.e. if f and g are well behaved with $f(x) = g(x) = 0$ for $|x| \geqslant \pi$, then, if $\hat{f}(m) = \hat{g}(m)$ for $m \in \mathbb{Z}$, it follows that $\hat{f}(\lambda) = \hat{g}(\lambda)$ for all $\lambda \in \mathbb{R}$).

(ii) By inversion, rescaling and translation, deduce that, if $F: \mathbb{R} \to \mathbb{C}$ is well behaved and $\hat{F}(\lambda) = 0$ for $|\lambda| \geqslant \pi/\omega$, then F is uniquely determined by its values on the sequence $a + n\omega$ $[n \in \mathbb{Z}]$ where a is any fixed number. (In engineering parlance we would say that F is *band limited* to $[-\omega/\pi, \omega/\pi]$. We might express the result of (ii) by saying that 'If F contains no frequencies higher than ω/π cycles per second, then it is completely determined by its values at a sequence of times spaced $1/\omega$ seconds apart'. The sampling theorem and the Heisenberg inequality are two fingers of the same hand. If we make $\hat{f}$ thin, we restrict f.)

The result above raises two questions.

(A) Can we sample less frequently and still have uniqueness?

(B) We know that F is determined in principle by $F(n\omega + a)$ $[n \in \mathbb{Z}]$. Can we determine F in practice?

The next part of this question answers (A).

(iii) Let $\pi > \varepsilon > 0$ be given. Let $g: \mathbb{R} \to \mathbb{R}$ be a twice continuously differentiable function such that $g(\lambda) = 0$ for $|\lambda| \geqslant \varepsilon$, $g(\lambda) \geqslant 0$ for all λ and $\int_{-\infty}^{\infty} g(\lambda)\, d\lambda = 1$. Show that, if we set $F(x) = (\sin \pi x)\hat{g}(x)$, then

(a) F is not identically zero.

(b) $\hat{F}(\lambda) = 0$ for $|\lambda| \geqslant \pi + \varepsilon$.

(c) $F(m) = 0$ for all $m \in \mathbb{Z}$.

Thus we cannot decrease the sampling rate below the value specified in (ii).

The lower bound is called the Nyquist sampling rate. (If you wish to see how I came to choose the counter example above, consider what happens as $\varepsilon \to 0+$.)

(iv) It is sometimes said that an entire library can be stored in the following manner. Translate the contents of the library into a string of zeros and ones. Writing the sequence as a terminating binary decimal 0.101... we get a rational number r between 0 and 1. Now cut a ruler to this exact length in metres. From measuring the length of the ruler, we can recover r and thus the library. In practice, using the best available equipment, how big a library (book, sentence, word, ...) could be stored?

In the real world equipment can only generate, transmit and receive in a certain band of frequencies. The broader the band used, the more it costs. Thus telephone systems do not carry the high and low frequencies but only the narrow band required to produce what custom has made an acceptable service for transmitting the human voice. If music is transmitted by telephone, it sounds very odd. Since sampling at a higher rate than the Nyquist rate produces, in principle, no more information about F, it is reasonable to conclude that the Nyquist rate is proportional to the rate at which information is transmitted, and we get the slogan

$$\frac{\text{rate of transmission of information}}{\text{band width of signal}} \leqslant \text{constant.}$$

Fibre optics, by allowing the use of high frequency electromagnetic waves, gives much higher band widths and therefore a much higher rate of transmission of information than previous methods of telecommunication.

In Question 7.2 we explained why very low frequency radio waves are used for communication with submerged submarines. By what we have just seen, such a system must have a low transmission rate of information. It is reported that in practice these systems would take more than a minute to transmit two letters of morse code. (Chapters 62, 65 and 66 look at this problem from a different angle.) It would appear that, because of the problem of communication, future submarine battles will be fought

> '... as on a darkling plane
> swept with confused alarms of struggle and fight
> Where ignorant armies clash by night'.

60.17 (i) We now turn to Problem (B) of the previous question. Can we determine F from a Nyquist sampling? In dealing with Fourier series, we looked at the Dirichlet kernel (see e.g. Chapter 18)

$$D_N(t) = (2\pi)^{-1} \sum_{n=-N}^{N} \exp(int) = \sin((N + 1/2)t)/(\sin t/2).$$

Here it turns out to be advantageous to look at the corresponding Dirichlet kernel for Fourier transform

$$\tilde{D}_R(t) = \frac{1}{2\pi}\int_{-R}^{R} \exp(i\lambda t)\, d\lambda.$$

We note (without trying to be too precise) that $\tilde{D}_\pi$ contains no frequencies λ with $|\lambda| > \pi$. Show that $\tilde{D}_\pi(t) = \sin \pi t/\pi t$ for $t \neq 0$ and that

(a) $\tilde{D}_\pi(0) = 1$,
(b) $\tilde{D}_\pi(m) = 0$ for all $m \in \mathbb{Z}$, $m \neq 0$.

Now suppose $F: \mathbb{R} \to \mathbb{C}$ is a well-behaved function with $\tilde{F}(\lambda) = 0$ for $|\lambda| > \pi$. Explain why (if everything is well behaved) we expect

$$G(t) = F(t) - \sum_{m=-\infty}^{\infty} F(m)\tilde{D}_\pi(t - m)$$

to give a well-behaved function G with $\hat{G}(\lambda) = 0$ for $|\lambda| > \pi$ and $G(m) = 0$ for all $m \in \mathbb{Z}$. Deduce that (if everything is well behaved)

$$F(t) = \sum_{m=-\infty}^{\infty} F(m)\tilde{D}_\pi(t - m).$$

(ii) It is thus not surprising that the formula written above should hold for some classes of F. What is more surprising is that, as we shall now see, it holds very generally. Suppose merely that $F, \hat{F} \in L^1 \cap C$ and that $\hat{F}(\lambda) = 0$ for $|\lambda| > \pi$. Let us write $f = \hat{F}$. Justify each step in the following chain of equalities and inequalities (you will need Schwarz's inequality and Parseval's equality for Fourier series as given in Theorem 34.2).

$$\begin{aligned}
&\left|F(t) - \sum_{n=-N}^{N} F(n)\tilde{D}_\pi(t-n)\right| \\
&\quad = (2\pi)^{-1}\left|\int_{-\pi}^{\pi}\left(f(\lambda)\exp(i\lambda t) - \frac{1}{2\pi}\sum_{n=-N}^{N}\hat{f}(-n)\exp(i\lambda(t-n))\right)d\lambda\right| \\
&\quad \leqslant (2\pi)^{-1}\int_{-\pi}^{\pi}\left|f(\lambda) - \frac{1}{2\pi}\sum_{n=-N}^{N}\hat{f}(n)\exp(i\lambda n)\right| d\lambda \\
&\quad \leqslant (2\pi)^{-1/2}\left(\int_{-\pi}^{\pi}\left|f(\lambda) - \frac{1}{2\pi}\sum_{n=-N}^{n}\hat{f}(n)\exp(i\lambda n)\right|^2 d\lambda\right)^{1/2} \\
&\quad \leqslant (2\pi)^{-1}\left(\sum_{|n|\geqslant N+1}|\hat{f}(n)|^2\right)^{1/2}.
\end{aligned}$$

(Note that, as in the proof of Poisson's formula, the key idea is to roll up $\mathbb{R}$ into $\mathbb{T}$.)

(iii) Conclude that, in general, if $F, \hat{F} \in L^1 \cap C$ and $\hat{F}(t) = 0$ for $|t| > a$, then

$$\sum_{n=-N}^{N} F(n\pi/a)\frac{\sin(at-\pi n)}{at-\pi n} \to F(t)$$

uniformly on $\mathbb{R}$ as $N \to \infty$. (The uniformity and the estimate which supports it are important since, in practice, we can only sample F finitely often.) Show also that, under the same hypotheses,

$$\sum_{n=-N}^{N} \frac{N+1-|n|}{N+1} F(n\pi/a)\frac{\sin(at-\pi n)}{at-\pi n} \to F(t)$$

uniformly on $\mathbb{R}$.

(Although the performance of a compact disc must satisfy very exigent listeners, the human ear is only sensitive to a limited band of frequencies. Thus, provided the sampling rate is high enough and sampling done with sufficient precision, the music to be recorded can be reduced to digital form from which the sound can later be recovered using a compact disc player. Digitising information has many advantages, one being that we can now use error correcting codes (another idea associated with the name of Shannon).)

The results of Questions 60.15 to 60.17 are often called 'uncertainty principles'. I have chosen to name them after the people who brought them to the centre of the intellectual stage, but the first person to write down Heisenberg's inequality may well have been Wiener in 1925 and the first person to publish the reconstruction formula obtained in Question 60.17 may well have been Whittaker in 1915 (see the interesting article 'Five short stories about the cardinal series' by Higgins).

60.18 (A Paley–Wiener type result) Another indication of how special 'band limited' functions are given by the following observations.

(i) Suppose $f: \mathbb{R} \to \mathbb{C}$ is continuous and $f(x) = 0$ for all $|x| > a$. Show, developing the result of Question 60.11 (iii), that $(\mathcal{F}f)(z) = \int_{-\infty}^{\infty} f(x)e^{-ixz}\,dx$ gives a well-defined analytic function $\mathcal{F}f: \mathbb{C} \to \mathbb{C}$ with $|\mathcal{F}f(z)| \leqslant Ae^{a\operatorname{Im} z}$ for some constant A.

(ii) Conversely suppose $f, \hat{f} \in L^1 \cap C$ and there exists an analytic function $\mathcal{F}f: \mathbb{C} \to \mathbb{C}$ such that

(a) $\mathcal{F}f(\lambda) = \hat{f}(\lambda)$ for all $\lambda \to \mathbb{R}$,
(b) $|\mathcal{F}f(z)| \leqslant Ae^{a(\operatorname{Im} z)}$ for all $z \in \mathbb{C}$,

where A, $a > 0$. By integrating $e^{ixz}\mathcal{F}f(z)$ round rectangles with vertices R, $R+iy$, $-R+iy$, $-R$ (take care with the choice of sign of y), show that $f(x) = 0$ for all $|x| > a$.

60.19 (The Kramers–Krönig relations)

(i) Suppose that $f: \mathbb{R} \to \mathbb{C}$ is once continuously differentiable and that $f(x) = 0$ for $x \leqslant \varepsilon$ and $x \geqslant R$ where $R > \varepsilon > 0$. Show that

$\mathscr{F}f(z) = \int_{-\infty}^{\infty} f(x)e^{ixz}\,dx$ defines an analytic function $\mathscr{F}f: \mathbb{C} \to \mathbb{C}$ such that $\mathscr{F}f(re^{i\theta}) \to 0$ uniformly for $\pi \geqslant \theta \geqslant 0$ as $r \to \infty$. (We use $\int_{-\infty}^{\infty} f(x)e^{ixz}\,dx$ rather than $\int_{-\infty}^{\infty} f(x)e^{-ixz}\,dx$ because this is the way the Kramers–Krönig relations are usually stated.)

(ii) Suppose that $f \in L^1 \cap C$ and that $f(x) = 0$ for $x \leqslant 0$. Show that we can find a sequence of functions $f_n: \mathbb{R} \to \mathbb{C}$ such that f_n is once continuously differentiable, $f_n(x) = 0$ for $x \leqslant n^{-1}$ and $x \geqslant n$ and $\int_{-\infty}^{\infty} |f(x) - f_n(x)|\,dx \to 0$. (Look at Appendix C, if you need a hint.) Show further that $\mathscr{F}f(z) = \int_{-\infty}^{\infty} f(x)e^{ixz}\,dx$ is well defined for $\operatorname{Im} z \geqslant 0$ and that $|\mathscr{F}f_n(z) - \mathscr{F}f(z)| \to 0$ uniformly on $\{z: \operatorname{Im} z \geqslant 0\}$.

(iii) Deduce that, if $f \in L^1 \cap C$ and $f(x) = 0$ for $x \leqslant 0$ (i.e. f is causal), then $\mathscr{F}f(z) = \int_{-\infty}^{\infty} f(x)e^{ixz}\,dx$ is a well-defined function $\mathscr{F}f: \{z: \operatorname{Im} z \geqslant 0\} \to \mathbb{C}$ such that

(a) $\mathscr{F}f$ is continuous on $\{z: \operatorname{Im} z \geqslant 0\}$,
(b) $\mathscr{F}f$ is analytic on $\{z: \operatorname{Im} z > 0\}$,
(c) $\mathscr{F}f(re^{i\theta}) \to 0$ uniformly for $\pi \geqslant \theta \geqslant 0$ as $r \to \infty$.

(You may wish to refer to Theorem 75.3 (iv) to obtain (b).)

(iv) Suppose $g: \{z: \operatorname{Im} z \geqslant 0\} \to \mathbb{C}$ is a function such that

(a) g is continuous on $\{z: \operatorname{Im} z \geqslant 0\}$,
(b) g is analytic on $\{z: \operatorname{Im} z > 0\}$,
(c) $g(re^{i\theta}) \to 0$ uniformly for $\pi \geqslant \theta \geqslant 0$ as $r \to \infty$.

By applying Cauchy's theorem to large semicircles, or otherwise, show that, if $y_0 > 0$,

$$g(x_0 + iy_0) = \frac{1}{2\pi i}\int_{-\infty}^{\infty} \frac{g(x)}{x - (x_0 + iy_0)}\,dx.$$

Explain why

$$0 = \frac{1}{2\pi i}\int_{-\infty}^{\infty} \frac{g(x)}{x - (x_0 + iy_0)^*}\,dx.$$

By combining the two results, show that

$$g(x_0 + iy_0) = \frac{1}{\pi i}\int_{-\infty}^{\infty} \frac{g(x)(x - x_0)}{(x - x_0)^2 + y_0^2}\,dx \quad \text{for all } y_0 > 0,\ x_0 \in \mathbb{R}.$$

Deduce that, if we write $g(x + iy) = u(x, y) + iv(x, y)$, with u and v real, then

$$u(x_0, y_0) = \frac{1}{\pi}\int_{-\infty}^{\infty} \frac{(x - x_0)v(x, 0)}{(x - x_0)^2 + y_0^2}\,dx$$

$$v(x_0, y_0) = -\frac{1}{\pi}\int_{-\infty}^{\infty} \frac{(x - x_0)u(x, 0)}{(x - x_0)^2 + y_0^2}\,dx.$$

These equations are, in some sense, an integral version of the Cauchy–Riemann equations.

(v) Continuing with the ideas of (iv), observe that, by the usual technique of using a small indentation in our contours, we can extend our results to the case $y_0 = 0$. Show that

$$g(x_0) = \frac{1}{\pi i} \lim_{\varepsilon \to 0+} \int_{|x-x_0| \geqslant \varepsilon} \frac{g(x)}{x - x_0}\, dx$$

and deduce that

$$u(x_0, 0) = \frac{1}{\pi} \lim_{\varepsilon \to 0+} \int_{|x-x_0| \geqslant \varepsilon} \frac{v(x, 0)}{x - x_0}\, dx,$$

$$v(x_0, 0) = -\frac{1}{\pi} \lim_{\varepsilon \to 0+} \int_{|x-x_0| \geqslant \varepsilon} \frac{u(x, 0)}{x - x_0}\, dx.$$

(vi) Now return to the case of the causal function $f \in L^1 \cap C$ with $f(t) = 0$ for $t < 0$. Show that

$$\operatorname{Re} \hat{f}(\lambda) = \frac{1}{\pi} \lim_{\varepsilon \to 0+} \int_{|\lambda-\mu| \geqslant \varepsilon} \frac{\operatorname{Im} \hat{f}(\mu)}{\mu - \lambda}\, d\mu,$$

$$\operatorname{Im} \hat{f}(\lambda) = -\frac{1}{\pi} \lim_{\varepsilon \to 0+} \int_{|\lambda-\mu| \geqslant \varepsilon} \frac{\operatorname{Re} \hat{f}(\mu)}{\mu - \lambda}\, d\mu,$$

for all $\lambda \in \mathbb{R}$. Explain the statement 'If F contains no negative frequencies, then F is determined by its real part'. If f is real, show that

$$\operatorname{Re} \hat{f}(\lambda) = \frac{2}{\pi} \lim_{\varepsilon \to 0+} \int_{\mu \geqslant 0, |\lambda-\mu| \geqslant \varepsilon, |\mu+\lambda| \geqslant \varepsilon} \frac{\mu \operatorname{Im} \hat{f}(\mu)}{\mu^2 - \lambda^2}\, d\mu.$$

(vii) Suppose that you know that $f \in L^1 \cap C$, that $f(x) = 0$ for $x \leqslant 0$ and that $\operatorname{Re} \hat{f}(\lambda) = a(a^2 + \lambda^2)^{-1}$ for some real $a > 0$. Use (vi) to find $\operatorname{Im} \hat{f}(\lambda)$ and hence recover f.

(viii) Give an example of an $f \in L^1 \cap C$ such that $\operatorname{Im} \hat{f}(\lambda) = 0$ for all λ, but f is not identically zero.

60.20 (The theta function) In Questions 38.6 to 38.10 we studied doubly periodic meromorphic functions. The simplest one that we constructed was $\sum(n + mi + z)^{-3}$ and we had to work quite hard to produce the general form. An alternative approach was found by Jacobi. Instead of seeking a doubly periodic meromorphic function with periods 1 and τ (with $\tau \notin \mathbb{R}$), let us seek an analytic function θ which has period 1 and 'quasiperiod' τ. More precisely, let us seek a meromorphic function θ such that $\theta(z + 1) = \theta(z)$ and $\theta(z + \tau) = f(z)\theta(z)$ for some simple non constant function f. What follows should be treated as exploratory and rigour should be ignored.

(i) Show that $f(z) = f(z + 1)$ for all z.

(ii) The simplest choices of non constant f which satisfy (i) are

$$f(z) = c\exp(2\pi iz) \text{ or } f(z) = c\exp(-2\pi iz).$$

In conformity to standard notation, we take $f(z) = c\exp(-2\pi iz)$ with $c \neq 0$.

(iii) Since θ has period 1, it is natural (see Question 38.4) to seek to expand it as a Fourier series $\theta(z) = \sum_{n=-\infty}^{\infty} A_n \exp(2\pi inz)$. Show that, taking f as suggested in the last sentence of (ii), this yields $A_{n+1} = c^{-1}A_n \exp(2\pi in\tau)$ for all n. Conclude that $A_n = c^{-n}\exp(\pi in(n-1)\tau)A_0$ for all n.

(iv) We are still at liberty to choose c to give a neat formula. Setting $q = \exp(\pi i\tau)$ and $c = q^{-1}$, we obtain $\theta(z) = \sum_{-\infty}^{\infty} q^{n^2}\exp(2\pi inz)$.

(v) Returning to normal standards of rigour, verify that, if $\operatorname{Im}\tau > \delta$ for some $\delta > 0$ and $q = \exp(\pi i\tau)$, then $\sum_{-\infty}^{\infty} q^{n^2}\exp(2\pi inz)$ converges to a function $\vartheta(z, \tau)$ for all $z \in \mathbb{C}$, the convergence being uniform on each band $|\operatorname{Im} z| < R$. Using standard theorems of complex variable (see Theorem 75.3 if necessary), verify that ϑ: $\{\tau \in \mathbb{C}: \operatorname{Im}\tau > 0\} \times \mathbb{C} \to \mathbb{C}$ is analytic in τ and in z. Show also that

$$\vartheta(z+1, \tau) = \vartheta(z, \tau),\ \vartheta(z+\tau, \tau) = q^{-1}\exp(-2\pi iz)\vartheta(z, \tau).$$

(vi) Show that $\vartheta((1+\tau)/2, \tau) = 0$ and $\vartheta(0, \tau) \neq 0$.

(vii) Show that, if we write

$$f(z) = \frac{\vartheta(z + (1+\tau)/2, \tau)\,\vartheta(z - (1-\tau)/2, \tau)}{\vartheta(z, \tau)^2}$$

then f is a doubly periodic meromorphic function with periods 1 and τ. By using (vi), or otherwise, show that f is non constant.

The reader may suspect that since, as we saw in Question 38.8, the doubly periodic meromorphic functions can only take a very restricted form, it should be possible to obtain all doubly periodic meromorphic functions from ϑ. (We shall prove this in the next question by a method which is so routine that the reader may well wish to try the proof without looking at the question.) The importance of this observation lies in the fact that, whereas sums like $\sum(m + ni + z)^{-3}$ converge exceedingly slowly, the convergence of

$$\sum_{-\infty}^{\infty} q^{n^2}\exp(2\pi inz)$$

will be extremely rapid for $|q|$ less than, and not too close to, 1. (In Question 60.22 we shall see that sometimes something can be done even then.)

60.21 In order to carry out the program just outlined, the main thing we need to know is the nature of the zeros of $\vartheta(z, \tau)$. To simplify notation, let us write $\vartheta(z) = \vartheta(z, \tau)$.

(i) Explain why we can find an a such that θ has no zeros on the parallelogram Λ consisting of the straight line C_1 from a to $a+1$, the straight line C_2

from $a+1$ to $a+1+\tau$, the straight line C_3 from $a+1+\tau$ to $a+\tau$ and the straight line C_4 from $a+\tau$ to a. Explain why the number N of zeros of ϑ within Λ is given by

$$2\pi i N = \sum_{j=1}^{4} \int_{C_j} \frac{\vartheta'(z)}{\vartheta(z)}\, dz$$

(see Appendix E, but only if necessary.) By using the formulae at the end of part (v) of Question 60.20 and matching the contribution from opposite sides of the parallelogram show that $N = 1$.

(ii) Conclude that $\vartheta(z, \tau)$ has a simple zero at $(1+\tau)/2$, but no other zeros in $\Gamma = \{z = \lambda + \mu\tau: 0 \leqslant \lambda, \mu < 1\}$.

(iii) Let $N \geqslant 2$ and let $\alpha_1, \alpha_2, \ldots, \alpha_N, \beta_1, \beta_2, \ldots, \beta_N \in \Gamma$ with $\sum_{j=1}^{N}\alpha_j - \sum_{j=1}^{N}\beta_j = 0$ and $\{\alpha_1, \alpha_2, \ldots, \alpha_N\} \cap \{\beta_1, \beta_2, \ldots, \beta_N\} = \varnothing$. Show that, if f is a doubly periodic meromorphic function with periods 1 and τ whose zeros within Γ are $\alpha_1, \alpha_2, \ldots, \alpha_N$ and whose poles are $\beta_1, \beta_2, \ldots, \beta_N$ (multiple zeros and poles given multiply), then

$$f(z) = C\prod_{j=1}^{N} \frac{\vartheta(z - \alpha_j - (1+\tau)/2, \tau)}{\vartheta(z - \beta_j - (1+\tau)/2, \tau)}$$

for some constant C. (Compare Questions 38.9 and 38.10.)

(iv) Show by the same method that, given $\alpha_1, \alpha_2, \ldots, \alpha_N, \beta_1, \beta_2, \ldots \beta_N \in \Gamma$ with $\sum_{j=1}\alpha_j - \sum_{j=1}^{N}\beta_j \in \{n + m\tau:\ n, m \in \mathbb{Z}\}$ and $\{\alpha_1, \alpha_2, \ldots, \alpha_N\} \cap \{\beta_1, \beta_2, \ldots \beta_N\} = \varnothing$, then we can find a doubly periodic meromorphic function with periods 1 and τ and zeros$\alpha_1, \alpha_2, \ldots, \alpha_N$ and poles $\beta_1, \beta_2, \ldots, \beta_N$ within Γ.

(v) Suppose g is a doubly periodic meromorphic function with periods τ_1 and τ_2. Show that we can find f, τ and τ_3 with $\operatorname{Im}\tau > 0$ and f a doubly periodic function with periods 1 and τ such that $g(z) = f(\tau_3 z)$.

60.22 (Jacobi's identity) Since I would like this question to be independent of the previous two questions, I start by restating the definition of the theta function

$$\vartheta(z, \tau) = \sum_{n=-\infty}^{\infty} \exp(\pi i n^2\tau + 2\pi i n z).$$

If you have done the previous two questions, you should check that this is identical with our previous definition. If you have not, you should check that $\vartheta(z, \tau)$ is well defined for all $z \in \mathbb{C}$ and all $\tau \in \mathbb{C}$ with $\operatorname{Im}\tau > 0$. Our object is to prove Jacobi's identity

$$\vartheta(z, \tau) = \sqrt{\left(\frac{i}{\tau}\right)} \exp\left(-\frac{\pi i z^2}{\tau}\right) \vartheta\left(\frac{z}{\tau}, -\frac{1}{\tau}\right) \quad [z \in \mathbb{C}, \operatorname{Im}\tau > 0],$$

where we set $\sqrt{(re^{i\theta})} = r^{1/2}e^{i\theta/2}$ for $r > 0, \pi > 0 > -\pi$.

(i) To do this fix $z \in \mathbb{C}$ and τ with $\operatorname{Im} \tau > 0$ and define $f: \mathbb{R} \to \mathbb{C}$ by $f(x) = \exp(\pi i \tau (x+z)^2)$. By verifying that the conditions set out in Question 60.4 (iii) hold, or otherwise, show that we can apply the Poisson summation formula to obtain

$$\sum_{n=-\infty}^{\infty} \exp(\pi i \tau (n+z)^2) = \sum_{n=-\infty}^{\infty} \sqrt{(i/\tau)} \exp(-\pi m^2 i \tau^{-1} + 2\pi i m z).$$

(If you need a hint, look at Lemma 50.2 (i). If you are lazy, prove the result for z real and τ purely imaginary and then use analytic continuation.)

(ii) By simple manipulation obtain Jacobi's identity.

(iii) Setting $z = 0$ and $t = -i\tau\pi$, obtain the particularly elegant formula

$$\sum_{n=-\infty}^{\infty} \exp(-n^2 \pi t) = \sqrt{\left(\frac{\pi}{t}\right)} \sum_{n=-\infty}^{\infty} \exp(-\pi n^2/t)$$

valid for $\operatorname{Re} t > 0$. Setting $\psi(t) = \sum_{n=-\infty}^{\infty} \exp(-n^2 \pi t)$, this gives the so called functional equation of the theta function

$$\psi(t) = \sqrt{(1/t)}\psi(1/t) \quad \text{for } \operatorname{Re} t > 0.$$

(iv) Give a rough estimate, with reasons, of the number of terms required to compute $\psi(t)$ to five significant figures when $t = 10^2$ and when $t = 10^{-1}$, using each of the expressions

$$\psi(t) = 1 + 2\sum_{n=1}^{\infty} \exp(-n^2 \pi t) \text{ and } \psi(t) = t^{-1/2}\left(1 + 2\sum_{n=1}^{\infty} \exp(-n^2 \pi t^{-1})\right).$$

If t is real and strictly positive, explain which expression you would use to compute $\psi(t)$ and show that, in every case, summing up to and including $n = 4$ gives substantially better than 30-figure accuracy.

(The functional equation of the theta function is not just an isolated formula. It appears in Riemann's celebrated derivation of the functional equation for the zeta function, which we give in Question 108.6 (iii). It can be used to derive the equally celebrated Gauss sum formula, which we discuss (though with different proofs) in Questions 97.2 to 97.5. For more information the reader can profitably consult Bellman's *A Brief Introduction to Theta Functions*. In all Bellman's books there are moments when the author forgets that the reader is unlikely to be as mathematically gifted or well informed as himself, but they remain a marvellous source of inspiration rather than organised knowledge.)

60.23 (A Wirtinger type inequality. The results of Questions 60.23 to 60.25 are required for Question 60.26.)

(i) Suppose that $u: [0, \pi/2] \to \mathbb{R}$ is once continuously differentiable with $f(0) = 0$. Show that $u(x) \cot x \to 0$ as $x \to 0$ and that

$$\int_0^{\pi/2} (u'(x)^2 - u(x))^2)\,dx = \int_0^{\pi/2} (u'(x)^2 - u(x)\cot x)^2\,dx.$$

Deduce that

$$\int_0^{\pi/2} (u(t))^2\,dt \leqslant \int_0^{\pi/2} (u'(t))^2\,dt$$

with equality if and only if $u(t) = C \sin t$ for some $C \in \mathbb{R}$.
(This proof comes from Hardy, Littlewood and Pólya's *Inequalities* and is suggested by another variational proof (see Sections 7.6 and 7.7 of that book). There are alternative treatments in Questions 36.1 and 39.6 (ii) which the reader may find more transparent.)

(ii) If $f: \mathbb{R} \to \mathbb{R}$ is once continuously differentiable, $a \leqslant b$, and either $f(a) = 0$ or $f(b) = 0$, show that

$$\int_a^b (f(t))^2\,dt \leqslant \frac{4(b-a)^2}{\pi^2} \int_a^b (f'(t))^2\,dt.$$

(iii) If $f: \mathbb{R} \to \mathbb{R}$ is once continuously differentiable, $a \leqslant b$, and either $f(a) = 0$ or $f(b) = 0$ show that

$$\int_a^b (f(t))^2\,dt \leqslant \frac{4(b-a)^2}{\pi^2} \int_a^b (f'(t))^2\,dt.$$

(iv) If $g: \mathbb{R} \to \mathbb{C}$ is once continuously differentiable, $a \leqslant c \leqslant b$ and $g(c) = 0$, show that

$$\int_a^b |f(t)|^2\,dt \leqslant \frac{4(b-a)^2}{\pi^2} \int_a^b |f'(t)|^2\,dt.$$

60.24 (Bernstein's inequality for band limited functions)

(i) By using the Parseval equality of Question 60.7(i) show that, if f, $\hat{f} \in L^1 \cap C$ then

$$\int_{-\infty}^{\infty} |f(t)|^2\,dt = \frac{1}{2\pi} \int_{-\infty}^{\infty} |\hat{f}(\lambda)|^2\,d|\lambda.$$

(This equality is also referred to as Parseval's equality.)

(ii) Suppose that F, $\hat{F} \in L^1 \cap C$ and that $\hat{F}(\lambda) = 0$ for $|\lambda| > \pi$. By differentiating under the integral sign in the inversion formula (you must justify this), or otherwise, show that F is differentiable and find a formula for F' in terms of $\hat{F}$. By using (i) show that

$$\int_{-\infty}^{\infty} |F'(t)|^2\,dt = \frac{1}{2\pi} \int_{-\pi}^{\pi} |\lambda \hat{F}(\lambda)|^2\,d\lambda.$$

Deduce (possibly using Parseval's equality again) that

$$\int_{-\infty}^{\infty} |F'(t)|^2\,dt \leqslant \pi^2 \int_{-\pi}^{\pi} |F(t)|^2\,dt.$$

(This is a much easier inequality than the Bernstein inequality for trigonometric polynomials which deals with the uniform and not the mean square norm.)

(iii) Show that the inequality just obtained is best possible.

(If you need a hint look at Question 60.15(iii).)

60.25$^{(-)}$ (The mean square triangle inequality)

(i) Suppose $f,\ g\colon \mathbb{R}\to\mathbb{C}$ are continuous. Explain why

$$\left(\int_{-X}^{X} |f(x)+g(x)|^2\,dx\right)^{1/2} \leqslant \left(\int_{-X}^{X} |f(x)|^2\,dx\right)^{1/2} + \left(\int_{-X}^{X} |g(x)|^2\,dx\right)$$

and deduce that if $\int_{-\infty}^{\infty}|f(x)|^2\,dx$ and $\int_{-\infty}^{\infty}|g(x)|^2\,dx$ converge then so does $\int_{-\infty}^{\infty}|g(x)-f(x)|^2\,dx$ and that

$$\left(\int_{-\infty}^{\infty} |f(x)+g(x)|^2\,dx\right) \leqslant \left(\int_{-\infty}^{\infty} |f(x)|^2\,dx\right)^{1/2} + \left(\int_{-\infty}^{\infty} |g(x)|^2\,dx\right)^{1/2}.$$

(ii) Conclude that, if we write $L^2\cap C$ for the set of continuous functions $f\colon \mathbb{R}\to\mathbb{C}$ such that $\int_{-\infty}^{\infty}|f(x)|^2\,dx$ converges, and set

$$\|f\|_2 = \left(\int_{-\infty}^{\infty} |f(x)|^2\,dx\right)^{1/2}$$

for all $f\in L^2\cap C$ then $(L^2\cap C, \|\ \|_2)$ is a normed vector space.

60.26 (Irregular sampling) The results of Question 60.17 look complete but from a practical point of view they are not. Consider a well-behaved function $F\colon \mathbb{R}\to\mathbb{C}$ with $\hat{F}(\lambda)=0$ for $|\lambda|>\pi$. In Question 60.17 we saw how to recover F from the values of $F(n)$ $[n\in\mathbb{Z}]$. However, we also saw, in Question 60.16 (iii) that if we only know $F(n)$ for $n\in\mathbb{Z}$, $n\neq 0$ (i.e. if one sample point is missing) then F ceases to be uniquely determined and so cannot be reconstructed.

In the real world we must expect the occasional sample point to go missing and we try to get round the problem by sampling more closely. (In compact discs this 'oversampling' corresponds, roughly speaking, to evaluating $F(n/4)$ for all $n\in\mathbb{Z}$.)

(i) Estimate the number of sample points required for a compact disc of 1 hour duration.

There still remains the problem of reconstructing F from an oversampled system with some points missing. More formally suppose we have a sequence of points x_j with and a δ with $1>\delta>0$ such that

$$0 < x_j - x_{j-1} \leqslant \delta$$

for all j and $x_{-j} \to -\infty$, $x_j \to \infty$ as $j \to \infty$, and suppose we know $F(x_j)$ for each j. How can we find F? It seems clear that we cannot hope to find a closed formula along the lines of the Whittaker–Shannon formula of Question 60.17. However it is possible to find an algorithm which (given sufficient computing power) will do the trick.

(The method that follows is due to Gröchenig (see bibliography). I had the pleasure of listening to the author's exposition at one of the NATO conferences organised by Jim Byrnes to introduce pure Fourier analysts to engineering Fourier analysts in surroundings so congenial as to cause both sides to drop their guard.) From the point of view of this book it has the disadvantage that, even if we start with well-behaved continuous functions, the method produces discontinuous functions (though of a rather simple kind; they are 'piecewise well behaved'). The reader should assume without proof that all the equalities and inequalities obtained in Questions 60.23 to 60.25 go over without problems. At the end she can, if she wishes, check (by e.g. approximating by well-behaved functions) that all the inequalities used work in the slightly extended form required but this is not the point of the exercise.

Gröchenig begins by introducing a map P given by

$$(Pf)(t) = \frac{1}{2\pi}\int_{-\pi}^{\pi} \hat{f}(\lambda)\exp(i\lambda t)\,dt$$

(ii) Show that, assuming everything is well behaved, P is linear, $\widehat{Pf}(\lambda) = 0$ for $|\lambda| > \pi$, $\widehat{Pf}(\lambda) = \hat{f}(\lambda)$ for $|\lambda| < \pi$ (so that, in some sense, Pf represents the effect of 'band limiting' f in the crudest possible way). Show also that

$$\|Pf\|_2 \leqslant \|f\|_2.$$

(Here and in the development that follows the required results will be found somewhere in Questions 60.23 to 60.25.)

Next he considers a rather crude attempt at reconstruction of the function F of the first paragraph from its sample values at the x_j considered in the paragraph following (i). Let $y_j = (x_{j+1} + x_j)/2$ and set $\chi_j(t) = 1$ if $y_{j-1} \leqslant t < y_j$, $\chi_j(t) = 0$ otherwise. We define Q by

$$(QF)(t) = \sum_{j=-\infty}^{\infty} F(x_j)\chi_j(t).$$

(iii) Show that Q is linear.

Finally he sets $AF = PQG$.

(iv) Explain why A is linear and $\widehat{AF}(\lambda) = 0$ for $|\lambda| > \pi$.

We develop the properties of A in the next question.

60.27 (Continuation of 60.26)

(i) On the assumption that everything is well behaved justify each step in the following calculation.

$$\begin{aligned}
\|F - AF\|_2^2 &= \|P(F - QF)\|_2^2 \\
&\leqslant \|F - QF\|_2^2 \\
&= \int_{-\infty}^{\infty} \left| \sum_{j=-\infty}^{\infty} (F(t) - F(x_j))\chi_j(t) \right|^2 dt \\
&= \sum_{j=-\infty}^{\infty} \int_{y_{j-1}}^{y_j} |(F(t) - F(x_j))\chi_j(t)|^2 \, dt \\
&\leqslant \sum_{j=-\infty}^{\infty} (y_j - y_{j-1})^2 \frac{4}{\pi^2} \int_{y_{j-1}}^{y_j} |F'(t)|^2 \, dt \\
&\leqslant \frac{\delta^2}{\pi^2} \sum_{j=-\infty}^{\infty} \int_{y_{j-1}}^{y_j} |F'(t)|^2 \, dt \\
&= \frac{\delta^2}{\pi^2} \|F'\|_2^2 \\
&\leqslant \delta^2 \|F\|_2^2.
\end{aligned}$$

(ii) To the uninitiated the result of (i) might appear a little disappointing, but anyone with any experience of Banach algebras or approximation methods will see that it is precisely what is required. Set $Bf = f - Af$. Explain why, if f is well behaved and $\hat{f}(\lambda) = 0$ for $|\lambda| > \pi$, then $\widehat{Bf}(\lambda) = 0$ for $|\lambda| > \pi$ and why $\|Bf\|_2 \leqslant \delta\|f\|$. Show that $\|B^n f\|_2 \leqslant \delta^n \|f\|$ and that (writing $B^0 F = F$)

$$\left\| F - \sum_{n=0}^{N} B^n AF \right\|_2 = \|B^{N+1} F\|_2 \leqslant \delta^{N+1} \|F\|_2.$$

Thus, writing $C_N F = \sum_{n=0}^{N} B^n AF$ we see that $\|C_N F - F\|_2 \to 0$ geometrically fast as $N \to \infty$.

It might be objected that convergence in mean square is not the same as pointwise convergence. This is true but for practical purposes mean square convergence (which corresponds to convergence in signal energy) is what we seek. Gröchnig's algorithm seems to me mathematically very elegant and it is certainly practicable. However it is only one of a family of competing algorithms (for details see his paper cited above) and it remains to see which one will win, or, indeed, if the march of technology will enable engineers to do without any of them.

63

A SECOND APPROACH TO HEAT FLOW IN A SEMI-INFINITE ROD

63.1$^{(-)}$ Comment on the following argument. Let $C(1)$ be the straight line contour from $-i-1$ to $-i+1$ and $C(2)$ the straight line contour from $i-1$ to $i+1$. If $f(z)=1/z$, then the linear change of variable $w=z+2i$ (for which $dw/dz=1$) yields

$$\int_{C(1)} f(z)\,dz = \int_{C(2)} f(w)\,dw.$$

Thus $\int_{-1}^{1} \frac{1}{x-i}\,dx = \int_{-1}^{1} \frac{1}{x+i}\,dx$ and so

$$0 = \int_{-1}^{1} \frac{1}{x-i}\,dx - \int_{-1}^{1} \frac{1}{x+i}\,dx = \int_{-1}^{1} \left(\frac{1}{x-i} - \frac{1}{x+i}\right) dx$$

$$= \int_{-1}^{1} \frac{2i}{x^2+1}\,dx = 2i\,[\tan^{-1} x]_{-1}^{1} = \pi i.$$

64

THE WAVE EQUATION

64.1$^{(-)}$ Consider the wave equation $\partial^2\phi/\partial t^2 = c^2\partial\phi/\partial x^2$ with initial conditions $\phi(x,0) = u(x)$, $\partial\phi(x, 0)/\partial t = v(x)$. By considering the relation between u, v and F, G in Theorem 64.1, or otherwise, find the conditions which must be imposed on u, v to give the following:

(i) a disturbance propagating in the positive direction only,

(ii) identical disturbances propagating in the positive and negative directions,

(iii) identical disturbances propagating in the positive and negative directions which have no region of overlap after a time T,

(iv) equal and opposite disturbances in the positive and negative directions (i.e. $\phi(x, t) = -\phi(-x, t)$).

64.2$^{(-)}$ Find the general solution of the equation

$$\frac{\partial^2\phi}{\partial x^2} - c^2\frac{\partial^2\phi}{\partial t^2} = xt + x^2.$$

71

THE CENTRAL LIMIT THEOREM II

71.1 (i) Recall from Question 5.2 that Bernstein associated with each continuous function $f: [0, 1] \to \mathbb{R}$ a sequence of polynomials

$$B_n(f, t) = \sum_{r=0}^{n} f(r/n) \binom{n}{r} t^r (1-t)^{n-r}.$$

Let $f_k(t) = (t - 1/2)^k$ with k a strictly positive integer. Show that

$$B_n(f_k, 1/2) = f_k(1/2) \quad \text{if } k \text{ is odd,}$$

and use de Moivre's theorem (or the general central limit theorem) to show that, when k is even, there exist $A_k > 0$ such that for each $\varepsilon > 0$

$$(A_k + \varepsilon)n^{-k/2} \geqslant |B_n(f_k, 1/2) - f_k(1/2)| \geqslant (A_k - \varepsilon)n^{-k/2},$$

for all sufficiently large n.

(ii) Let M be a positive integer and $\eta > 0$. Show that we can find a $\delta > 0$ and an M times continuously differentiable function $g: [0, 1] \to \mathbb{R}$ such that $|g^{(r)}(t)| < \eta$ for all $t \in [0, 1]$ and all $0 \leqslant r \leqslant M$ but $g(t) = (t - 1/2)^2$ for all $|t - 1/2| < \delta$. Show that

$$n|B_n(g, 1/2) - g(1/2)| \to A_2 \quad \text{for } n \to \infty.$$

(Thus the convergence of Bernstein polynomials may be sure, but it is also slow.)

72
STABILITY AND CONTROL

72.1 This continues Question 21.1 and so starts with a part (iii).

(iii) Show that, if $\theta = 2n\pi + \eta$, with η and $\dot{\eta}$ small, equation $(*)$ becomes approximately

$$\ddot{\eta} = -b\eta - a\dot{\eta}. \qquad (**)$$

Conclude by looking at the solutions of $(**)$ that we expect $\theta(t) \to 2n\pi$.

(iv) What happens if $\theta = (2n+1)\pi + \eta$ with η and $\dot{\eta}$ small? Comment on the relation of parts (ii), (iii) and (iv).

(v) Let $y(t) = \theta(t)$, $x(t) = \dot{\theta}(t)$. Sketch the behaviour of $(x(t), y(t))$ in a typical case when $\dot{\theta}(0)$ is large.

73

INSTABILITY

73.1 Suppose that we wish to solve the equation

$$\dot{\mathbf{x}}(t) + A\mathbf{x}(t) = \mathbf{c}(t)$$

with $\dot{\mathbf{x}}: \mathbb{R} \to \mathbb{R}^n$, A a constant $n \times n$ matrix and $\mathbf{c}: \mathbb{R} \to \mathbb{R}^n$. Formally we might wish to say that

$$\frac{d}{dt}((\exp(At)\mathbf{x}(t)) = \exp(At)\mathbf{c}(t)$$

and so

$$\mathbf{x}(t) = \exp(-At)\left(\mathbf{x}(0) + \int_0^t \exp(As)\mathbf{c}(s)ds\right).$$

If challenged as to what $\exp(At)$ actually meant, we might take refuge in the further formalism

$$\exp(At) = \sum_{k=0}^{\infty} (At)^k/k!.$$

In this question we justify these formal ideas.

(i) If A is an $n \times n$ real matrix with coefficients a_{ij}, let us write $\|A\| = \max_{i,j}|a_{ij}|$. Show that, if A and B are $n \times n$ real matrices and $\lambda \in \mathbb{R}$, we have $\|\lambda A\| = |\lambda|\|A\|$, $\|A + B\| \leqslant \|A\| + \|B\|$ and that $\|A\| = 0$ if and only if $A = 0$. Show that $\|AB\| \leqslant n\|A\|\,\|B\|$.

(ii) Write $E_r(A) = \sum_{k=0}^{r} A^k/k!$. Show that if $s \geqslant r \geqslant 0$

$$\|E_r(A) - E_s(A)\| \leqslant \sum_{k=r+1}^{s} (n\|A\|)^k/k!$$

and, by considering coefficients, deduce that there is a real $n \times n$ matrix $\exp A$ such that $\|E_r(A) - \exp A\| \to 0$. Show that, in fact,

$$\|E_r(A) - \exp A\| \leqslant \sum_{k=r+1}^{\infty} (n\|A\|)^k/k!$$

and conclude that $\|E_r(A) - \exp A\| \to 0$ uniformly as $r \to \infty$ for all $\|A\| \leqslant K$ where K if any fixed constant.

(iii) We say that an $n \times n$ matrix $B(t)$ whose coefficients $b_{ij}(t)$ are functions of t is differentiable if each b_{ij} is. We write $dB(t)/dt$ for the matrix with coefficients $db_{ij}(t)/dt$. Show carefully that, if A is a constant $n \times n$ matrix, then $\exp At$ is differentiable with $d(\exp At)/dt = (\exp At)A$.

(iv) By considering coefficients, show carefully that

$$d(\exp(At)\exp(-At))/dt = 0$$

and deduce that $\exp(At)\exp(-At) = I$. Conclude, in particular, that $\exp A$ is always invertible with $(\exp A)^{-1} = \exp(-A)$.

(v) By considering coefficients, prove that the computations made formally at the beginning of this question are correct and we have indeed

$$\mathbf{x}(t) = \exp(-At)\left(\mathbf{x}(0) + \int_0^t \exp(As)\mathbf{c}(s)ds\right).$$

(vi) Deduce that, if there exist constants K_1, K_2 such that $\max_{1\leqslant i\leqslant n}|c_i(s)| \leqslant K_1\exp(K_2 t)$ for $t \geqslant 0$, then there exist constants K_3, K_4 such that $\max_{1\leqslant i\leqslant n}|x_i(t)| \leqslant K_3\exp(K_4 t)$ for $t \geqslant 0$.

(vii) By observing that $d\mathbf{x}^{(r)}(t)/dt = \mathbf{x}^{(r+1)}(t)$, show that equation $(*)$ on the first page of Chapter 73 may be rewritten as

$$\dot{\mathbf{z}}(t) + A\mathbf{z}(t) = \mathbf{g}(t)$$

with $\mathbf{z}: \mathbb{R} \to \mathbb{R}^{kn}$, A a constant $kn \times kn$ matrix, g a continuous function $\mathbb{R} \to \mathbb{R}^{kn}$ where $x_j(t) = z_{(k-1)n+j}(t)\,[1 \leqslant j \leqslant n]$ and $g_j(t) = c_j(t)\,[1 \leqslant j \leqslant n]$, $g_j(t) = 0$ otherwise. State precisely and prove the statement made about the growth of solutions of $(*)$.

73.2 Objects like $\exp(At)$ form a marvellous tool for advanced investigations. However, we must be careful not to assume that everything we wish to be true is thereby made true. One of Banach's favourite sayings was 'Hope is the mother of fools'.

(i) Show that, if A is a real $n \times n$ matrix with $\|A\| \leqslant n^{-1}$, then

$$\|\exp A - (I + A + A^2/2)\| \leqslant \sum_{k=3}^{\infty} (n\|A\|)^k/k!$$

$$\leqslant 6^{-1}\sum_{k=3}^{\infty} (n\|A\|)^k/4^{k-3} \leqslant (n\|A\|)^3.$$

(ii) Show that, if A, B, C, D are real $n \times n$ matrices, then

$$\begin{aligned}\|AB - CD\| &\leq \|AB - CB\| + \|CB - CD\| \\ &\leq n(\|A - C\|\,\|B\| + \|C\|\,\|B - D\|).\end{aligned}$$

(iii) Show that there is a $\delta_0 > 0$ and a constant K_1 (both depending on A and B) such that

$$\|\exp(tA)\exp(tB) - (I + tA + t^2A^2/2)(I + tB + t^2B^2/2)\| \leq K_1|t|^3$$

for $|t| < \delta_0$. Deduce that there is a constant K_2 (depending on A and B) such that

$$\|\exp(tA)\exp(tB) - (I + t(A + B) + \tfrac{1}{2}t^2(A^2 + 2AB + B^2))\| \leq K_2|t|^3$$

for $|t| < \delta_0$.

(iv) By interchanging the role of A and B, or otherwise, show that there exist $\delta_1 > 0$ and K_3 (depending on A and B) such that

$$\|(\exp(tA)\exp(tB) - \exp(tB)\exp(tA)) - t^2(AB - BA)\| \leq K_3|t|^3$$

for all $|t| < \delta_1$.

(v) Deduce that, if $n \geq 2$, there exist real $n \times n$ matrices C and D with $\exp C \exp D \neq \exp D \exp C$. Is it true that $\exp A \exp B = \exp(A + B)$ for all real $n \times n$ matrices A and B?

(vi) Show that, if A and B are $n \times n$ matrices with differentiable coefficients, then

$$\frac{d}{dx}(AB) = \left(\frac{dA}{dx}\right)B + A\left(\frac{dB}{dx}\right).$$

By giving a proof or counter example, establish whether it is always true that $d(A^2)/dx = 2A\,dA/dx$.

73.3 This question assumes a little more knowledge of matrix algebra than the previous two. All matrices will be $n \times n$ matrices with complex coefficients.

(i) Check briefly that the results on real matrices in the previous two questions apply to complex matrices.

(ii) Let A, B be $n \times n$ matrices which commute (i.e. $AB = BA$). By bounding

$$\left\|\left(\sum_{k=0}^{r} A^k/k!\right)\left(\sum_{k=0}^{r} B^k/k!\right) - \sum_{k=0}^{r}(A + B)^k/k!\right\|,$$

or otherwise, show that $\exp A \exp B = \exp(A + B)$.

(iii) If $A = (a_{ij})$, we write $A^* = (a_{ji}^*)$ with a^* the complex conjugate of a. Show that $(\exp A)^* = \exp A^*$.

(iv) If C is invertible, show that $C(\exp A)C^{-1} = \exp(CAC^{-1})$.

(v) Recall that we say that A is diagonalisable if we can find C invertible such that $CAC^{-1} = D$ a diagonal matrix. Show that, if A is diagonalisable, $\det(\exp A) = e^{\operatorname{tr} A}$ (where $\operatorname{tr} A = \sum_{i=1}^{n} a_{ii}$).

(vi) Recall the (not entirely trivial) theorem that, given any matrix A, we can find a sequence A_n of diagonalisable matrices with $\|A - A_n\| \to 0$. Show that, if A is any $n \times n$ matrix, we have $\det(\exp A) = e^{\operatorname{tr} A}$.

(vii) Show that there is a matrix B with $\exp B = A$ if and only if A is invertible.

(viii) Let $A = \begin{pmatrix} 0 & -1 \\ 1 & 0 \end{pmatrix}$. Show that $A^4 = I$ and so, in particular, A is invertible. Choose B such that $\exp B = A$, and set $C = 4B$. Show that $C \neq C^*$ but $\exp C = I$. Find all diagonal matrices D with $\exp(iD) = I$.

(ix) Recall that we say that U is unitary if $UU^* = I$. Show that, if A is self adjoint (i.e. $A = A^*$), then $\exp iA$ is unitary. Show that, if $\exp(i\delta A)$ is unitary for all real δ with $\delta > 0$, then A is self adjoint. Is it true that, if $\exp(iA)$ is unitary, then A is self adjoint?

(x) Recall the non trivial theorem that, if B is normal (i.e. if $BB^* = B^*B$) so, in particular, if B is unitary, we can find a unitary matrix W with WBW^* diagonal. Show that, if U is unitary, there exists a self adjoint matrix A with $\exp(iA) = U$. What does this result reduce to if $n = 1$? Is A unique?

(xi) We say that U is special unitary if U is unitary and $\det U = 1$. Show that U is special unitary if and only if we can write $U = \exp(iA)$, where A belongs to a set of matrices to be identified.

73.4 (Volterra's predator–prey equations)

(i) Animal populations may vary substantially in size from year to year. Famous examples include the locust, and the pairs of populations whalers and whales, ladybirds and aphids. Volterra sought simple mathematical models which might mimic such behaviour. For example consider a population of $P(t)$ predators (lynx, say) which live off a population of $N(t)$ herbivores (hares, say). Without being too critical, explain why the following pair of equations might give a suitable model

$$\left.\begin{aligned} \dot{N}(t) &= aN(t) - \alpha N(t)P(t) \\ \dot{P}(t) &= -bP(t) + \beta N(t)P(t). \end{aligned}\right\} \qquad (*)$$

where $a, b, \alpha, \beta > 0$. (Remember that animals are born, die, eat and are eaten.)

(ii) Show that there are unique $N_1, P_1 > 0$ such that $N(t) = N_1$, $P(t) = P_1$ represents a solution of $(*)$. Suppose that we seek solutions of the form $N(t) = N_1 + n(t)$, $P(t) = P_1 + p(t)$ with $n(t)$, $p(t)$ small. Find what happens to $(*)$ if we ignore terms of the form $n(t)p(t)$ as being negligibly small. Show that this procedure leads to solutions in which $n(t)$ and $p(t)$ and so $N(t)$ and $P(t)$ oscillate with period $2\pi(ab)^{-1/2}$.

(iii) Suppose a natural (or man made) disaster suddenly reduces the populations to $N(0) = \varepsilon N_1$, $P(0) = \varepsilon P_1$ with ε small. What will happen (at least in the short term)?

(iv) There are, as Volterra knew, biological populations whose size seems to vary periodically, but the result of (ii), although suggestive, is no more than this. We cannot expect to observe small fluctuations in biological populations; it is hard enough to be confident to be confident of the reality of large ones. However, Volterra's equations have a very curious mathematical property. Show, using $(*)$, that, if $N(t), P(t) > 0$, then

$$\beta \dot{N}(t) + \alpha \dot{P}(t) = a\beta N(t) - \alpha b P(t)$$

and

$$b\frac{\dot{N}(t)}{N(t)} + a\frac{\dot{P}(t)}{P(t)} = a\beta N(t) - \alpha b P(t).$$

By eliminating between these equations and integrating, show that

$$N(t)^b P(t)^a \exp(-\beta N(t)) \exp(-\alpha P(t)) = H \qquad (**)$$

where H is a constant.

(v) Show that, if $(N(0), P(0)) \neq (N_1, P_1)$ (i.e. we do not start in the equilibrium position), then $(N(t), P(t))$ will lie on a path Γ_H given by $(**)$. Sketch Γ_H. Show that $\dot{N}(t)^2 + \dot{P}(t)^2 \geqslant \delta_H$ with $\delta_H > 0$ (there is no need to calculate δ_H) and conclude that $(N(t), P(t))$ will move round Γ_H repeating periodically. (A reasonable argument is all that is required. Use only as much rigour as you can easily supply.)

(vi) Sketch Γ_H for various values of H. What is the effect of making $N(0)$ or $P(0)$ small?

(It is clear that the account of the solution of Volterra's equation just given tells us much more about the special structure of $(*)$ than about natural populations. It does, none the less, raise the possibility that biological systems could exhibit large-scale oscillations for a long period of time. Kolmogorov showed that more general models

$$\dot{N}(t) = f(N(t), T(t)), \quad \dot{P}(t) = g(N(t), T(t)),$$

in which f and g are not precisely specified but have certain natural properties one would expect in the context, will also exhibit large-scale oscillation. (For an introduction see Hirsch and Smale's *Differential Equations, Dynamical Systems and Linear Algebra*. Chapter 12 deals with predator and prey systems and Exercise 4 of that chapter gives a version of Kolmogorov's theorem.)

Experiments have produced test tube ecologies in which two species of bacteria form the predator and prey and for which Volterra–Kolmogorov oscillation actually occurs. It should be noted however that, in spite of the apparent generality of Kolmogorov's result, it appears that great care must be taken in setting up the experiment. It is an open question, certainly biologically and, I think, mathematically, whether large-scale oscillations of this type are more or less likely in more complicated systems.

M. L. Zeeman (see bibliography) has completed a very nice investigation of the case of three competing species. Interestingly the Volterra equations now produce oscillations for a range of values.)

73.5 It is quite interesting to see what happens when we try using Fourier transforms to solve the equation

$$y''(t) + 2by'(t) + \lambda_0^2 y(t) = f(t), \qquad (*)$$

where f is a real continuous function with $f(t) = 0$ for $|t| > T$, say. We shall suppose that λ_0 and b are real with $\lambda_0^2 > b^2 > 0$.

(i) Without bothering too much about rigour (but noting that we implicitly assume that $y, y', y'' \in L^1 \cap C$) show, by taking Fourier transforms in (*), that $y = f * g$ with $\hat{g}(\lambda) = (-\lambda^2 + 2ib\lambda + \lambda_0^2)^{-1}$. Conclude that $y = f * g$ with

$$g(t) = \frac{1}{2\pi} \int_{-\infty}^{\infty} \frac{e^{i\lambda t}}{(-\lambda^2 + 2ib\lambda + \lambda_0^2)}\, d\lambda. \qquad (**)$$

(ii) By carefully using contour integration, or otherwise, show that, if g is defined by (**), then

(I) If $b < 0$, then

$$g(t) = 0 \qquad \text{for } t \geqslant 0,$$

$$g(t) = e^{-bt} \frac{\sin (\lambda_0^2 - b^2)^{1/2} t}{(\lambda_0^2 - b^2)^{1/2}} \qquad \text{for } t \leqslant 0.$$

(II) If $b > 0$, then

$$g(t) = e^{-bt} \frac{\sin (\lambda_0^2 - b^2)^{1/2} t}{(\lambda_0^2 - b^2)^{1/2}} \qquad \text{for } t \geqslant 0$$

$$g(t) = 0 \qquad \text{for } t \leqslant 0.$$

Whichever method you use, you must make clear why it gives different answers for $b > 0$ and for $b < 0$. (Here $(\lambda_0^2 - b^2)^{1/2}$ is the positive square root of $\lambda_0^2 - b^2$.)

(iii) Verify directly that, if g is as in (iii), then $f * g$ is a bounded solution of (*). Give a rough sketch of $f * g$ in the two cases $b > 0$ and $b < 0$. Explain why $f * g$ is the only bounded solution of (*).

(iv) Show that we obtain similar results if $b^2 > \lambda_0^2 > 0$

(v) Explain clearly where and how the argument of parts (ii) and (iii) fails when $b = 0$, $\lambda_0^2 > 0$.

(vi) If $b=0$, $\lambda_0^2>0$ does (*) always have a bounded solution? Does (*) always have a solution with $|y(t)|\to 0$ as $|t|\to\infty$. (You may find it helpful to forget about Fourier transforms and think first about solving

$$y''(t)+\lambda_0^2 y(t)=f(t)$$

for $-2T\leqslant t\leqslant 2T$ with initial conditions $y(-2T)=y'(-2T)=0$.)

74

THE LAPLACE TRANSFORM

74.1 (Partial fractions) We work over $\mathbb{C}$.

(i) Show that, if P is a polynomial of degree n, we can write $P(z)/(z-\lambda) = Q(z) + A/(z-\lambda)$ where Q is polynomial of degree $n-1$ and $A = P(\lambda)$.

(ii) Hence, or otherwise, show that if P is a polynomial of degree n, the complex numbers $\lambda(1), \lambda(2), \ldots, \lambda(k)$ are distinct and $n(1), n(2), \ldots, n(k)$ are integers with $n(j) \geqslant 1$ $[1 \leqslant j \leqslant k]$, then we can find a_{jl} $[1 \leqslant l \leqslant n(j), 1 \leqslant j \leqslant k]$ and a polynomial Q, which is zero if $n < \sum_{j=1}^{k} n(j)$ and of degree $n - \sum_{j=1}^{k} n_j$ otherwise, such that

$$\frac{P(z)}{\prod_{j=1}^{k}(z-\lambda(j))^{n(j)}} = Q(z) + \sum_{j=1}^{k}\sum_{l=1}^{n(j)} \frac{\alpha_{jl}}{(z-\lambda(j))^{l}} \qquad (*)$$

(iii) Suppose $\lambda(1), \lambda(2), \ldots, \lambda(k)$ are distinct, $n(1), n(2), \ldots, n(k)$ are integers with $n(j) \geqslant 1$, $a_{jl} \in \mathbb{C}$ $[1 \leqslant l \leqslant n(j), 1 \leqslant j \leqslant k]$ and Q is a polynomial. Write $f(z) = Q(z) + \sum_{j=1}^{k} \sum_{l=1}^{n(j)} \alpha_{jl}(z-\lambda(j))^{-l}$. By examining the behaviour of $(z-\lambda(j))^{r} f(z)$ as $z \to \lambda(j)$ and the behaviour of $f(z)$ as $z \to \infty$, or otherwise, show that, if $f(z) = 0$ for all $z \in \mathbb{C}$, then $a_{jl} = 0$ for all $1 \leqslant l \leqslant n(j)$, $1 \leqslant j \leqslant k$ and $Q = 0$. Hence show that the α_{jl} and Q in part (ii) are uniquely determined by $(*)$.

(iv) Let us return to the situation in (ii). Suppose $n < \sum_{j=1}^{k} n(j)$ so $Q = 0$ and $(*)$ becomes

$$P(z) = \sum_{j=1}^{k}\sum_{l=1}^{n(j)} \alpha_{jl}\left(\prod_{r \neq j}(z-\lambda(r))^{n(r)}\right)(z-\lambda(j))^{n(j)-l}. \qquad (**)$$

By setting $z = \lambda(j)$, obtain

$$\alpha_{jn(j)} = P(\lambda(j))\left(\prod_{r \neq j}(\lambda(j)-\lambda(r))^{n(r)}\right)^{-1}$$

the so-called 'cover up rule'. There are more complicated formulae for $\alpha_{jn(j)-1}$ and so on, but these are rarely used. Here is a procedure for obtaining (*).

(A) By algebraic long division obtain

$$\frac{P(z)}{\prod_{j=1}^{k}(z-\lambda(j))^{n(j)}} = Q(z) + \frac{R(z)}{\prod_{j=1}^{k}(z-\lambda(j))^{n(j)}}$$

where the degree of R is less than $\sum_{j=1}^{k} n(j)$.

(B) We know have

$$\frac{R(z)}{\prod_{j=1}^{k}(z-\lambda(j))^{n(j)}} = \sum_{j=1}^{k}\sum_{l=1}^{n(j)} \frac{\alpha_{jl}}{(z-\lambda(j)))^{l}}$$

so that

$$R(z) = \sum_{j=1}^{k}\sum_{l=1}^{n(j)} \alpha_{jl}\left(\prod_{r\neq j}(z-\lambda(r))^{n(r)}\right)(z-\lambda(j))^{n(j)-l}. \quad (***)$$

(C) Determine the $\alpha_{jn(j)}$ by the cover up rule applied to (***).

(D) Determine the remaining coefficients of powers of z in (***).

(E) Check your working by verifying that (*) holds with your calculated Q by a_{jl} for some simple choice of $z \notin \{\lambda(1), \lambda(2), \ldots, \lambda(k)\}$.

(If the reader ever needs to take an advanced English school leaving exam from the 1960s, there are several further little tricks she should know, but otherwise the procedure should be adequate for most circumstances.)

(v) Prove the so-called Heaviside expansion theorem. Suppose P, Q are polynomials with the degree of P less than Q and suppose the zeros $\lambda(1)$, $\lambda(2)$, $\ldots$, $\lambda(k)$ of Q are distinct. Then if $u \in \mathscr{E}$ and $(\mathscr{L}u)(z) = P(z)/Q(z)$ for $\operatorname{Re} z > a$, it follows that

$$u(t) = \sum_{j=1}^{k} (P(\lambda(j))/Q'(\lambda(j)))e^{\lambda(j)t} \text{ for } t \geq 0.$$

(vi) By integrating an equation of the form

$$\frac{1}{1+x^3} = \frac{A}{1+x} + \frac{a+bx}{1-x+x^2},$$

or otherwise, show that

$$1 - \frac{1}{4.8} + \frac{1}{7.8^2} - \frac{1}{10.8^3} + \ldots = \frac{1}{3}(\log 3 + \pi/\sqrt{3}).$$

74.2 (i) Let $\omega = \exp 2\pi i/N$. Show that $\Pi_{k=1}^{N-1}(x - \omega^k) = \sum_{r=0}^{N-1} x^r$.
(ii) Find the partial fraction expansion of $(1 - x^N)^{-1}$.
(iii) If $a \in \mathbb{C}$ and $a \neq 0$, show that

$$1/a + 1/a^* = 2\,\mathrm{Re}\, a/|a|^2.$$

(iv) Show that there exists a constant $A > 0$ such that if $\pi \geq \theta \geq 0$, then $|re^{i\theta} - 1| \geq A\theta$ for all real r.
(v) Using (ii) and allowing $N \to \infty$, show carefully that

$$\frac{1}{e^s - 1} = \frac{1}{s} + \sum_{n=1}^{\infty} \frac{s}{s^2 + 4n^2\pi^2} \quad \text{for all } s \neq 0.$$

(Remark. To show that $\sum_{n=1}^{\infty} a(n, N) \to \sum_{n=1}^{\infty} \alpha(n)$ as $N \to \infty$ we can sometimes proceed by showing (i) that, given $\varepsilon > 0$, there exists an M with $\sum_{n=M}^{\infty} |a(n, N)| < \varepsilon$ for all N and (ii) that $\alpha(n, N) \to \alpha(n)$ as $N \to \infty$.)

74.3 (Convolution)
(i) Suppose $f \in \mathscr{E}_a$, $g \in \mathscr{E}_b$ and f and g are continuous. Show that, if $c > \max(a, b)$, we have $f * g \in \mathscr{E}_c$ and, by modifying the proof of Theorem 51.7 (iv), or otherwise, show that

$$\mathscr{L}(f * g)(z) = (\mathscr{L}f)(z)(\mathscr{L}g)(z) \quad \text{for all } \mathrm{Re}\, z > \max(a, b).$$

(ii) Let $e_{\lambda,b}(t) = (T_b e_\lambda)(t)H(t)$ in the notation of Lemma 74.6. Show that, if $f \in \mathscr{E}_a$, then we can write

$$f = F + \sum_{i=1}^{n} \mu(i) e_{\lambda,b(i)}(t),$$

where $F \in \mathscr{E}_a$, F is continuous, $\mu(i) \in \mathbb{C}$, $b(i) \geq 0$ $[i = 1, \ldots, n]$ and λ is real with $\lambda < a$.
(iii) Suppose $f \in \mathscr{E}_a$ and f is continuous. Show that, if λ is real and $\lambda < a$, $b \leq 0$, then $e_{\lambda,b} * f$ is a continuous function, $e_{\lambda,b} * f \in \mathscr{E}_c$ for all $c > a$ and (by direct computation, or otherwise) show that

$$\mathscr{L}(e_{\lambda,b} * f)(z) = (\mathscr{L}e_{\lambda,b})(z)(\mathscr{L}f)(z) \quad \text{for all } \mathrm{Re}\, z > a.$$

Show also that, if $b(1), b(2) \geq 0$, then $e_{\lambda,b(1)} * e_{\lambda,b(2)}$ is a continuous function, $e_{\lambda,b(1)} * e_{\lambda,b(2)} \subset \mathscr{E}_a$ and

$$\mathscr{L}(e_{\lambda(1),b} * e_{\lambda(2),b})(z) = (\mathscr{L}e_{\lambda(1),b})(z)(\mathscr{L}e_{\lambda(2),b})(z) \quad \text{for all } \mathrm{Re}\, z > a.$$

(iv) Combine the results above to show that, if $f, g \in \mathscr{E}_a$, then $f * g \in \mathscr{E}_c$ for all $c > a$, $f * g$ is continuous and

$$\mathscr{L}(f * g)(z) = (\mathscr{L}f)(z)(\mathscr{L}g)(z) \quad \text{for all } \mathrm{Re}\, z > a.$$

(v) Solve the equation

$$f(x) = xe^{-x} + \int_0^x (x-u)e^{-(x-u)} f(u)\,du \quad [x \geqslant 0]$$

where $f \in \mathscr{E}$.

(vi) Show that if $ab + b^2 > 0$, the solution of the integral equation

$$f(t) + a\int_0^t f(s)\sin b(t-s)\,ds = g(t) \ [t \geqslant 0],$$

where $f, g \in \mathscr{E}$ and $a,b \in \mathbb{R}$ is

$$f(t) = g(t) - \frac{ab}{c}\int_0^t g(s)\sin c(t-s)\,ds,$$

where c is the positive square root of $(ab + b^2)$. Obtain the corresponding formulae for $ab + b^2 < 0$ and $ab + b^2 = 0$.

74.4$^{(-)}$ (i) (The initial value formula) If $f \in \mathscr{E}$, show that $z\mathscr{L}f(z) \to f(0+)$ as $\operatorname{Re} z \to \infty$.

(ii) If $f \in \mathscr{E}$ for all $a > 0$ and $\int_0^\infty |f(t)|\,dt < \infty$, show that $\mathscr{L}f(x) \to \int_0^\infty f(t)\,dt$ as $x \to 0$ through positive real values of x.

(iii) (The final value formula) If

$$f, f' \in \mathscr{E}_a \text{ for all } a > 0 \text{ and } \int_0^\infty |f'(t)|\,dt < \infty,$$

show that $\lim_{t\to\infty} f(t)$ exists and that $x\mathscr{L}f(x) \to \lim_{t\to\infty} f(t)$ as $x \to 0$ through positive real values of x.

(iv) Which of the two results (i) and (iii) do you think is likely to be more useful in practice, and why?

74.5 Laplace transforms can be used in much the same way as we used Fourier transforms in Chapter 55 to solve partial differential equations. In this question I suggest that the reader concentrates on using the Laplace transform as a exploratory tool and does not bother about rigour.

(i) Consider the heat equation

$$\frac{\partial \phi}{\partial t}(x,t) = K\frac{\partial^2 \phi}{\partial x^2}(x,t) \text{ for } x,t > 0 \qquad (*)$$

subject to the boundary conditions

$$\phi(x,t) \to 0 \text{ as } t \to 0+ \text{ for all } x \to 0,$$

$$\phi(x,t) \to 1 \text{ as } x \to 0+ \text{ for all } t > 0,$$

$$\phi(x,t) \to 0 \text{ as } x \to \infty \text{ for all } t > 0.$$

Verify that this problem is solved by taking $\phi = \phi_0$ with $\phi_0(x, t) = E(x/(2Kt)^{1/2})$ where $E(s) = 2/(2\pi)^{1/2}\int_s^\infty \exp(-u^2/2)\,du$.

(ii) By taking the Laplace transform of equation (*) with respect to t find $\mathcal{L}_2\phi_0(x,z)$ the Laplace transform of $\phi_0(x, \ )$. (Since you are not being rigorous, you may assume, without proof, that $\mathcal{L}_2\phi_0(x,z)$ behaves well as $x \to \infty$.)

(iii) Now consider the heat equation

$$\frac{\partial \phi}{\partial t}(x,t) = K\frac{\partial^2 \phi}{\partial x^2}(x,t) \text{ for } R > x > -R,\ t > 0$$

subject to the boundary conditions

$$\phi(x,t) \to 1 \text{ as } t \to 0+ \text{ for all } R > x > -R,$$

$$\phi(x,t) \to 0 \text{ as } x \to R- \text{ for all } t > 0,$$

$$\phi(x,t) \to 0 \text{ as } x \to -R+ \text{ for all } t > 0,$$

Show that

$$\mathcal{L}_2\phi(x,z) = \frac{1}{z} - \frac{\cosh(x(z/K)^{1/2})}{z\cosh(R(z/K)^{1/2})}$$

for $\operatorname{Re} z > 0$. (Comment briefly on how you define $(z/K)^{1/2}$.) By expanding the right hand side deduce that

$$\phi(x,t) = 1 + \sum_{n=1}^{\infty}(-1)^n\left\{E\left(\frac{(2n-1)R - x}{(2Kt)^{1/2}}\right) + E\left(\frac{(2n-1)R + x}{(2Kt)^{1/2}}\right)\right\}.$$

(If you have done Question 57.3, you should check that this solution is, after trivial modifications, the same as that obtained in that question.)

74.6 (A Theorem of I. Schur on making change) It is hard to find questions which test mathematical ability independent of mathematical knowledge. One question which, in my opinion, comes close to being such a test is the following.

(i) In how many ways can you make up 50 pence using only 1 pence, 2 pence, 5 pence and 10 pence pieces? (The reader may replace the British pence by her favourite currency: cents, francs or whatever.)

Here is another related problem.

(ii) In the game of rugby you can score 3 or 5 points at a time. Which positive integers can not occur as total scores? Prove your result. (Non-British readers are informed that rugby is a ruffians' game played by gentlemen whilst soccer is a gentlemen's game played by ruffians.)

Schur showed that, given an endless supply of coins in integer denominations $a_1, a_2, \ldots, a_m$ such that $a_1, a_2, \ldots, a_m$ have highest common factor 1, there exists an integer N such that every sum of money $n \geqslant N$ can be made up using only those coins. You may wish to try your hand at proving this before examining the proof sketched in the next question.

74.7 (A proof of Schur's theorem) The reader may have sensed in Question 74.1 a certain lack of enthusiasm for partial fractions on my part. Since writing out that question I have looked at the book *Generating Functionology* by H. S. Wilf and have substantially modified my attitude. Here is a beautiful proof of Schur's theorem taken from that book which uses partial fractions in an essential manner. We use the notation of the last paragraph of Question 74.6 and take $z \in \mathbb{C}$ with $|z| < 1$ throughout.

(i) Suppose that $c(r)$ is the number of ways of making up r pence using coins of the given denominations. By comparing coefficients of z^r show that (if $|z| < 1$)

$$\sum_{r=0}^{\infty} c(r)z^r = \prod_{j=1}^{m}\left(\sum_{k=0}^{\infty} z^{ka_j}\right).$$

Deduce that, if we write $H(z) = \sum_{r=0}^{\infty} c(r)z^r$, then

$$H(z) = \prod_{j=1}^{m}(1 - z^{a_j})^{-1}.$$

(ii) Show that

$$H(z) = \prod_{l=0}^{L}(\lambda(l) - z)^{-n(l)}$$

where the $\lambda(l)$ are distinct roots of unity and the $n(l)$ are strictly positive integers. (You should not attempt to find these quantities explicity.) Show that one of the $\lambda(l)$ takes the value 1. For definiteness take $\lambda(0) = 1$. Show that $n(0) = m > n(l)$ for all $1 \leq l \leq L$. (Remember that the denominations are coprime.)

(iii) Use Question 74.1 to show that there exist $\alpha_{lk} \in \mathbb{C}$ such that

$$H(z) = \sum_{l=0}^{L}\sum_{k=1}^{n(l)} \frac{\alpha_{lk}}{(\lambda(l) - z)^k}.$$

Explain why $(1 - z^{a_j})/(1 - z) \to a_j$ as $z \to 1$ and, by considering the behaviour of $H(z)(1 - z)^m$ as $z \to 1$, show that

$$\alpha_{0m} = (a_1 a_2 \ldots a_m)^{-1}.$$

(iv) Show that

$$\frac{1}{(\lambda(l) - z)^k} = \frac{1}{\lambda(l)^k}\sum_{r=0}^{\infty}\binom{k + r - 1}{r}\lambda(l)^{-r}z^r,$$

and use part (iii) to obtain an expression for $c(r)$ in terms of r, the $\lambda(l)$ and the α_{lk}.

(v) Show, using the expression just obtained, that

$$c(r)r^{1-m} \to \frac{\alpha_{0m}}{(m-1)!}$$

as $r \to \infty$.

(vi) Estimate the number of ways that you can make up 5 000 000 pence using only 1 pence, 2 pence, 5 pence, 10 pence, 20 pence, 50 pence and 100 pence pieces.

(vii) Explain why (v) proves Schur's theorem.

75

DEEPER PROPERTIES

75.1 An immediate use for Theorem 75.3 (iv) occurs if we develop complex variable theory leaving the introduction of power series to last.

(i) Recall the definition of the radius of convergence.

(ii) By using $\sum_{n=0}^{\infty} z^n$ as an example, show that a power series may not coverge uniformly within its radius of convergence.

(iii) Show however that, if $\sum_{n=0}^{\infty} a_n z^n$ has radius of convergence R and $\rho < R$, then $\sum_{n=0}^{N} a_n z^n$ converges uniformly to $\sum_{n=0}^{\infty} a_n z^n$ for $|z| \leqslant \rho$.

(iv) Conclude that $\sum_{n=0}^{\infty} a_n z^n$ is analytic on $\{z: |z| < R\}$ and find its derivative.

(Compare Question 48.2.)

75.2 (Uniqueness via complex variable)

(i) Suppose $f: \mathbb{R} \to \mathbb{C}$ is Riemann integrable with $\int_{-\infty}^{\infty} |f(x)|\, dx < \infty$. Show, in the manner of our proof of Theorem 75.2, that, if $t \in \mathbb{R}$ is fixed, the formula $F(z) = \int_{-\infty}^{\infty} e^{-ixz} f(x)\, dx$ defines a function F: $(z \in \mathbb{C}: \operatorname{Im} z \geqslant 0\} \to \mathbb{C}$ such that F is continuous on $\{z \in \mathbb{C}: \operatorname{Im} z \geqslant 0\}$ and analytic on $\{z \in \mathbb{C}: \operatorname{Re} z > 0\}$. In the same way, observe that $G(z) = \int_{t}^{\infty} e^{-ixz} f(x)\, dx$ defines a continuous function on $\{z \in \mathbb{C}: \operatorname{Im} z \leqslant 0\}$ which is analytic on $\{z \in \mathbb{C}: \operatorname{Im} z < 0\}$.

(ii) Suppose now that, in addition, $\hat{f}(\lambda) = 0$ for all $\lambda \in \mathbb{R}$. Show that if we define $H: \mathbb{C} \to \mathbb{C}$ by $H(z) = F(z)$ for $\operatorname{Im} z \geqslant 0$, $H(z) = G(z)$ for $\operatorname{Im} z < 0$, then H is continuous. Now show successively that, by Cauchy's theorem,

(a) $\int_{\Delta} H(z)\, dz = 0$ for any triangular contour Δ in $\{z: \operatorname{Im} z > 0\}$ so, by continuity,

(b) $\int_{\Delta} H(z)\, dz = 0$ for any triangular contour Δ in $\{z: \operatorname{Im} z \geqslant 0\}$, and

(b)′ $\int_{\Delta} H(z)\, dz\, 0$ for any triangular contour Δ in $\{z: \operatorname{Im} z \leqslant 0\}$ whence, by a decomposition argument,

(c) $\int_{\Delta} H(z)\, dx = 0$ for any triangular contour Δ.

Now recall that Morera's theorem (Theorem 75.3 (iii)) remains true, even if we restrict ourselves to triangular contours, and deduce that H is everywhere analytic. (You may recognise an argument used to establish the reflection principle for analytic continuation.)

(iii) Use Liouville's theorem to show that H is constant and then, by considering the behaviour of $H(iy)$ with y real as $y \to \infty$, show that $H = 0$. Conclude that $\int_{-\infty}^{t} e^{-ixz} f(x)\, dx = 0$ for all $t \in \mathbb{R}$.

(iv) (a) Show that, if f is continuous, $f = 0$.

(b) Is it true for general f satisfying the hypotheses of (i) and (ii) that $f(t) = 0$ for all t? Give a proof or a counter example.

(c) Show that, if f satisfies the hypotheses of (i) and (ii), then $f(t) = 0$ at all points t where f is continuous. (This is a stronger version of Theorem 61.5.) Hence obtain Theorem 74.5.

75.3 (Euler's formula (see Questions 18.7, 18.8 and 60.5) via complex variable)

(i) Suppose $|z| < R < 2R < N$. Show that $|2z/(z^2 - n^2)| \leq 4Rn^{-2}$ for all $n > N$. Conclude that $\sum_{n=N}^{\infty} 2z/(z^2 - n^2)$ converges uniformly on $|z| < R$. Hence show that $z^{-1} + \sum_{n=1}^{\infty} 2z/(z^2 - n^2)$ defines a function analytic on $\{z: |z| < R,\ z \notin \mathbb{Z}\}$ with simple poles at points m with m integral and $|m| < R$. What are the residues at these points? By allowing R to vary, show that $f(z) = z^{-1} + \sum_{n>0} 2z/(z^2 - n^2)$ defines a meromorphic function on $\mathbb{C}$ with poles at integer points.

(ii) For fixed non integral z show that $f(z) = \lim_{N\to\infty} \sum_{n=-N}^{N} (z - n)^{-1}$ and deduce that $f(z + 1) = f(z)$ i.e. f is periodic with period 1.

(iii) Show that $g(z) = \pi \cot \pi z$ defines a meromorphic function on $\mathbb{C}$ with poles at integer points. What are the residues at these points? Show also that g is periodic with period 1.

(iv) Use the information obtained so far to show that $f - g$ is a meromorphic function on $\mathbb{C}$ with removable singularities at the integer points. Hence conclude that $f - g$ can be extended to an analytic function h which is periodic with period 1.

(v) Show that $\sum_{n=1}^{\infty} a/(a^2 + n^2) \leq \int_0^{\infty} a/(a^2 + x^2)\, dx = \pi/2$. By obtaining an upper bound for $|2z/(z^2 - n^2)|$, or otherwise, show that $|f(z)|$ is uniformly bounded in the region $|\operatorname{Re} z| \leq 1/2$, $|\operatorname{Im} z| \geq 10$. Show also that $g(z)$ is uniformly bounded in the same region.

(vi) Hence, using the idea of the last paragraph of the proof of Theorem 38.1, or otherwise, show that h is uniformly bounded everywhere and conclude, using Liouville's theorem, that h is constant. By examining the behaviour of $h(z)$ as $z \to 0$, show that $h(0) = 0$ and so $h = 0$ i.e. $f(z) = g(z)$ for all $z \notin \mathbb{Z}$.

(vii) By expanding $2z/(z^2 - n^2)$ in powers of z and interchanging sums (you must justify both operations) show that

$$f(z) = z^{-1} - 2\sum_{m=1}^{\infty}\left(\sum_{n=1}^{\infty} n^{-2m}\right) z^{2m-1} \text{ for } |z| < 1.$$

By writing $g(z) = \pi i(1 + 2(e^{2\pi iz} - 1)^{-1})$, expanding in powers and equating coefficients, deduce that

$$\sum_{m=1}^{\infty} n^{-2m} = (-1)^{m+1}2^{2m-1}b_{2m-1}\pi^{2m}/(2m-1)!$$

where the b_j are defined by the formula

$$(e^y - 1)^{-1} = y^{-1} - 2^{-1} + \sum_{n=1}^{\infty} b_n y^n/n! \quad \text{for } y \notin 0.$$

75.4 Example 75.8 is not quite conclusive. However, only a little sharpening is required to produce a complete result.

(i) Let $S = \{iy: y > 0, y \text{ rational}\}$. Since S is countable, we can write $S = \{z(1), z(2), \ldots\}$ with $z(1) = iy(1), z(2) = iy(2), \ldots$ distinct. Let $\varepsilon(1) = 1$. Show that we can define $x(1), x(2), \ldots$ and $\varepsilon(2), \varepsilon(3), \ldots$ inductively in such a way that $2^{-j} \geqslant x(j) > 0$, $2^{-j+1} \geqslant \varepsilon(j) > 0$ for all $j \geqslant 1$ and writing $w(j) = x(j) + iy(j)$ we have

(a) $$\left|\sum_{j=1}^{n} \frac{\varepsilon(j)}{2i}\left(\frac{1}{w(n) - z(j)} - \frac{1}{w(n) - z(j)^*}\right)\right| \geqslant 2^n,$$

(b) $$\left|\frac{\varepsilon(n)}{2i}\left(\frac{1}{w(j) - z(j)} - \frac{1}{w(j) - z(n)^*}\right)\right| \geqslant 2^{-n} \quad \text{for } 1 \leqslant j \leqslant n-1$$

(ii) Show that $\sum_{j=1}^{\infty}\varepsilon(j)\sin y(j)t$ converges uniformly on $\mathbb{R}$ to a continuous $F \in \mathscr{E}_0$ with

$$\mathscr{L}F(z) = \sum_{j=1}^{\infty} \frac{\varepsilon(j)}{2i}\left(\frac{1}{z - z(j)} - \frac{1}{z - z(j)^*}\right)$$

for all $\operatorname{Re} z > 0$.

(iii) Show that $|\mathscr{L}F(w(j))| \to \infty$ and $|\mathscr{L}F(w(j)^*)| \to \infty$ as $j \to \infty$. Show also that, given any $y \in \mathbb{R}$ with $y \geqslant 0$, we can find $j(n) \to \infty$ such that $w(j(n)) \to iy$. Conclude that, given any $y \in \mathbb{R}$ we can find $a(j) \in \mathbb{C}$ with $\operatorname{Re} a(j) > 0$, $a(j) \to iy$ and $\mathscr{L}F(a(j)) \to \infty$ as $j \to \infty$.

(iv) Show that if $y \in \mathbb{R}$, $\delta > 0$, there does not exist a meromorphic function g defined on the disc $\{z: |z - y| < \delta\}$ such that $\mathscr{L}F(z) = g(z)$ for $|z - y| < \delta$, $\operatorname{Re} z > 0$.

75.5 In the next few questions we forget about Laplace transforms and concentrate on the problem of extending analytic functions. If $D = \{z: |z| < 1\}$ and $f: D \to \mathbb{C}$ is analytic, we say that $\{z: |z| = 1\}$ is a natural boundary of f if, for any $|w| = 1$ and $\delta > 0$, there is no analytic function $g: \{z: |z - w| < \delta\} \to \mathbb{C}$ with $g(z) = f(z)$ for $|z| < 1, |z - w| < \delta$.

(i) Let $f(z) = \sum_{n=0}^{\infty} z^{2^n}$ for $|z| < 1$. If $\omega = \exp(q2^{-m}\pi i)$ for positive integers

q and m, show that $|f(r\omega)| \to \infty$ as $r \to 1-$. Conclude that $\{z: |z| = 1\}$ is a natural boundary for f.

(ii) Let $f(z) = \sum_{n=1}^{\infty} n^{-2} z^{2^n}$ for $|z| < 1$. Show that there is a bounded continuous function $\tilde{f}: \{z: |z| \leq 1\} \to \mathbb{C}$ with $\tilde{f}(z) = f(z)$ for $|z| < 1$. By considering f', show that, none the less, $\{z: |z| = 1\}$ is a natural boundary for f.

(iii) Construct an analytic function $f: D \to \mathbb{C}$ such that $|f(z)|, |f'(z)|, \ldots, |f^{(k)}(z)| \leq M$ for all $|z| < 1$ and some M yet $\{z: |z| = 1\}$ is a natural boundary for f.

(iv) Show that $|z| = 1$ is not a natural boundary for $f(z) = \sum_{n=0}^{\infty} z^n$ (look at $(1 - z)^{-1}$).

(v) Which of the following have $\{z: |z| = 1\}$ as a natural boundary? Give reasons.

(a) $\sum_{n=0}^{\infty} nz^n$.
(b) $\sum_{n=0}^{\infty} z^{n!}$.
(c) $\sum_{n=0}^{\infty} (n!)^{-3} z^{n!}$.

(vi) Let us say that a power series $\sum_{n=0}^{\infty} a_n z^n$ with radius of convergence R has natural boundary $\{z: |z| = R\}$ if $\sum_{n=0}^{\infty} a_n (Rz)^n$ has $\{z: |z| = 1\}$ as a natural boundary. Suppose $\sum_{j=0}^{\infty} a_j z^j$ has natural boundary $\{z: |z| = R_1\}$ and $\sum_{j=0}^{\infty} b_j z^j$ has natural boundary $\{z: |z| = R_2\}$. Let $c_j = a_j + b_j$. Which of the following statements are always true, always false, sometimes true and sometimes false, and why?

(a) If $R_1 = R_2$, $\sum_{j=0}^{\infty} c_j z^j$ has $\{z: |z| = R_1\}$ as natural boundary.
(b) If $R_1 \neq R_2$, $\sum_{j=0}^{\infty} c_j z^j$ has $\{z: |z| = \min(R_1, R_2)\}$ as natural boundary.

75.6 (i) Let Ω_1 be an open set in $\mathbb{C}$ with closure $\mathrm{cl}\,\Omega_1$. Set $\Omega_2 = \mathbb{C} \backslash \mathrm{cl}\,\Omega_1$. Suppose $\Omega_1, \Omega_2 \neq \emptyset$. Let

$$T = \Omega_1 \cap \{x + iy: x, y \text{ rational}\}.$$

Since T is countable, we can take $T = \{z(1), z(2), \ldots\}$. Show that, if we choose $\varepsilon(j) > 0$ suitably, then $f(z) = \sum_{j=1}^{\infty} \varepsilon(j)(z - z(j))^{-1}$ gives a well defined analytic function $f: \Omega_2 \to \mathbb{C}$ such that, if $w \in \mathbb{C}$, $\delta > 0$ satisfy $N(w, \delta) \cap \Omega_1 \neq \emptyset$, $N(w, \delta) \cap \Omega_2 \neq \emptyset$ (where $N(w, \delta) = \{z | z: - w| < \delta\}$), then there does not exist an analytic function $g: N(w, \delta) \to \mathbb{C}$ with $g(z) = f(z)$ for $z \in N(w, \delta) \cap \Omega_2$.

(ii) Show that, if we choose $\varepsilon(j)$ suitably in (i), we can ensure that $|f(z)| \leq 1$ for all $z \in \Omega_2$. (Remember 75.5 (ii).) Why does this not contradict Liouville's theorem that a bounded analytic function is constant?

(iii) Is the following statement true or false, and why? If Ω is an open subset of $\mathbb{C}$ and $\Omega \neq \emptyset$, there exists an analytic function $f: \Omega \to \mathbb{C}$ such that, if $w \in \mathbb{C}$, $\delta > 0$ satisfy $N(w, \delta) \cap \Omega \neq \emptyset$, $N(w, \delta) \backslash \Omega \neq \emptyset$, there does not exist an analytic function $g: N(w, \delta) \to \mathbb{C}$ with $g(z) = f(z)$ for $z \in N(w, \delta) \cap \Omega$.

75.7 (i) Let $L_1 = \{iy: y \geqslant 0\}$. Construct an analytic function $f: \mathbb{C}\backslash L_1 \to \mathbb{C}$ such that, for each $w \in L_1$, we can find $w_n \in \mathbb{C}\backslash L_1$ such that $w_n \to w$ and $|f(w_n)| \to \infty$ as $n \to \infty$.

(ii) Let $L_2 = \{y: y \leqslant 0\}$. Define $F: \mathbb{C}\backslash L_2 \to \mathbb{C}$ by $F(r^2 e^{2i\theta}) = f(re^{i\theta})$ for $r > 0$, $\pi/2 > \theta > -\pi/2$. Explain why F is analytic. Suppose $w \in L_2$ and $|w| > \delta > 0$. Show that, on the one hand, there exists an analytic function $G_1: N(w, \delta) \to \mathbb{C}$ such that $G_1(z) = F(z)$ for $|z - w| < \delta$, $\operatorname{Im} z < 0$ but, on the other hand, there exists no analytic function $G_2: N(w, \delta) \to \mathbb{C}$ such that $G_2(z) = F(z)$ for $|z - w| < \delta$, $\operatorname{Im} z > 0$.
(This explains why we make no attempt to define the concept of a natural boundary in cases more general than the disc.)

75.8 (Three warnings)

(i) Show that $\sum_{n=1}^{\infty} z^n/((1 + z^n)(1 + z^{n+1}))$ converges to $z(1 - z^2)^{-1}$ for $|z| < 1$ but to $(z^2 - 1)^{-1}$ for $|z| > 1$. Why does this not contradict Theorem 75.6?

(ii) Evaluate $\phi(z) = \sum_{r=1}^{\infty} z^r(z - 1)/((z^r - 1)(z^{r+1} - 1))$ for $|z| \neq 1$. If f, $g: \mathbb{C} \to \mathbb{C}$ are given show that, if

$$\psi(z) = (z - 1)^{-1}(zg(z) - f(z)) + (f(z) - g(z))\phi(z),$$

then $\psi(z) = f(z)$ for $|z| < 1$, $\psi(z) = g(z)$ for $|z| > 1$.

(iii) Show that, if α is rational, then $\sum_{n=1}^{\infty} 1/(2^n(z - e^{2\pi i n\alpha}))$ defines a meromorphic function on $\mathbb{C}$, but that, if α is irrational, it defines an analytic function f_1 on $\{z: |z| > 1\}$ and an analytic function f_2 on $\{z: |z| < 1\}$ both having $\{z: |z| = 1\}$ as a natural boundary.

75.9 In Question 75.3 (i) we produced an f with $\{z: |z| = 1\}$ as natural boundary by considering a power series of a very special form. It is, however, reminiscent of the kind of series we used to obtain nowhere differentiable functions in Chapter 11 and the potential theory counter example of Hadamard in Chapter 30. Following Question 11.2, let us call a sequence of integers $0 \leqslant \lambda(1) < \lambda(2) < \ldots$ lacunary if there exists a $q > 1$ and $N \geqslant 1$ such that $\lambda(n + 1) \geqslant q\lambda(n)$ for all $n \geqslant N$. In his celebrated gap theorem, Hadamard showed that, for such a sequence, if $\sum_{n=1}^{\infty} a_n z^{\lambda(n)}$ has radius of convergence 1, then $\{z: |z| = 1\}$ is a natural boundary. In the next two questions we present a proof due to Mordell.

To start, we need a result which is important in its own right. Suppose $\sum_{n=0}^{\infty} a_n z^n$ is a power series with radius of convergence at least 1. If $|w| = 1$, we say that w is a regular point if there exists a $\delta(w) > 0$ and an analytic function $g_w: N(w, \delta) \to \mathbb{C}$ with $g_w(z) = \sum_{n=0}^{\infty} a_n z^n$ for $|z| < 1$, $|z - w| < \delta(w)$.

(i) Suppose every w with $|w| = 1$ is regular. Show that there exists an $\varepsilon > 0$ such that for each $|w| = 1$ there exists $G_w: N(w, \varepsilon) \to \mathbb{C}$ with $G_w(z) = \sum_{n=0}^{\infty} a_n z^n$ for $|z| < 1$, $|z - w| < \varepsilon$.

(Suppose not. Then we can find w_n with $|w_n| = 1$ such that there exists no $G: N(w_n, 1/n) \to \mathbb{C}$ analytic with $G(z) = \sum_{n=0}^{\infty} a_n z^n$ for $|z| < 1$, $|z - w_n| < 1/n$. Show that there exists a w with $|w| = 1$ and $n(j) \to \infty$ such that $w_{n(j)} \to w$ and examine w.)

(ii) With the notation of (i), show that the following definition of $G: N(0, 1 + \varepsilon/8) \to \mathbb{C}$ is unambiguous. If $|z| < 1$, set $G(z) = \sum_{n=0}^{\infty} a_n z^n$. If $|z| < 1 + \varepsilon/8$, $|w| = 1$ and $|z - w| < \varepsilon/8$, set $G(z) = G_w(z)$. Show that G is analytic. Explain why G has a Taylor series in $\{z: |z| < 1 + \varepsilon/8\}$ and find it.

(iii) Conclude that, if $\sum_{n=0}^{\infty} a_n z^n$ has radius of convergence exactly 1, then there exists a w with $|w| = 1$ which is not a regular point.

75.10 (Mordell's proof of the Hadamard gap theorem) Let $q > 1$ and let $1 \leqslant \lambda(1) < \lambda(2) < \ldots$ be a sequence of integers with $\lambda(n+1) \geqslant q\lambda(n)$ for all $n \geqslant 1$. Suppose further that $\sum_{n=0}^{\infty} a_n z^{\lambda(n)}$ has radius of convergence at least 1. Choose p a positive integer such that $p(q-1) > 1$. (Initially you may wish to consider the simplest case when $q > 2$ and we can take $p = 1$.)

(i) Show that $|\frac{1}{2}(w^p + w^{p+1})| < 1$, if $|w| \leqslant 1$, $w \neq 1$.

(ii) Show that, if $|w| < 1$, we may define $\phi(w) = \sum_{n=0}^{\infty} a_n (\frac{1}{2}(w^p + w^{p+1}))^{\lambda(n)}$ and that $\phi(w) = \sum_{n=0}^{\infty} b_n w^n$ where the coefficients b_n are to be found explicitly. Explain why, if $\sum_{n=0}^{\infty} b_n w^n$ converges and $z = \frac{1}{2}(w^p + w^{p+1})$, then $\sum_{n=0}^{\infty} a_n z^{\lambda(n)}$ converges. (Be careful.)

(iii) Show, using (i), that every point z with $|z| = 1$, $z \neq 1$ is regular for $\sum_{n=0}^{\infty} b_n z^n$. Conclude that, if 1 is a regular point for $\sum_{n=1}^{\infty} a_n z^{\lambda(n)}$, then every point z with $|z| = 1$ is regular for $\sum_{n=0}^{\infty} b_n z^n$ and so, using Question 75.9, $\sum_{n=0}^{\infty} b_n z^n$ has radius of convergence greater than 1. Deduce, by a choosing z suitably in the last sentence of (ii), that $\sum_{n=0}^{\infty} a_n z^{\lambda(n)}$ converges for some z with $|z| > 1$ i.e. that $\sum_{n=0}^{\infty} a_n z^{\lambda(n)}$ has radius of convergence greater than 1.

(iv) Deduce that, $\sum_{n=0}^{\infty} a_n z^{\lambda(n)}$ has radius of convergence exactly 1, then 1 is not a regular point.

(v) Explain why this means that, if $\sum_{n=0}^{\infty} a_n z^{\lambda(n)}$ has radius of convergence exactly 1, no point z with $|z| = 1$ can be regular, i.e. $\{z: |z| = 1\}$ is a natural boundary.

(vi) Where does the proof above break down when applied to $\sum_{n=0}^{\infty} (-1)^n z^n$?

75.11 (Kolmogorov's zero-one law) Consider the space s of real sequences $\mathbf{x} = (x_0, x_1, x_2, \ldots)$ and well-behaved functions $f, g: s \to [0, 1]$. Suppose $X_0, Y_0, X_1, Y_1, \ldots$ are independent random variables such that X_j and Y_j are identically distributed. Let us write $Z_j^{(N)} = X_j$ for $0 \leqslant j \leqslant N$ and $Z_j^{(N)} = Y_j$ for $j \geqslant N + 1$. If the notion of expected value behaves reasonably, we should have

$$\mathbb{E}(f(\mathbf{X})g(\mathbf{Z}^{(N)})) \to \mathbb{E}(f(\mathbf{X})g(\mathbf{X})) \quad \text{as } N \to \infty.$$

Let us say that the function f is asymptotic if $f(\mathbf{x}) = f(\mathbf{y})$ whenever there exists an m such that $x_j = y_j$ for $j \geqslant m$.

(i) Consider the following functions $f_k\colon s \to [0, 1]$.

(I) $f_1(\mathbf{x}) = 1$ if $|x_j| \leqslant 1$ for all j, $f_1(\mathbf{x}) = 0$ otherwise.
(II) $f_2(\mathbf{x}) = 1$ if $\lim_{j\to\infty} x_j$ exists, $f_2(\mathbf{x}) = 0$ otherwise.
(III) $f_3(\mathbf{x}) = 1$ if $\sup_{j\geqslant 0}|x_j| < \infty$, $f_3(\mathbf{x}) = 0$ otherwise.

State, with reasons, which of these f_k are asymptotic and which are not.

(ii) If g is asymptotic, explain why

$$\mathbb{E}(f(\mathbf{X})g(\mathbf{Z}^{(N)})) = \mathbb{E}(f(\mathbf{X})g(\mathbf{Y})) = \mathbb{E}f(\mathbf{X})\mathbb{E}g(\mathbf{Y}).$$

Deduce that

$$\mathbb{E}(f(\mathbf{X})g(\mathbf{X})) = \mathbb{E}f(\mathbf{X})\mathbb{E}g(\mathbf{X}).$$

(iii) Let Q be a well-behaved subset of s. We say that Q is asymptotic if, whenever $\mathbf{x}$ and $\mathbf{y}$ are sequences such that $x_j = y_j$ for all $j \geqslant m$ for some $m \geqslant 0$, then either $\mathbf{x} \in Q$ and $\mathbf{y} \in Q$ or $\mathbf{x} \notin Q$ and $\mathbf{y} \notin Q$. Let $f(\mathbf{x}) = 1$ if $\mathbf{x} \in Q$, $f(\mathbf{x}) = 0$ otherwise. Show that

$$\mathbb{E}(f(\mathbf{X})(1 - f(\mathbf{X}))) = \mathbb{E}f(\mathbf{X})\mathbb{E}(1 - f(\mathbf{X}))$$

and deduce that

$$0 = Pr(\mathbf{X} \in Q)(1 - Pr(\mathbf{X} \in Q)).$$

Conclude that, if Q is asymptotic, then either $Pr(\mathbf{X} \in Q) = 0$ or $Pr(\mathbf{X} \in Q) = 1$.

(iv) Let $X_0, X_1, X_2, \ldots$ be a sequence of independent random variables. Which of the following statements is always true and which may be false? (Give an explanation or produce a counter example.)

(a) $Pr(\sum|X_n|^2$ diverges) has value 0 or 1.
(b) $Pr(X_n = 0$ for infinitely many values of $n)$ has value 0 or 1.
(c) $Pr(X_n$ tends to a limit) has value 0 or 1.
(d) $Pr(X_0X_n$ tends to a limit) has value 0 or 1.
(e) $Pr(\sup_{n>0}|X_n| \leqslant 1)$ has value 0 or 1.
(f) $Pr(\limsup_{n\to\infty}|X_n| \leqslant 1)$ has value 0 or 1.

(I have done my best to make Kolmogorov's zero-one law plausible but I would be surprised if the arguments above leave the reader fully convinced. At best, I have shown that, if the laws of probability apply in a particularly nice way to sequences, then the zero-one law holds. Kolmogorov showed that, by choosing a suitable axiomatic basis for probability, it was possible, after further quite hard work, to put probability structure on our independent sequences $X_0, X_1, \ldots$ in such a way that the argument above is valid. It is fashionable in some quarters to pour scorn on the axiomatic method. Such a view may be just, but its proponents would be more convincing if, instead of knocking down straw men, they explained how we could develop modern probability theory on a convincing but non axiomatic basis.)

75.12 Let $\sum_{n=0}^{\infty} a_n z^n$ be a power series with radius of convergence 1. Let $X_0, X_1, \ldots$ be a sequence of independent, identically distributed random variables each with the uniform distribution on $[0, 1)$. Set $W_k = \exp(2\pi i X_k)$. We claim that, with probability 1, $\sum_{n=0}^{\infty} a_n W_n z^n$ has the circle $\{z \in \mathbb{C}: |z| = 1\}$ as natural boundary. Borel was the first mathematician to realise that a result of this form should be true. The result presented here is due to Steinhaus. The beautiful proof using Kolmogorov's zero-one law (see Question 75.11) is taken from Kahane *Some Random Series of Functions*.

(i) By recalling the standard proof of the existence of the radius of convergence, or otherwise, show that, if $w_0, w_1, \ldots$ satisfy $|w_k| = 1$ for all k, then $\sum_{n=0}^{\infty} a_n w_n z^n$ has radius of convergence 1.

(ii) Suppose $w_0, w_1, \ldots$ satisfy $|w_k| = 1$ for all k. By modifying Question 75.10, show that, if N is an integer with $N \geqslant 1$, then there exists an integer r with $N - 1 \geqslant r \geqslant 0$ which has the following property. If $\Delta(r, N) = \{z \in \mathbb{C}: |z - \exp(2\pi i r/N)| \leqslant 10N^{-1}\}$, then there does not exist an analytic function $f_{rN}: \Delta(r, N) \to \mathbb{C}$ with $f_{rN}(z) = \sum_{n=0}^{\infty} a_n w_n z^n$ for all $z \in \Delta(r, N)$, $|z| < 1$.

(iii) Returning to our random series $\sum_{n=0}^{\infty} a_n W_n z^n$, let $E(r, N)$ be the event that there exists an analytic function $f_{rN}: \Delta(r, N) \to \mathbb{C}$ with $f_{rN}(z) = \sum_{n=0}^{\infty} a_n W_n z^n$ for all $z \in \Delta(r, N)$, $|z| < 1$. Explain why

(a) $Pr(E(r, N)) = Pr(E(0, N))$ for $0 \leqslant r \leqslant N - 1$,
(b) $Pr(E(0, N))$ has value 0 or 1.

If $Pr(E(0, N)) = 1$, show that $Pr(\bigcup_{r=0}^{N-1} E(r, N)) = 1$ and use (ii) to obtain a contradiction. Deduce that $Pr(E(r, N)) = 0$ for all $0 \leqslant r \leqslant N - 1$ and all N. We conclude that, with probability 1, given any $N \geqslant 1$ and any $0 \leqslant r \leqslant N - 1$, there does not exist an $f: \Delta(r, N) \to \mathbb{C}$ with $f(z) = \sum_{n=0}^{\infty} a_n W_n z^n$ for all $z \in \Delta(r, N)$, $|z| < 1$.

(iv) If $|z_0| = 1$ and $\delta > 0$, show that there exist $N \geqslant 1$ and $0 \leqslant r \leqslant N - 1$ such that

$$\{z: |z - z_0| < \delta\} \subseteq \Delta(r, N).$$

Deduce that, with probability 1, $\sum_{n=0}^{\infty} a_n W_n z^n$ has $\{z: |z| = 1\}$ as natural boundary.

(v) Let $a_n = (n + 1)^{-\log(n+1)}$. Show that $\sum_{n=0}^{\infty} a_n z^n$ has radius of convergence 1 and, by using (iv), show that there exist $b_n \in \mathbb{C}$ with $|b_n| = a_n$ such that $\sum_{n=0}^{\infty} b_n z^n$ has natural boundary $\{z: |z| = 1\}$. Let $g(z) = \sum_{n=0}^{\infty} b_n z^n$ for $|z| < 1$. Show by simple estimates that $g^{(r)}$ (the rth derivative of g) is bounded for each

r. More exactly, show that there exist $M_0, M_1, M_2, \ldots$ such that $|g^{(r)}(z)| \leqslant M_r$ for each $|z| < 1$ and each $r \geqslant 0$.
(There is a family resemblance between 75.10 and 75.12. Experience shows that results on random series and lacunary series tend to parallel one another.)

75.13 (i) Recall that $t^z = e^{z \log t}$ for t real and positive and $z \in \mathbb{C}$. Show that $\int_0^\infty t^z e^{-t}\,dt$ converges for all $z \in \mathbb{C}$ with $\operatorname{Re} z > -1$.
Let us write

$$F(z) = \int_0^\infty t^z e^{-t}\,dt \quad \text{for } \operatorname{Re} z > -1.$$

(ii) Show, using the kind of proof of Theorem 75.2, or otherwise, that F is analytic.
(iii) Show that, if x is real and $x > 0$, then $F(x) = xF(x-1)$. By looking at the zeros of $F(z) - zF(z-1)$ and using Theorem 75.6, show that $F(z) = zF(z-1)$ for all $\operatorname{Re} z > 0$.
(iv) Show that the conditions $G(z) = F(z)$ for $\operatorname{Re} z > -1$, $G(z-1) = z^{-1} G(z)$ for $z \notin \{0, -1, 2, \ldots\}$ define, unambiguously, an analytic function $G: \mathbb{C}\backslash\{-1, -2, \ldots\} \to \mathbb{C}$. Show that $-1, -2, \ldots$ are simple poles and find the residues at these points.
(v) We have thus found a meromorphic function $G: \mathbb{C} \to \mathbb{C}$ satisfying $G(z) = zG(z-1)$. What is the value of $G(0)$? What is the value of $G(n)$ for n a positive integer? We write $z! = G(z)$. (The other common notation is $\Gamma(z) = (z-1)!$; Γ is called the gamma function.)

75.14 (i) For what real values of a does the integral $\int_{-\infty}^{\infty} e^{ax}/(1+e^x)\,dx$ converge?
(ii) By considering the contour integral $\int_{C(R)} e^{ax}/(1+e^x)\,dx$ round a rectangle $C(R)$ with vertices $-R, R, R+2\pi i, -R+2\pi i$, and allowing $R \to \infty$, evaluate $\int_{-\infty}^{\infty} e^{ax}/(1+e^x)\,dx$ when it converges.
(iii) Deduce the value of $\int_0^\infty t^b/(1+t)\,dt$ for those real values of b for which the integral converges.
(iv) Show that, if x is real and $0 < x < 1$,

$$x!(-x)! = 4\int_0^\infty \int_0^\infty e^{-(u^2+v^2)} u^{2x+1} v^{-2x+1}\,du\,dv$$

and (recalling the ideas of Lemma 48.10, if necessary, and then making a further change of variable) deduce that

$$x!(-x)! = \int_0^\infty \frac{t^x}{(1+t)}\,dt = \pi x \operatorname{cosec} \pi x.$$

(v) By considering the zeros of a meromorphic function, deduce that $z!(-z)! = \pi z \operatorname{cosec} \pi z$ for all $z \in \mathbb{C}$ with z not an integer.
(vi) Show that $z!$ has no zeros in the complex plane.

(vii) Show that $(-1/2)! = \pi^{1/2}$. What is the value of $(n - 1/2)!$ for general integer n? Show that, if n is a non negative integer

$$(2n)! = 2^{2n} n!(n - 1/2)!\pi^{-1/2}.$$

(This is a special case of the duplication formula proved in Question 75.16 (vi).)

75.15 In Question 75.13 we defined $z!$ and saw that it was a natural extension of the factorial $n!$. Is it the only natural extension?

(i) Let $a \geqslant -1$. Suppose that $\gamma: \{z: \operatorname{Re} z > a\} \to \mathbb{C}$ is analytic with $\gamma(z) = z\gamma(z-1)$ for $\operatorname{Re} z > a + 1$. By considering $\gamma(z)/z!$, or otherwise, show that there exists a periodic analytic function $\omega: \mathbb{C} \to \mathbb{C}$ with period 1 such that $\gamma(z) = \omega(z)z!$. State and prove the (very easy) converse.

(ii) The fact that $z!$ is defined on $\mathbb{C}$ does not seem to make the task of finding a uniqueness criterion any easier, so let us confine ourselves to the real axis and ask what extra conditions on an $f: (0, \infty) \to \mathbb{R}$ with $f(x) = xf(x-1)$ for $x > 1$, $f(1) = 1$ will ensure that $f(x) = x!$. Our first thought is convexity (see Question 32.4 (i)). Show that, writing $g(x) = x!$, we have $g''(x) > K$ for all $x > 0$ and some $K > 0$.

(iii) Consider a twice continuously differentiable function $h: (0, \infty) \to \mathbb{R}$ such that $h(x) = h(x-1)$ for $x > 1$, $h(x) = 1$ for $|x - 1| < 1/4$. Show that, provided ε is small enough $d^2(g(x) + \varepsilon h(x))/dx^2 > K/2$ for $x > 0$. Conclude that convexity is not enough to give uniqueness.

(iv) We say that $f: (0, \infty) \to \mathbb{R}$ is log convex if $f(x) > 0$ for all $x > 0$ and $\log f$ is convex. If f is twice differentiable and strictly positive, find a condition involving first and second derivatives of f which ensures that f is log convex. Use this to show that $x!$ is log convex.

(v) (The following result will be used in the next question.) Show that, if $f: (0, \infty) \to \mathbb{R}$ is convex, then

$$\frac{f(x) - f(a)}{x - a} \geqslant \frac{f(y) - f(a)}{y - a} \quad \text{if } x \geqslant y \; [x, y \neq a].$$

(*Hint.* Draw a picture and convert the picture into algebra.)

75.16 Artin wrote a charming monograph called *The Gamma Function*. The following is a typical result.

Suppose $f: (0, \infty) \to \mathbb{R}$ satisfies the following three conditions

(1) $f(x) = xf(x-1)$ for $x > 1$,
(2) f is log convex,
(3) $f(1) = 1$.

Then $f(x) = x!$.

Here is the proof he gives. (This approach to the factorial function $x!$ comes from Bohr and Mollerup.)

(i) Use induction to show that $f(n) = n!$. Use condition (2) together with Question 75.15 (v) to show that, if $0 < x \leqslant 1$ and n is an integer with $n \geqslant 2$,

$$\frac{\log f(-1+n) - \log f(n)}{(-1+n) - n} \leqslant \frac{\log f(x+n) - \log f(n)}{(x+n) - n}$$

$$\leqslant \frac{\log f(1+n) - \log f(n)}{(1+n) - n}$$

and deduce that

$$\log n \leqslant \frac{\log f(x+n) - \log(n!)}{x} \leqslant \log(n+1)$$

Conclude that

$$n^x n! \leqslant f(x+n) \leqslant (n+1)^x n!.$$

(ii) Use the last inequalities of (i) together with condition (1) to obtain

$$\frac{n^x n!}{(x+1)\ldots(x+n)} \leqslant f(x) \leqslant \frac{(n+1)^x n!}{(x+1)\ldots(x+n)}$$

and hence deduce that

$$\frac{n}{x+n} f(x) \leqslant \frac{n^x n!}{(x+1)\ldots(x+n)} \leqslant f(x).$$

Conclude that

$$\frac{n^x n!}{(x+1)\ldots(x+n)} \to f(x) \quad \text{as } n \to \infty$$

for all $0 < x \leqslant 1$.

(iii) Show more generally that

$$\frac{n^x n!}{(x+1)\ldots(x+n)} \to f(x) \quad \text{as } n \to \infty$$

for all $x > 0$.

(iv) Observe that $x!$ obeys conditions (1), (2) and (3) and conclude, setting out your reasoning carefully, that $f(x) = x!$ as required. Deduce also Gauss's formula

$$x! = \lim_{n\to\infty} \frac{n^x n!}{(x+1)\ldots(x+n)} \quad \text{for all } x > 0.$$

(v) By using the relation $x! = x(x-1)!$ show that the formula

$$\frac{n^x n!}{(x+1)\ldots(x+n)} \to x! \quad \text{as } n \to \infty$$

holds for all real x with $x \neq -1, -2, \ldots$.
(In fact it holds for complex values as well.)

(vi) Use the result of (v) to show that

$$\frac{(n!)^2 2^{2n}}{(2n)!n^{1/2}} \to \frac{2^{2x}x!(x-1/2)!}{(2x)!} \quad \text{as } n \to \infty$$

for all real x with $x \neq -\frac{1}{2}, -1, -\frac{3}{2}, -2, \ldots$ (and so $2^{2x}x!(x-1/2)!/(2x)!$ is constant). By setting $x = 0$ and using Question 75.14 (vii), obtain

$$(2x)! = 2^{2x}x!(x - \tfrac{1}{2})!\pi^{1/2}$$

for all real x with $x \neq -\frac{1}{2}, -1, -\frac{3}{2}, -2, \ldots$. Explain why this gives the 'duplication formula'

$$(2z)! = 2^{2z}z!(z - \tfrac{1}{2})!\,\pi^{1/2}$$

for all $z \in \mathbb{C}\backslash\{-n/2: n \in \mathbb{Z}, n \geqslant 0\}$.

75.17 (The Beta function)

(i) We start by generalising Lemma 74.6 (v). If α is real and $\alpha \geqslant 0$, let us write $\tau_\alpha(t) = t^\alpha H(t)$ $[t \in \mathbb{R}]$ so $\tau_\alpha \in \mathscr{C}$. By making a simple change of variable, show that $\mathscr{L}\tau_\alpha(z) = \alpha! z^{-\alpha-1}$ for $\operatorname{Re} z > 0$.

(ii) Suppose α, β are real with $\alpha, \beta > 0$. By considering the convolution formula $\mathscr{L}(\tau_a * \tau_\beta)(z) = (\mathscr{L}\tau_\alpha)(z)(\mathscr{L}\tau_\beta)(z)$ of Question 74.3 and making an appropriate change of variable, show that

$$\alpha!\beta! = (\alpha + \beta + 1)! \int_0^1 t^\alpha(1-t)^\beta\, dt.$$

(iii) Show that $\int_0^1 t^\alpha(1-t)^\beta\, dt$ is a well-defined integral for all $\alpha, \beta \in \mathbb{C}$ with $\operatorname{Re}\alpha, \operatorname{Re}\beta > -1$. By first considering $\alpha!\beta!/(\alpha+\beta+1)!$ as an analytic function of α and then as an analytic function of β, show that

$$\frac{\alpha!\beta!}{(\alpha+\beta+1)!} = \int_0^1 t^\alpha(1-t)^\beta\, dt \quad \text{for all } \operatorname{Re}\alpha, \operatorname{Re}\beta > -1.$$

(iv) By setting $\alpha = \beta = -1/2$ obtain another proof that $(-1/2)! = \pi^{1/2}$.

(v) Obtain the result of (iii) by imitating the proofs in Question 75.14 (iv) and (v).

(The function

$$\frac{\Gamma(\alpha)\Gamma(\beta)}{\Gamma(\alpha+\beta)} = \frac{(\alpha-1)!(\beta-1)!}{(\alpha+\beta-1)!} = B(\alpha, \beta)$$

is called Euler's Beta function. We have shown that, if $\operatorname{Re}\alpha, \operatorname{Re}\beta > 0$, then $B(\alpha,\beta) = \int_0^1 t^{\alpha-1}(1-t)^{\beta-1}\, dt$.)

(vi) Show that, if $\operatorname{Re}\alpha, \operatorname{Re}\beta > 0$, then

$$2\int_0^{\pi/2} \sin^{2\alpha-1}\theta \cos^{2\beta-1}\theta\, d\theta = B(\alpha, \beta).$$

Write down the value of the integral when α, β are positive integers (note that you will have to distinguish between even and odd integers).

(vii) Suppose $\gamma: \{z: \operatorname{Re} z > 0\} \to \mathbb{C}$ is an analytic function satisfying $\gamma(\alpha)\gamma(\beta) = \gamma(\alpha+\beta)f(\alpha,\beta)$ for all $\operatorname{Re}\alpha$, $\operatorname{Re}\beta > 0$ and some function $f: \{z: \operatorname{Re} z > 0\}^2 \to \mathbb{C}$. Show that, if $\gamma(z_0) = 0$ for some z_0 with $\operatorname{Re} z_0 > 0$, we can find $z_n \to z_0$ with $\gamma(z_n) = 0$ and deduce that γ is identically zero. Conclude that, if $\gamma \neq 0$, then γ cannot have a zero.

75.18 (i) Let $a, b, c, \alpha, \beta, \gamma$ be strictly positive real numbers and let

$$\Sigma = \{(x, y, z) \in \mathbb{R}: x^2/a^2 + y^2/b^2 + z^2/c^2 \leqslant 1, x, y, z \geqslant 0\}$$

and

$$I = \iint_\Sigma x^{\alpha-1} y^{\beta-1} z^{\gamma-1}\, dx\, dy\, dz.$$

By making appropriate changes of variable, show that

$$\begin{aligned} I &= a^\alpha b^\beta c^\gamma \int_0^1 \int_0^{1-u} \int_0^{1-u-v} u^{(\alpha/2-1)} v^{(\beta/2-1)} w^{(\gamma/2-1)}\, dw\, dv\, du \\ &= \frac{a^\alpha b^\beta c^\gamma}{4\gamma} \int_0^1 u^{(\alpha/2-1)} (1-u)^{(\beta+\gamma)/2} B(\beta/2, \gamma/2+1)\, du \\ &= \frac{a^\alpha b^\beta c^\gamma}{\alpha\beta\gamma} \frac{\Gamma(\alpha/2)\Gamma(\beta/2)\Gamma(\gamma/2)}{\Gamma((\alpha+\beta+\gamma)/2)} \\ &= \frac{a^\alpha b^\beta c^\gamma}{\alpha\beta\gamma} \frac{(\alpha/2-1)!(\beta/2-1)!(\gamma/2-1)!}{((\alpha+\beta+\gamma)/2-1)!}. \end{aligned}$$

(ii) Extend the formula of (i) to higher dimensions and check your result by obtaining the final result of Question 48.3 (the volume V_n of the n-dimensional sphere).

(iii) Suppose p, q, r are strictly positive real numbers,

$$\Sigma(p, q, r) = \{(x, y, z \in \mathbb{R}^3: (x/a)^p + (y/b)^q + (z/c)^r \leqslant 1, x, y, z \geqslant 0\}$$

and

$$I(p, q, r) = \iint_{\Sigma(p,q,r)} x^{\alpha-1} y^{\beta-1} z^{\gamma-1}\, dx\, dy\, dz.$$

Find the value of $I(p, q, r)$ in a form similar to that given in (i).

(iv) Let $X_1, X_2, \ldots, X_{n-1}$ be independent, identically distributed random variables, each uniformly distributed on [0, 1]. Let Y_1 be the smallest, Y_2 the second smallest, etc. Show that

$$Pr(Y_{k-1} \leqslant \lambda) = \frac{1}{B(k, n-k)} \int_0^\lambda x^{k-1}(1-x)^{n-k-1}\, dx$$

for all $0 \leqslant \lambda \leqslant 1$ and deduce that Y_{k-1} has density function

$$B_{k,n-k}(x) = \frac{1}{B(k, n-k)} x^{k-1}(1-x)^{n-k-1} \quad \text{for } 0 \leqslant x \leqslant 1.$$

(v) More generally, if $\alpha, \beta > 0$, a random variable X taking values on $[0, 1]$ with probability density

$$B_{\alpha,\beta}(x) = \frac{1}{B(\alpha, \beta)} x^{\alpha-1}(1-x)^{\beta-1}$$

is called a Beta random variable. Find its nth moment $\mathbb{E}(X^n)$ for all integer $n \geqslant 1$.
(The Beta distribution occurs in several places in statistics. Karl Pearson was particularly convinced of its importance and his last major work was the tabulation of the incomplete Beta function

$$Pr(X < t) = \frac{1}{B(\alpha, \beta)} \int_0^t x^{\alpha-1}(1-x)^{\beta-1} dx.$$

The tables had a quarter of a million entries. One of his assistants was the future eminent group theorist Philip Hall but, after a few weeks of computation, Hall left the project declaring that, as far as he was concerned, the incomplete Beta function could remain forever incomplete.

At the time these exercises are being compiled the theoretical physicists' favourite theory of everything is superstring theory. The following quotation is taken from the introduction of Michio Kaku's *Introduction to Superstrings*. 'Superstring theory . . . has a bizarre history beginning with the purely accidental discovery of the quantum theory in 1968 by G. Veneziano and M. Suzuki. Thumbing through old math books, they stumbled by chance on the Beta function, written down . . . by mathematician Leonhard Euler. To their amazement, they discovered that the Beta function satisfied almost all the stringent requirements of the scattering matrix describing particle interactions.')

75.19 (Fractional integrals. This question makes use of ideas from the previous question.)

(i) Throughout this question $f: [0, \infty) \to \mathbb{R}$ will be continuous and, except where we specify otherwise, we take $f(0) = 0$. By direct computation, or better (from the point of view of understanding where the idea comes from), by reinterpreting Taylor's theorem in Question 4.3 (iii), show that, if $n \geqslant 1$,

$$(I_n f)(t) = \int_0^t \frac{(t-s)^{n-1}}{(n-1)!} f(s)\, ds$$

defines an n times continuously differentiable function $I_n f\colon [0, \infty) \to \mathbb{R}$ with nth derivative f.

(ii) More generally, if a is a complex number with $\operatorname{Re} a > 0$, let us define $I_a f(t)$ for $t \geqslant 0$ by

$$(I_a f)(t) = \int_0^t \frac{(t-s)^{a-1}}{(a-1)!} f(s)\, ds$$

and set $(I_a f)(0) = 0$. Show that $I_a f\colon [0, \infty) \to \mathbb{R}$ is a well-defined continuous function. Show further (by the usual argument of obtaining the integral as the uniform limit of analytic functions, or otherwise) that, if t is fixed, $(I_a f)(t)$ is an analytic function of a.

(iii) If $\operatorname{Re} a > 1$, show how to write $I_a f$ in terms of $\tau_{\alpha-1}$ and f. Show that if $\operatorname{Re} a$, $\operatorname{Re} b > 1$, then $I_a I_b f = I_{a+b} f$. By using an argument based on analytic continuation, or by a direct argument involving integrals of unbounded functions, or otherwise, show that, in fact,

$$I_a I_b f = I_{a+b} f \quad \text{for all } \operatorname{Re} a, \operatorname{Re} b > 0.$$

(iv) Show that $I_{a+1} f$ is differentiable with

$$\frac{d}{dt}(I_{a+1} f)(t) = I_a f(t) \quad \text{for all } t \geqslant 0 \text{ when } \operatorname{Re} a > 0.$$

(As usual, the result is easy for $\operatorname{Re} a \geqslant 1$ but requires a slightly harder argument based e.g. on uniform convergence for $1 > \operatorname{Re} a > 0$. You may wish to deal with the point $t = 0$ separately.)

(v) By a simple modification of our standard kernel argument, show that

$$I_a f(t) \to f(t) \quad \text{as } a \to 0+ \text{ for all } t \geqslant 0.$$

(vi) Suppose we drop the condition $f(0) = 0$. By writing $f = f(0) + f_0$, or otherwise, investigate which of the results above remain true.

(vii) Suppose now that f is k times continuously differentiable with $f(0) = f^{(1)}(0) = \ldots = f^{(k)}(0) = 0$. Show that we can extend the definition of $I_a f$ is a consistent manner to all $a \in \mathbb{C}$ with $\operatorname{Re} a > -k$ by setting

$$I_a f = I_{a+k} f^{(k)}.$$

Show further that

(1) $I_{-r} f = f^{(r)}$ for r an integer, $0 \geqslant r \geqslant k-1$;
(2) if t is fixed, $I_a f(t)$ is an analytic function of a for $\operatorname{Re} a > -k$;
(3) $I_a I_b f = I_{a+b} f$ for $\operatorname{Re} a$, $\operatorname{Re} b$, $\operatorname{Re}(a+b) > -k$;
(4) $I_a f(t) \to f^{(k)}(t)$ as $a \to -k$ from above for all $t \geqslant 0$.

(This treatment, which is due to Riemann, is satisfactory in many ways, but leaves certain question unanswered. Is the definition of fractional integration the only natural one in some sense like that which in Question 75.16 makes the gamma function the only natural extension of the factorial? If g is many times differentiable and $g(0) = g'(0) = g''(0) = 0$, then we know that $g'''(x) = 0$ for all $x \geq 0$ implies $g = 0$, but, if we omit the conditions $g(0) = g'(0) = g''(0) = 0$, we know that the most general form of solution of $g'''(x) = 0$ is $g(x) = Ax^2 + Bx + C$. If we remove the restriction $f^{(r)}(0) = 0$ for $0 \leq r \leq k$, what should be the most general form of solution for $(I_{-\alpha}f)(x) = 0$ for all $x \geq 0$?)

(viii) Suppose $g: [0, \infty] \to \mathbb{R}$ is a continuous function with $g(0) = 0$ and suppose $0 < \operatorname{Re} \mu < 1$. Show that Abel's equation

$$\int_0^t \frac{h(s)}{(t-s)^{\mu}} \, ds = g(t) \quad [t \geq 0]$$

has the unique continuous solution h, with $h(0) = 0$, given by

$$h(t) = \frac{\sin \pi\mu}{\pi} \frac{d}{dt} \int_0^t \frac{g(s)}{(t-s)^{1-\mu}} \, ds.$$

What happens if we remove the condition $g(0) = 0$?

75.20 (i) By integrating e^{iz}/z^{α} round a suitable contour, show that, if α is real and $0 < \alpha < 1$, then

$$\int_0^{\infty} \frac{\sin x}{x^{\alpha}} \, dx = \frac{\pi}{2} \frac{1}{(\alpha - 1)!} \frac{1}{\sin(\pi\alpha/2)}.$$

By integrating $(e^{iz} - 1)/z^{\alpha}$ round a suitable contour, or otherwise, show that the formula continues to hold if $1 < \alpha < 2$. Show, finally, by another suitable contour integration, or otherwise that the result is true when $\alpha = 1$.

(ii) If b is real and $b > 0$, show that, if α is real and $0 < \alpha < 2$, then

$$\int_0^{\infty} \frac{\sin bx}{x^{\alpha}} \, dx = \frac{\pi}{2} \frac{b^{\alpha-1}}{(\alpha - 1)!} \frac{1}{\sin(\pi\alpha/2)}.$$

What happens if b is real and $b < 0$? What happens if $b = 0$? If b is real and $b \neq 0$, does the integral converge for any other real values of α?

(iii) For which values of real b and real α does $\int_0^{\infty} (\cos bx/x^{\alpha}) \, dx$ converge, and what is its value when it does converge?

(iv) For which real $\alpha, \beta, > 0$ does $\int_0^{\infty} (\sin x^{\beta}/x^{\alpha}) \, dx$ converge, and what is its value when it does converge? For which real α and real θ does $\int_0^{\infty} (\sin(x + \theta)/x^{\alpha}) \, dx$ converge and what is its value when it does converge?

(v) (This lies a little deeper than (iv).) Suppose b is real and $b > 0$, but that we now allow α to be complex. For which values of α does $\int_0^{\infty} (\sin bx/x^{\alpha}) \, dx$ converge, and what is its value when it does converge? Suppose that we take α real with $\alpha > 0$. Does the integral converge for any b which is not real?

75.21 (Failure of high-order Newton–Cotes quadrature methods. You should do Questions 41.2, 41.3 and have some familarity with the contents of Questions 75.13 to 75.16 before tackling this question.)

(i) With the notation of Question 41.3 (ii), show that

$$A(2n+1, 4n+1) = \frac{1}{(2n)((2n)!)^2}\int_0^{4n} (t-2n)^{-1}\prod_{k=0}^{4n}(t-k)\,dt$$
$$= \frac{\lambda_n}{2n}\binom{4n}{2n}$$

where

$$\lambda_n = \int_0^{4n} \frac{t!(t-2n)^{-1}}{(t-4n-1)!(4n)!}\,dt.$$

(ii) By using Question 75.14 (v), show that

$$\lambda_n = \frac{1}{\pi}\int_0^{4n} \frac{t!(4n-t)!}{(4n)!}\frac{\sin \pi t}{(t-2n)}\,dt$$
$$= -\frac{2}{\pi}\int_0^{2n} \frac{t!(4n-t)!}{(4n)!}\frac{\sin \pi t}{(2n-t)}\,dt$$

(iii) Using the fact proved in Question 75.15 (iv) that the factorial function is log convex, show that $t!(4n-t)!$ is a decreasing function of t on the interval $[0, 2n]$. By using elementary inequalities for sine, deduce that, if $n \geqslant 1$,

$$\left|\int_1^3 \frac{t!(4n-t)!}{(4n)!}\frac{\sin \pi t}{2n-t}\,dt\right| \leqslant \frac{1!(4n-1)!}{(4n)!}\frac{2}{2n-3},$$

and

$$\left|\int_3^{2n} \frac{t!(4n-t)!}{(4n)!}\frac{\sin \pi t}{2n-t}\,dt\right| \leqslant \frac{3!(4n-3)!}{(4n)!}\pi(2n-3),$$

so

$$\left|\int_1^{2n} \frac{t!(4n-t)!}{(4n)!}\frac{\sin \pi t}{2n-t}\,dt\right| \leqslant \frac{4}{n^2}.$$

Show also that

$$\int_0^1 \frac{t!(4n-t)!}{4n!}\frac{\sin \pi t}{2n-t}\,dt \geqslant \frac{1}{16n}\frac{(1/2)!(4n-1/2)!}{(4n)!}$$

(iv) By using the duplication formula of Question 75.16 (vi) and Stirling's formula (see e.g. Question 18.16), show that, provided n is sufficiently large,

$$\frac{(1/2)!(4n-1/2)!}{(4n)!} \geqslant \frac{2^{-29}}{n^{1/2}}$$

(feel free to replace 2^{-29} by any suitable positive number). Deduce that there exists an n_0 such that

$$\int_0^{2n} \frac{t!(4n-t)!\sin \pi t}{(4n)!}\,dt \geqslant \frac{2^{-30}}{n^{3/2}}$$

for all $n \geqslant n_0$.

(v) Conclude, using Stirling's formula again, that

$$A(2n+1, 4n+1) < -2^{4n}/n^4$$

for all sufficiently large n. (Thus, not merely does $-A(2n+1, 4n+1)$ tend to infinity, but it does so explosively fast. The paper of Pólya indicated in Question 41.3 contains a more refined calculation from which the calculations above are adapted.)

75.22 (Watson's lemma. This question is independent of the previous ones.)

(i) Suppose $f \in \mathscr{E}$. Graph e^{xt} for large x and state what aspects of the behaviour of f you expect to control the behaviour of $\mathscr{L}f(x)$ for large x.

(ii) Suppose $f \in \mathscr{E}$ with $|f(t)| \leqslant Me^{at}$ for all $t \in \mathbb{R}$. Suppose further that there exist A, $\varepsilon > 0$ and an integer $n \geqslant 0$ such that $|f(t)| \leqslant At^n$ for $0 \leqslant t \leqslant \varepsilon$. Show that, if x is positive and $x > a$, then

$$\left|\int_0^{\varepsilon} f(t)e^{-xt}\,dt\right| \leqslant Ax^{-n-1}n!, \quad \left|\int_{\varepsilon}^{\infty} f(t)e^{-xt}\,dt\right| \leqslant \frac{Me^{\varepsilon(a-x)}}{(x-a)}.$$

Deduce that we can find an x_0 such that $|x^n\mathscr{L}f(x)| \leqslant 2A/n!$ for all $x \geqslant x_0$.

(iii) Suppose $f \in \mathscr{E}$ and f is n times differentiable on the right at 0 (i.e. $f(0+), f'(0+), \ldots, f^{(n)}(0+)$ exist, where $f(0+) = \lim_{t\to 0+} f(t)$, $f^{(r+1)}(0+) = \lim_{h\to 0+}(f^{(r)}(h) - f^{(r)}(0+))/h$). Use Taylor's theorem (as presented e.g. in Question 4.2 (vi)) to show that

$$x^n\left(\mathscr{L}f(x) - \sum_{r=0}^{n} f^{(r)}(0+)x^{-r-1}\right) \to 0 \quad \text{as } x \to \infty.$$

(iv) Let $f\colon \mathbb{R} \to \mathbb{R}$ be a continuous function such that $f(t) = 0$ for $t \leqslant 1$ and for $t \geqslant 2$, but $f(3/2) \neq 0$. Show that $x^n\mathscr{L}f(x) \to 0$ as $x \to \infty$ for all n, but that $\mathscr{L}f(x)$ is not identically zero on any interval. Observe that, in particular, there are arbitrarily large x such that

$$\sum_{r=0}^{n} f^{(r)}(0)x^{-r} \not\to \mathscr{L}f(x) \quad \text{as } n \to \infty.$$

(v) Modify the arguments of (ii) and (iii) to show that, if f satisfies the hypotheses of (ii) and θ is fixed with $\pi/2 > \theta \geqslant 0$, then

$$R^N(\mathscr{L}f(Re^{i\phi}) - \sum_{r=0}^{N} f^{(r)}(0)(Re^{-i\phi})^{-r-1}) \to 0$$

uniformly as $R \to \infty$ for $|\phi| \leq \theta$. More succinctly, we say that

$$z^N\left(\mathcal{L}f(z) - \sum_{r=0}^{N} f^{(r)}(0)z^{-r-1}\right) \to 0$$

uniformly as $|z| \to \infty$ with $|\arg z| \leq \theta$.

75.23 (An extended Watson's lemma. This question requires the information contained in Question 75.13.)

(i) Generalise the arguments of the previous question to show that, if $\lambda(0)$, $\lambda(1), \ldots, \lambda(n)$ are real numbers with $-1 < \lambda(0) < \lambda(1) < \ldots < \lambda(n) < \lambda$ and $f \in \mathscr{E}$ is such that

$$t^{-\lambda}\left(f(t) - \sum_{r=0}^{n} a_r t^{\lambda(r)}\right) \to 0 \quad \text{as } t \to 0+$$

for some $a_0, a_1, \ldots, a_n \in \mathbb{C}$, then, for any fixed θ with $\pi/2 > \theta \geq 0$,

$$z^{\lambda}\left(\mathcal{L}f(z) - \sum_{r=0}^{n} \lambda(r)!a_r z^{-\lambda(r)-1}\right) \to 0$$

uniformly as $|z| \to \infty$ with $|\arg z| \leq \theta$. (Note that you must specify which definition of $z^{-\lambda(r)-1}$ you adopt.)

(ii) Suppose $g: \mathbb{R} \to \mathbb{R}$ is an infinitely differentiable bounded function. By writing

$$\int_{-\infty}^{\infty} g(t) \exp(-xt^2)\, dt = \int_0^{\infty} g(t) \exp(-xt^2)\, dt + \int_{-\infty}^{0} g(t) \exp(-xt^2)\, dt,$$

making simple changes of variable and applying (i), show that

$$x^{n-1/2}\left|\int_{-\infty}^{\infty} g(t) \exp(-xt^2)\, dt - \sum_{r=0}^{n} \frac{g^{(2r)}(0)}{(2r)!}(r - 1/2)!x^{-r-1/2}\right| \to 0$$

as $x \to \infty$. Why can we improve this, without further work, to

$$x^{\kappa}\left|\int_{-\infty}^{\infty} g(t) \exp(-xt^2)\, dt - \sum_{r=0}^{n} \frac{g^{(2r)}(0)}{(2r)!}(r - 1/2)!x^{-r-1/2}\right| \to 0$$

as $x \to \infty$ for all κ with $n + 3/2 > \kappa$?

(iii) Suppose $\alpha, \beta > 0$ and $g: [-\alpha, \beta] \to \mathbb{R}$ is a continuous function which is $2n$ times differentiable at 0. Show that

$$x^{n+1/2}\left|\int_{-\alpha}^{\beta} g(t) \exp(-xt^2)\, dt - \sum_{r=0}^{n} \frac{g^{(2r)}(0)}{(2r)!}(r - 1/2)!x^{-r-1/2}\right| \to 0$$

as $x \to \infty$ Why is the sum $\sum_{r=0}^{n} (g^{(2r)}(0)/(2r)!)(r - 1/2)!x^{-r-1/2}$ independent of α and β?

(iv) If g is as in (iii), obtain a similar result to the one given there, but involving $\int_{-\alpha}^{\beta} g(t)\exp(-xt^3)\,dt$.
(A more impressive application of Watson's lemma occurs in Question 95.6.)

75.24 (The pendulum and the tautichrone. This question requires a smattering of elementary mechanics. The second part makes use of Question 75.19.)

(i) Consider a simple pendulum consisting of a particle (the bob) of mass m attached to one end of a rod length l and negligible mass whose other end is attached to a fixed point. If θ is the angular displacement of the rod from the vertical, show, using conservation of energy, that

$$l\left(\frac{d\theta}{dt}\right)^2 = 2g(\cos\theta - \cos\alpha),$$

where α is the greatest value of θ. Deduce that the period $T(\alpha)$ is given by

$$2\left(\frac{l}{g}\right)^{1/2}\int_0^{\alpha}(\sin^2(\alpha/2) - \sin^2(\theta/2))^{-1/2}\,d\theta.$$

(Throughout this question all square roots will be taken positive.)

By making the substitution $\sin(\theta/2) = \sin(\alpha/2)\sin\psi$, and expanding the integrand in powers of $\sin(\alpha/2)$, prove that the period is approximately

$$2\pi\left(\frac{l}{g}\right)^{1/2}(1 + \alpha^2/16).$$

Observe that to a high order of accuracy the period $T(\alpha)$ is independent of the 'length of the swing' α, provided that α is kept small. This is why (as Galileo observed empirically) the swing of the pendulum can be used for timekeeping.

(ii) The great physicist and mathematician Huygens combined the theoretical and practical to a remarkable extent. He both worked out the mathematical theory of the pendulum (including all the results in this question) and built the first pendulum clock. (Until the coming of the atomic clock the pendulum clock in its various refinements was the most accurate time keeper available. A walk through the clock gallery of a big scientific museum is simultaneously soothing and awe inspiring.) The result of (i) shows that, for the standard pendulum, the period is not completely independent of amplitude. Huygens asked himself whether, if the bob was constrained to move in some different, non-circular path, if might be possible to remove all amplitude dependence. Such a path he called a tautochrone (from the Greek *tauto*, the same, and *chronos*, time). He solved this problem and we shall give a general method of Abel for solving more general problems of this nature.

Consider a particle of mass m moving in an xy plane with Ox axis vertically upward and Oy axis horizontal. This unusual choice of coordinates reflects the fact that it is height and not horizontal distance which is the key variable.

Suppose the particle is constrained (e.g. by being made to slide along a frictionless wire) to move along a well-behaved curve $y = f(x)$ passing through the origin $(0, 0)$. Let $s(x)$ be length of the curve from $(0, 0)$ to $(x, f(x))$ so that

$$s(x) = \int_0^x (1 + f'(u)^2)^{1/2}\, du.$$

By using conservation of energy, or otherwise, show that if the particle is initially at $(X, f(X))$ the time taken to reach the origin is

$$T(X) = \left(\frac{l}{2g}\right)^{1/2} \int_0^X \frac{s'(u)}{(X-u)^{1/2}}\, du.$$

Thus, given the initial height X, we can compute the time $T(X)$ of descent. 'Abel's mechanical problem' is to reverse the process, i.e. given $T(X)$ for all X to find f.

Using Question 75.19 (viii), show that

$$s'(x) = \frac{d}{dx}\frac{(2g)^{1/2}}{\pi}\int_0^x \frac{T(u)}{(x-u)^{1/2}}\, du.$$

(iii) We now specialise back to Huygen's original problem. Here we want $T(X) = T_0$ a constant independent of X. Show that

$$s'(x) = \left(\frac{b}{x}\right)^{1/2}$$

where $b = 2gT_0^2/\pi^2$. By recalling the definition of s, deduce that the curve is given by

$$\frac{dy}{dx} = \left(\frac{b-x}{x}\right)^{1/2}.$$

By making the substitution $x = b\sin^2\phi$ show that curve has the parametric form

$$x = \frac{b}{2}(1 - \cos 2\phi) \quad\text{and}\quad y = \frac{b}{2}(2\phi + \sin 2\phi).$$

By reparameterising obtain the form

$$x = a(1 - \cos\theta) \quad\text{and}\quad y = a(\theta + \sin\theta).$$

(iv) Show that this describes the path through $(0, 0)$ traced out by a fixed point on a circle of radius a rolling along the line $x = 2a$ with its centre following the line $x = a$. This curve is called a cycloid.

Hugens' first pendulum clock was constructed so that the bob moved in a cycloidal arc. However, it turned out to be easier to improve accuracy for the simple pendulum by using small amplitudes (halving the amplitude reduces the dependence by a factor of 4) and and using mechanisms to ensure that the amplitude of the swing remained essentially constant. As sometimes happens, the most elegant solution was not the most practical.

76

POLES AND STABILITY

76.1 (The first infinite product) By studying the methods of Archimedes for computing π the great sixteenth-century mathematician Vieta was led to an idea which I now sketch in twentieth-century form.

(i) Prove by induction, or otherwise, that

$$\sin x = 2^m \sin 2^{-m} x \prod_{n=1}^{m} \cos 2^{-n} x$$

for all $m \geqslant 1$ and all $x \in \mathbb{R}$.

(ii) Deduce that $x \prod_{n=1}^{m} \cos 2^{-n} x \to \sin x$ as $m \to \infty$ for all x. Is the convergence uniform?

(iii) By setting $x = \pi/2$ show that, in an obvious notation,

$$\frac{2}{\pi} = \left(\frac{1}{2}\right)^{1/2} \left(\frac{1}{2} + \frac{1}{2}\left(\frac{1}{2}\right)^{1/2}\right)^{1/2} \left(\frac{1}{2} + \frac{1}{2}\left(\frac{1}{2} + \frac{1}{2}\left(\frac{1}{2}\right)^{1/2}\right)^{1/2}\right)^{1/2} \cdots$$

(iv) Use a hand calculator to investigate the way the expression on the right approaches $2/\pi$ as we take more terms.

76.2 In the good old days (all days when old are good), a first course in analysis would cover infinite products as well as sums.

(i) If $a \geqslant 0$, show that $\exp a \geqslant 1 + a$.

(ii) If $a_j \geqslant 0$, for $j \geqslant 1$, show that

$$\exp\left(\sum_{j=1}^{n} a_j\right) \geqslant \prod_{j=1}^{n}\left(1 + a_j\right) \geqslant \sum_{j=1}^{n} a_j.$$

Deduce that $\prod_{j=1}^{n}(1 + a_j)$ tends to a limit (which we shall call $\prod_{j=1}^{\infty}(1 + a_j)$) as $n \to \infty$ if and only if $\sum_{j=1}^{n} a_j$ does.

(iii) If $b_j \in \mathbb{C}$, show that

$$\prod_{j=1}^{n}(1 + |b_j|) - 1 \geqslant \left| \prod_{j=1}^{n}(1 + b_j) - 1 \right|.$$

(iv) If $a_j \in \mathbb{C}$ and $\sum_{j=1}^{\infty} a_j$ is absolutely convergent, show that

$$\left(\exp\left(\sum_{j=m}^{n}|a_j|\right)-1\right)\prod_{j=1}^{\infty}(1+|a_j|)\geqslant\left|\prod_{j=1}^{n}((1+a_j)-\prod_{j=1}^{m}(1+a_j)\right|$$

for all $n \geqslant m \geqslant 0$. Deduce that $\prod_{j=1}^{n}(1+a_j)$ tends to a limit which we shall call $\prod_{j=1}^{\infty}(1+a_j)$.

(v) Suppose that Ω is an open set in $\mathbb{C}$, $f_j: \Omega \to \mathbb{C}$ is analytic and $\sum_{j=1}^{\infty}|f_j(z)|$converges uniformly on Ω. Show that $g(z)=\prod_{j=1}^{\infty}(1+f_j(z))$ gives a well-defined analytic function $g: \Omega \to \mathbb{C}$.

(vi) Suppose $a, b \in \mathbb{C}$ and $(1+a)(1+b)=1$. If $|a|<1/2$, show that $|b|<2|a|$.

(vii) Suppose $a_j, b_j \in \mathbb{C}$ and $(1+a_j)(1+b_j)=1$ for each j. Show that, if $\sum_{j=1}^{\infty} a_j$ is absolutely convergent, so is $\sum_{j=1}^{\infty} b_j$ and, further,

$$\prod_{j=1}^{\infty}(1+a_j)\prod_{j=1}^{\infty}(1+b_j)=1$$

(the argument is easy, but take care to get it right). Deduce that, if $a_j \in \mathbb{C}$, $a_j \neq -1$ for all j and $\sum_{j=1}^{\infty} a_j$ is absolutely convergent, then $\prod_{j=1}^{\infty}(1+a_j)\neq 0$.

(viii) Suppose that $a_j>0$ for all j and $\sum_{j=1}^{\infty} a_j$ converges. Show that, if we set $R_n=\sum_{j=n+1}^{\infty} a_j$, then $\sum_{n=1}^{\infty} a_n/R_n$ diverges.

(ix) Using part (v) show that, if R is fixed, then $\prod_{j=1}^{\infty}(1-z^2/n^2)$ is a well-defined analytic function on $\{z: |z|>R\}$. Deduce that

$$g(z)=z\prod_{j=1}^{\infty}(1-z^2/n^2)$$

is a well-defined analytic function on $\mathbb{C}$.

(x) (The last two parts of this question depend on earlier questions, but an alternative derivation of the result proved will be given in Question 76.8.) In Question 38.15(iii) we saw that, if $1/2>\delta>0$ is fixed, then

$$x^{-1}-\sum_{1\leqslant n\leqslant N} 2x/(n^2-x^2)\to\pi\cot\pi x$$

uniformly as $N\to\infty$ for all real $x\in[\delta, 1-\delta]$. By integrating, show that

$$\log\left(x\prod_{n=1}^{N}(n^2-x^2)\right)-\log\left(1/2\prod_{n=1}^{N}(n^2-(1/2)^2)\right)\to\log\sin\pi x$$

for all $x\in[\delta, 1-\delta]$ and conclude that

$$\log g(x)-\log g(1/2)=\log\sin\pi x$$

for all $x\in[\delta, 1-\delta]$ with $1/2>\delta>0$.

(xi) Deduce that $g(x)/g(1/2)=\sin\pi x$ for $0<x<1$ and, by considering what happens as $x\to 0+$, show that $g(x)=\pi^{-1}\sin\pi x$ for all $0<x<1$. Deduce that $g(z)=\pi^{-1}\sin\pi z$ for all $z\in\mathbb{C}$, i.e.

$$\sin \pi z = \pi z \prod_{n=1}^{\infty} (1 - z^2/n^2)$$

for all $z \in \mathbb{C}$.

76.3 We shall see in Chapter 105 and onwards that infinite products are an important tool in number theory. They are also very useful in studying the zeros of analytic functions. Here is a beautiful proof of Weierstrass that, given $a_n \to \infty$, we can find an analytic function with zeros at, and only at, the points a_n. (Moreover we can demand that, if w occurs exactly r times in the sequence (a_n), then w will be a zero of order r).

(i) Suppose $a_n \in \mathbb{C}$ and $a_n \not\to \infty$ as $n \to \infty$. What can we say about an analytic function $f: \mathbb{C} \to \mathbb{C}$ with $f(a_n) = 0$ for all n, and why?

(ii) Let $E_k(z) = (1 - z)\exp(z + z^2/2 + \ldots z^k/k)$. Show that E_k has a simple zero at 1 and no other zeros. Show also that, if $\delta > 0$ is fixed, $E_k(z) \to 1$ uniformly as $k \to \infty$ for $|z| \leqslant 1 - \delta$.

(iii) Suppose $a_n \neq 0$ for all n and that $a_n \to \infty$. By reordering if necessary, we may suppose that $|a_1| \leqslant |a_2| \leqslant \ldots$ Show that we can find $k(n)$ such that $|E_{k(n)}(z/a_n) - 1| \leqslant 2^{-n}$ for all $|z| < |a_n|/2$.

(iv) Let $R > 0$. Choose N such that $|a_N| > 2R$. Show, using the results of Question 76.2, that $\prod_{n=N}^{\infty} E_{k(n)}(z/a_n)$ is a well-defined, nowhere zero, analytic function on $|z| < R$. Deduce that $g(z) = \prod_{n=1}^{\infty} E_{k(n)}(z/a_n)$ is a well-defined analytic function on $|z| < R$ and that, if $|w| < R$, then w is a zero of g of order m if and only if w occurs in the sequence $a_1, a_2, \ldots$ exactly m times.

(v) Deduce that, if $a_n \to \infty$, we can find $F: \mathbb{C} \to \mathbb{C}$ analytic such $w \in \mathbb{C}$ is a zero of F of order m if and only if w occurs in the sequence $a_1, a_2, \ldots$ exactly m times.

(vi) Suppose that $H: \mathbb{C} \to \mathbb{C}$ is an analytic function such that $w \in \mathbb{C}$ is a zero of H of order m if and only if w occurs in the sequence $a_1, a_2, \ldots$ exactly m times. By showing that H/F has removable singularities at $a_1, a_2, \ldots$, or otherwise, show that we can write $H(z) = h(z)F(z)$ where $h: \mathbb{C} \to \mathbb{C}$ is analytic and nowhere zero. The expression $h(z)\prod_{n=1}^{\infty} E_{k(n)}(z/a_n)$ is called a Weierstrass factorisation of H.

(vii) Let $D = \{z: |z| < 1\}$. Show that, if $|a_n| \to 1$, we can find $F: D \to \mathbb{C}$ analytic such that w is a zero of F of order m if and only if w occurs in the sequence $a_1, a_2, \ldots$ exactly m times.

(viii) In (vii) let $a_{2^n - r} = (1 - 2^{-n})\exp(2^{n-1}\pi i r)$ for all $0 \leqslant r \leqslant 2^n - 1$, $n \geqslant 0$. Show that, with the definition of Question 75.5, $\{z: |z| = 1\}$ is a natural boundary of F.

76.4 Suppose Ω is an open set in $\mathbb{C}$, $f: \Omega \to \mathbb{C}$ is an analytic function and $\sum_{j=1}^{\infty} |f_j(z) - 1|$ converges uniformly on every disc $D(w, \delta) = \{z: |z - w| < \delta\}$ with $D(w, \delta) \subseteq \Omega$. Suppose further that $f_j(z) \neq 0$ for all $z \in \Omega$ and $j \geqslant 1$.

(i) Show that, if $D(w, 2\delta) \subseteq \Omega$, then there exists an $\eta > 0$ such that $|\prod_{j=1}^{\infty} f_j(z)| > \eta$ for all $|z - w| \leq \delta$.

(ii) Let $D(w, 2\delta) \subseteq \Omega$. Show that, writing

$$g_n(z) = \prod_{j=1}^{n} f_j(z), \quad g(z) = \prod_{j=1}^{\infty} f_j(z),$$

we have $g_n(z) \to g(z)$, $1/g_n(z) \to 1/g(z)$ and $g_n'(z) \to g'(z)$ uniformly on $D(w, \delta)$. Hence deduce that $\sum_{j=1}^{\infty} f_j'(z)/f_j(z)$ converges uniformly to $g'(z)/g(z)$ on $D(w, \delta)$.

(iii) Conclude that $g'(z)/g(z) = \sum_{j=1}^{\infty} f_j'(z)/f_j(z)$ for all $z \in \Omega$.

(iv) If $b \in \mathbb{C}$ and $|1 - b| < 1/2$, show that we can find an $a \in \mathbb{C}$ with $|a| < 2|b|^{-1}$ such that $\exp a = b$.

(v) Suppose now that $\Omega = \{z \colon \operatorname{Re} z < \alpha\}$ and $w(0) \in \Omega$. Show that we can find $a_j \in \mathbb{C}$ with $\sum_{j=1}^{\infty} a_j$ convergent and $\exp a_j = f_j(w(0))$.

(vi) Define

$$\lg f_j(w) = a_j + \int_{C(w)} \frac{f_j'(z)}{f_j(z)} dz,$$

where $C(w)$ is the straight line from $w(0)$ to w. Show (e.g. by imitating the standard proof of Morera's theorem) that $\lg f_j \colon \Omega \to \mathbb{C}$ is analytic with $(\lg f_j)' = f_j'/f_j$. By considering the derivative of $(\exp(\lg f_j))/f_j$, or otherwise, show that $(\exp(\lg f_j(z))) = f_j(z)$ for all $z \in \Omega$.

(vii) Set

$$\lg g(w) = \sum_{j=1}^{\infty} a_j + \int_{C(w)} \frac{g'(z)}{g(z)} dz$$

with $C(w)$ as before. As before, $\lg g \colon \Omega \to \mathbb{C}$ is analytic with $(\lg g)' = g'/g$ and $(\exp(\lg g(z))) = g(z)$ for all $z \in \Omega$. Show that $\sum_{j=1}^{\infty} \lg f_j(z)$ converges to $\lg g(z)$.

76.5 In Question 31.1 we showed that, if $a_j \in \mathbb{C}$ and $|a_j| < 1$ for each j, then $a_1, a_2, \ldots$ can be the zeros of some non constant bounded analytic function $F \colon \{z \colon |z| < 1\} \to \mathbb{C}$ (multiple zeros counted multiply) only if $\sum_{j=1}^{\infty}(1 - |a_j|)$ converges.

(i) Let $\Omega = \{z \colon \operatorname{Re} z > 1\}$ and $D = \{z \colon |z| < 1\}$. Let $T \colon \Omega \to \mathbb{C}$ be given by $Tz = 1/z$ and $S \colon \mathbb{C} \to \mathbb{C}$ by $Sz = 2z - 1$. Show that $ST \colon \Omega \to D$ is a conformal map. Suppose $b_j \in \Omega$ and $|b_j| \to \infty$ as $j \to \infty$. Show that $\sum_{j=1}^{\infty} -\log|STb_j|$ converges if and only if $\sum_{j=1}^{\infty} |b_j|^{-1}$ converges.

(ii) Suppose $F \colon \Omega \to \mathbb{C}$ is an analytic bounded function which has zeros at $b_1, b_2, \ldots$ where $\sum_{j=1}^{\infty} |b_j|^{-1}$ diverges. Show that $F = 0$.

(iii) Let $a > d$. Suppose that $f \in \mathscr{E}_a$ and that $\mathscr{L}f$ has zeros at $b_1, b_2, \ldots$ where $\operatorname{Re} b_j > d$ and $\sum_{j=1}^{\infty} |b_j|^{-1}$ diverges. Show that $f = 0$.

(iv) Deduce Lerch's theorem (Theorem 75.10).

76.6 (A converse to Jensen's result of Question 31.1)
(i) If $0<|a|<1$, set

$$B(a, z) = \frac{|a|}{a}\left(\frac{a - z}{1 - a^* z}\right) \text{ for } |z|<1.$$

Using Question 31.1 (iii), show that $|B(a, z)|<1$ for $|z|<1$. Set $B_0(z) = z$.
(ii) If $0<|a|<1$, show that

$$1 - B(a, z) = \frac{(a + |a| z)}{(1 - a^* z)a}(1 - |a|) \text{ for } z \neq a^{*-1}.$$

(iii) Show that, if $|a_j|<1$ for all j and $\sum_{j=1}^{\infty}(1-|a_j|)$ converges, then $\sum_{j=1}^{\infty}|1 - B(a_j, z)|$ converges uniformly for $|z|<R$ whenever $R<1$.
(iv) Conclude that, under these hypotheses, $g(z) = \prod_{j=1}^{\infty} B(a_j, z)$ yields a well defined analytic function $g: D \to \mathbb{C}$ with $|g(z)|<1$ for $|z|<1$ and zeros precisely at $a_1, a_2, \ldots$ (multiple zeros appearing multiply). We call g the Blaschke product.
(v) Show that we can find a_j with $0<|a_j|<1$, $\sum(1-|a_j|)$ convergent such that $\prod_{j=1}^{n}(a_j - z)/(1 - a_j^* z)$ fails to converge on D as $n \to \infty$. (Look at $z = 0$.)
(vi) Show that we can find $|a_j|<1$ such that $\sum_{j=1}^{\infty}(1-|a_j|)$ converges and, given any $\theta \in \mathbb{R}$, we can find a subsequence $a_{n(k)} \to e^{i\theta}$. Let g be the function of part (iv). Show that with the definition of Question 75.5 $\{z: |z| = 1\}$ is the natural boundary of g.

76.7 (i) Let $f: \mathbb{C} \to \mathbb{C}$ be analytic and suppose $f(0) \neq 0$. Let A be the set of zeros of f. By rescaling the equality of Question 31.1 (iv), show that

$$\log|f(0)| = \frac{1}{2\pi}\int_0^{2\pi} \log|f(Re^{i\theta})|\,d\theta + \sum_{w \in A, |w|<R} \log(|w|/R)$$

for all $R>0$ such that $\{w \in A: |w| = R\} = \varnothing$.
(ii) Show that, if $g: \mathbb{C} \to \mathbb{C}$ is an analytic function such that $g(n) = 0$ for all $n \in \mathbb{Z}$ and $|g(z)| \leqslant Ke^{k|z|}$ for all $z \in \mathbb{C}$ and some $K>0$, $1>k \geqslant 0$, then $g = 0$. (*Hint.* Recall Stirling's formula (see e.g. Question 18.16 or Questions 76.9 and 76.10 below).) Can we replace the condition '$1>k\geqslant 0$' by '$1\geqslant k\geqslant 0$'? Give a proof or a counter example.
(iii) By modifying the ideas above, show that, if $a \in \mathbb{R}$ and $g: \{z: \operatorname{Re} z > a\} \to \mathbb{C}$ is an analytic function such that $g(n) = 0$ for all $n \in \mathbb{Z}$ with $n>a$, and $|g(z)| \leqslant Ke^{k|z|}$ for all $z \in \mathbb{C}$, $\operatorname{Re} z>a$ and some $K>0$, $1>k\geqslant 0$, then $g = 0$.
(iv) Deduce Lerch's theorem (Theorem 75.10).

76.8 The object of this question is to obtain the formulae

$$\sin \pi z = \pi z \prod_{n=1}^{\infty}(1 - z^2/n^2), \quad \cos \pi z = \prod_{n=1}^{\infty}(1 - z^2/(n - 1/2)^2)$$

in a different way from that used in Question 76.2.

(i) Show that the infinite products

$$\prod_{n=1}^{\infty}(1 - z^2/n^2) \text{ and } \prod_{n=1}^{\infty}(1 - z^2/(n - 1/2)^2)$$

converge uniformly in $|z| < R$ for any fixed R. Show that, if

$$\phi(z) = \pi z \prod_{n=1}^{\infty}(1 - z^2/n^2), \quad \psi(z) = \prod_{n=1}^{\infty}(1 - z^2/(n - 1/2)^2),$$

ϕ and ψ are well-defined analytic functions on $\mathbb{C}$. Explain why ϕ has zeros at the points $z = n$ with n an integer and nowhere else. (Recall Question 76.2(vii).) Where are the zeros of ψ?

(ii) By writing $\phi(z) = \lim_{N\to\infty} \pi z \prod_{|m|\leq N, m\neq 0}(1 + z/m)$, or otherwise, show that $\phi(z + 1) = -\phi(z)$ and ϕ is periodic with period 2. Show also that $\phi(-z) = -\phi(z)$, $\psi(z + 1) = -\psi(z)$ and $\psi(-z) = \psi(z)$ for all z.

(iii) Prove along the same lines that $\phi(2z) = 2\phi(z)\psi(z)$ and $\phi(1/2)\psi(z) = \phi(z + 1/2)$. Conclude that $\phi(2z)\phi(1/2) = 2\phi(z)\phi(z + 1/2)$.

(iv) Show that $g(z) = \pi\phi(z)/\sin \pi z$ is analytic, except for removable singularities at integer points, and so g can be extended to an analytic function. What is the value of $g(0)$? What is the value of $g'(0)$?

(v) For the next two parts of the question, we restrict our functions to have domain the real line. Show that $\phi(x) > 0$ for $0 < x < 1$ and deduce, using the formulae of (ii), that the restricted function $g: \mathbb{R} \to \mathbb{R}$ is a strictly positive, continuous function with period 1.

(vi) Set $f(x) = d(\log g(x))/dx$. Show, using the formula concluding (iii), that

$$f(2x) = \tfrac{1}{2}(f(x) + f(x + 1/2)) \quad \text{for all } x \in \mathbb{R}.$$

Explain why f has a maximum m, say, attained at x_0, say. Show from the formula for $f(2x)$ that $f(x_0/2) = m$, and so, by induction, that $m = f(2^{-n}x_0)$ for all $n \geq 1$. Deduce, by continuity, that $f(0) = m$ and so $f(0) \geq f(x)$ for all $x \in \mathbb{R}$. Show similarly that $f(x) \leq f(0)$ for all $x \in \mathbb{R}$ and so f is constant. Deduce that $g(x) = g(0)$ for all $x \in \mathbb{R}$.

(vii) We now return to the complex plane. Explain why the fact that $g(z) = g(0)$ for all real z implies that $g(z) = g(0)$ for all z and deduce that $\phi(z) = \sin \pi z$ as required. Show, using (iii), that $\psi(z) = \cos \pi z$.

(viii) By setting $z = 1/2$ in the formula $\phi(z) = \sin \pi z$, obtain the result $\prod_{r=1}^{n}(1 - 1/4r^2) \to 2/\pi$ as $n \to \infty$ proved in Question 18.15(iv).

(Euler was led to the formulae of these questions as generalisations of the Wallis formula.)

(ix) Show carefully that

$$\pi z \prod_{n=1}^{\infty}\left(1 - \frac{z^2}{n^2}\right) = \pi\left(z - z^3 \sum_{n=1}^{\infty}\frac{1}{n^2} + z^3\varepsilon(z)\right),$$

where $\varepsilon(z)\to 0$ as $z\to 0$. Hence deduce that $\sum_{n=1}^{\infty} n^{-2}=\pi^2/6$.

(x) If you have done Question 75.16(v), use Gauss's formula to obtain

$$\frac{1}{x!}\frac{1}{(-x)!}=\frac{\sin \pi x}{\pi x}$$

for all real x with x not an integer. Deduce that

$$z!(-z)! = \pi z \operatorname{cosec} \pi z$$

for all $z\in\mathbb{C}$ with z not an integer. (Thus we have a new proof of Question 75.14(v).)

76.9 (Another proof of Stirling's formula) In the December 1967 issue of the *American Mathematical Monthly*, Feller gave a very neat proof of Stirling's formula based on the infinite product expression for sine.

(i) Suppose $f\colon [0,a]\to\mathbb{R}$ is continuous and there exist $A>B>0$ such that $At\geqslant f(t)\geqslant Bt$ for all $t\in[0,a]$. Show that $\int_0^a \log f(t)\,dt$ exists.

(ii) Write $a_k=\frac{1}{2}\log k-\int_{k-1/2}^{k}\log x\,dx$, $b_k=\int_k^{k+1/2}\log x\,dx-\frac{1}{2}\log k$. Show that

$$a_1-b_1+a_2-b_2+\ldots+a_{n-1}-b_{n-1}+a_n=$$

$$\log n!-\tfrac{1}{2}\log n-\int_{1/2}^{n}\log x\,dx.$$

Show also that $a_k>b_k>a_{k+1}>0$ and that $a_k\to 0$ as $k\to\infty$. Draw a diagram showing $\log x$ and rectangles with height $\log k$ and base the interval $[k-1/2, k+1/2]$ to illustrate your argument.

(iii) Explain why $\sum_{k=1}^{\infty}(a_k-b_k)$ converges to a limit S, say, and by evaluating $\int_0^n \log x\,dx$, by integration by parts, or otherwise, show that

$$\log n!-(n+1/2)\log n+n\to S-\int_0^{1/2}\log x\,dx=C,\text{ say.}$$

(iv) Show that $a_k=-\int_0^{1/2}\log(1-t/k)\,dt$ and, by finding a similar expression for b_k, show that

$$a_k-b_k=-\int_0^{1/2}\log\left(1-\frac{t^2}{k^2}\right)dt.$$

Deduce that

$$S=-\int_0^{1/2}\log\left(\prod_{k=1}^{\infty}\left(1-\frac{t^2}{k^2}\right)\right)dt$$

and so

$$C=-\int_0^{1/2}\log\left(\frac{\sin\pi t}{\pi}\right)dt.$$

(v) Using the formula $\sin 2\pi t = 2 \sin \pi t \cos \pi t$ and appropriate symmetries, show that

$$\int_0^{1/2} \log \sin \pi t \, dt = \int_0^{1/2} \log \sin 2\pi t \, dt = \tfrac{1}{2} \log 2 + 2 \int_0^{1/2} \log \sin \pi t \, dt.$$

Deduce that $\int_0^{1/2} \log \sin \pi t \, dt = -\frac{1}{2} \log 2$ and so $C = \frac{1}{2} \log 2\pi$.

(vi) Conclude that $\dfrac{e^n n!}{n^{n+1/2}} \to (2\pi)^{1/2}$.

76.10 (Yet another proof of Stirling's formula. This is taken from a 1929 Cambridge examination paper.)

(i) Show that, if $x > -1$, there exists a θ with $0 < \theta < 1$ such that $\log(1 + x) = x - x^2/2(1 + \theta x)^2$.

(ii) If $t, a > 0$, show that the equation

$$t^2 = x - a - a \log(x/a)$$

has two real roots x_1, x_2 with $x_1 < x_2$. Show that we can find $-1 < \theta_1 < 0 < \theta_2 < 1$ such that

$$\frac{a}{a - x_1} + \theta_1 = \frac{a}{x_2 - a} + \theta_2 = \frac{1}{t}\sqrt{\left(\frac{a}{2}\right)}.$$

(iii) By making a substitution based on the equation $t^2 = x - a - a \log(x/a)$, show that, if $a > 0$,

$$a! = \int_0^\infty x^a e^{-x} \, dx = 2\left(\frac{a}{e}\right)^a \int_0^\infty e^{-t^2} \left(\frac{x_1(t)}{a - x_1(t)} + \frac{x_2(t)}{x_2(t) - a}\right) t \, dt,$$

where $x_1 = x_1(t)$, $x_2 = x_2(t)$ are defined as in (ii).

(iv) By using the last line of (ii), deduce that

$$\frac{a! e^a}{a^{a+1/2}} \to \sqrt{(2\pi)} \quad \text{as } a \to \infty.$$

76.11 Show that the infinite product $\prod_{n=1}^\infty (1 - z/n)e^{z/n}$ converges to limit $g(z)$, say, for all $z \in \mathbb{C}$. Show also that g is analytic with

$$g'(z) = g(z) \sum_{n=1}^\infty \left(\frac{1}{z - n} + \frac{1}{n}\right).$$

Deduce that

$$g(z + 1) = -zg(z)A \tag{*}$$

for some constant A. By considering a particular value of z, or otherwise, show that A is real and positive and that

$$\sum_{n=1}^{N} \frac{1}{n} - \log N \to \log A$$

as $N \to \infty$.

Thus we have another proof that Euler's constant

$$\gamma = \lim_{N\to\infty} \left(\sum_{n=1}^{N} \frac{1}{n} - \log N\right)$$

exists and (*) above takes the more elegant form

$$g(z+1) = -zg(z)e^{\gamma}.$$

77

A SIMPLE TIME DELAY EQUATION

77.1 (i) Suppose $g : \mathbb{R} \to \mathbb{R}$ is a bounded continuous function such that $g(t) = 0$ for $t \leqslant 0$. Without worrying too much about rigour, find the Laplace transform of the function $f : \mathbb{R} \to \mathbb{R}$ such that $f(t) = 0$ for $t \leqslant 0$ and

$$f'(t) + f(t) - f(t-1) = g(t) \text{ for all } t.$$

Deduce that

$$f(t) = \int_0^t R(t-s)g(s)\,ds$$

where $R(t) = \sum_{n=0}^{\infty} e^{n-t}((t-n)^n/n!)H(t-n)$ and H is the Heaveside function (see the definition before Lemma 74.6).

(ii) Show rigorously that f is indeed a solution of the stated problem, and (e.g. by imitating the proof of Lemma 77.1(i)) that it is the unique solution.

(iii) Obtain a similar explicit solution to the problem (*) discussed in Chapter 77.

77.2 (i) Find the number of roots of the equation $z^4 - z^3 + 6z^2 - 2z + 3 = 0$ in each of the four quadrants of the complex plane.

(ii) Consider the polynomial $p(z) = z^4 + 6z^3 + cz^2 + dz + e$ with real coefficients. Suppose that p has no real zeros. Show that, if none of the zeros of $z^4 + cz^2 + e$ are real, then p has one root in each of four quadrants of the complex plane.

77.3 (Do not worry too much about rigour in this question.)

(i) Laguerre's differential equation is

$$tL_n'' + (1-t)L_n' + nL_n(t) = 0, \qquad (*)$$

where n is a positive integer. Recalling the formula for $\mathscr{L}(f')$ and $\mathscr{L}(g)$, where $g(t) = tf(t)$, take the Laplace transform of (*) and solve the resulting differential equation to obtain

$$(\mathscr{L}L_n)(z) = C_n(z-1)^n/z^{n+1} \qquad (**)$$

where C_n is an arbitrary constant.

(ii) Is there anything about the form of (*) which could explain why (**) can contain an arbitrary constant? (You are not asked to go any further than making a suggestion.)

(iii) Show that the $L_n(t)$ given by (**) is a polynomial of degree n (for $t \geqslant 0$).

(iv) By taking Laplace transforms of both sides show that

$$\frac{e^{-xt/(1-x)}}{1-x} = \sum_{n=0}^{\infty} L_n(t)x^n$$

for a suitable choice of C_n.

77.4 (Causal black boxes) This question and the ones after it continue Questions 60.12 to 60.14 which the reader is supposed to have done. Write $\mathscr{A}$ for the set of continuous functions $f: \mathbb{R} \to \mathbb{R}$ such that $f(t) = 0$ for $t < 0$. If $a \geqslant 0$ define $S_a : \mathscr{A} \to \mathscr{A}$ by $S_a f(t) = f(t-a)$. Consider a function $T: \mathscr{A} \to \mathscr{A}$.

(i) Suppose that T satisfies the following conditions

(A) $TS_a f = S_a Tf$ for all $f \in \mathscr{A}$, $a \geqslant 0$.

(B) $T(\lambda_1 f + \lambda_2 f_2) = \lambda_1 Tf_1 + \lambda_s Tf_2$ for all $f_1, f_2 \in \mathscr{A}$, $\lambda_1, \lambda_2 \in \mathbb{R}$.

(C) Given $R > 0$, we can find a $\kappa(R) > 0$ such that, if $|f(t)| \leqslant 1$ for $0 \leqslant t \leqslant R$, then $|(Tf)(t)| \leqslant \kappa(R)$ for all $0 \leqslant t \leqslant R$.

(D) If Δ_n is defined as in Question 60.12 then, for each fixed $R \geqslant 0$, $T(n\Delta_n)(t) \to K(t)$ uniformly for $0 \leqslant t \leqslant R$.

Explain why $K \in \mathscr{A}$. Show by modifying the arguments of Question 60.12 that

$$(Tf)(t) = (f * K)(t) \text{ for all } t \in \mathbb{R}$$

where $f * K(t) = \int_{-\infty}^{\infty} f(t-u)K(u)\,du = \int_0^t f(t-u)K(u)\,du$ if $t \geqslant 0$ and $f * K(t) = 0$ if $t \leqslant 0$.

(ii) Prove conversely that, if $K \in \mathscr{A}$, then $Tf = f * K$ defines a function $T: \mathscr{A} \to \mathscr{A}$ satisfying conditions (A), (B), (C), (D). We call T a 'causal black box function' and K the 'transfer function of T'.

(iii) Discuss briefly the differences between the black box of question and that of Question 60.12. In order to apply Laplace transform techniques, we must restrict the generality of our causal black box a little. Let us call a causal black box a Laplace black box if

(E) We can find a, $A \in \mathbb{R}$ with $A > 0$ such that, if $f \in \mathscr{A}$ and $\int_{-\infty}^{\infty} |f(t)|\,dt \leqslant 1$, then $|Tf(t)| \leqslant Ae^{at}$ for all $t \in \mathbb{R}$.

Show that a causal black box is a Laplace black box if and only if its transfer function $K \in \mathscr{E}$.

(iv) Show that, if T is a Laplace black box, then, if $f \in \mathscr{A} \cap \mathscr{E}$, it follows that $Tf \in \mathscr{A} \cap \mathscr{E}$. (The reader who feels that condition (E) is merely a mathematical convenience should ask herself how a causal black box which is not Laplace would behave. She may also feed that we should not restrict ourselves to considering f and K in $\mathscr{A} \cap \mathscr{E} = \mathscr{E} \cap C = \{g \in \mathscr{E}: g \text{ continuous}\}$, but should allow $f, K \in \mathscr{E}$. There is, in fact, no problem in extending our discussion to functions with simple discontinuities and the reader is invited to provide the extension if she wishes.)

77.5 (A simple feedback system) In Question 60.14(v) we put together two black boxes to form a feedback system, but the scope of our investigation was limited by the fact that the result might not be a black box. Now let us consider the same system, but with the Laplace black boxes of the previous question.

(i) Explain why this reduces to investigating the equation

$$f_1 = k_1 * (f + \alpha k_2 * f_1)$$

where k_1, k_2, $f \in \mathscr{A}$. The first two quesions which arise are whether this equation actually has a solution $f_1 \in \mathscr{A}$ and whether, if it exists, it is unique. Show, by taking Laplace transforms, that, if f_1 exists, it is unique.

(ii) Show that, if the equation

$$g = h + w * g$$

where h, $w \in \mathscr{A}$, always has a solution $g \in \mathscr{A}$, then so does the equation in (i). It is natural to seek a solution by successive approximation

$$g_1 = h, g_2 = h + w * g_1, \ldots, g_n = h + w * g_{n-1}$$

and this we proceed to do.

(iii) Let $w: \mathbb{R} \to \mathbb{R}$ be continuous with $w(t) = 0$ for $t \leqslant 0$ and $|w(t)| \leqslant Ae^{at}$ for $t \geqslant 0$. Show by induction, or otherwise, that if we set $w_1(t) = w(t)$ and $w_{n+1} = w_n * w$ then w_n is continuous, $w_n(t) = 0$ for $t \leqslant 0$ and $|w_{n+1}(t)| \leqslant A^{n+1} t^n e^{at}/n!$. Conclude that

$$\sum_{n=1}^{N} w_n(t) \to W(t) \text{ uniformly on } (-\infty, R) \text{ as } N \to \infty$$

for each $R > 0$, where $W: \mathbb{R} \to \mathbb{R}$ is a continuous function with $W(t) = 0$ for $t \leqslant 0$ and $|W(t)| \leqslant Ae^{(A+a)t}$ for $t \geqslant 0$. Show also that $w * W = W - w$.

(iv) Show that the equation of (ii) always has the solution $g = h + W * h$. Show that there exists a $k \in \mathscr{A}$ such that the equation of (i) always has the solution $f_1 = k * f$. By taking Laplace transforms show that k is unique and there exists a b such that

$$(\mathscr{L}k)(z) = \frac{(\mathscr{L}k)(z)}{1 - \alpha(\mathscr{L}k_1)(z)(\mathscr{L}k_2)(z)}$$

for $\operatorname{Re} z > b$.

(v) In this part of the question you should assume that $\mathcal{L}k_1$ and $\mathcal{L}k_2$ are rational functions (i.e. the quotient of two polynomials) and that $\mathcal{L}k_1(z), \mathcal{L}k_2(z) \to 0$ as $|z| \to \infty$. Which of the following statements are true and which false, and why?

(a) If boxes 1 and 2 are stable, the system must be stable.
(b) If box 2 is unstable then, in practice, the system will be unstable.
(c) If box 1 is stable, then there exists a $\delta > 0$ such that, if $|\alpha| < \delta$, the system is stable.

77.6 (The Nyquist criterion) By a series of happy ideas Nyquist transformed the theory of Question 77.5 into a practical tool.

(i) Suppose $k \in L^1 \cap C$ and $k(t) = 0$ for $t \leqslant 0$. Suppose $\lambda \in \mathbb{R}$, $E_\lambda(t) = 0$ for $t \leqslant 0$, $E_\lambda(t) = \exp(i\lambda t)$ for $t \geqslant \delta$ and E_λ is continuous. Show that $\exp(-i\lambda t)(k * E_\lambda(t)) \to \hat{k}(\lambda)$ as $t \to \infty$. Explain how, given a Laplace black box with well behaved transfer function k, you would determine $\hat{k}(\lambda)$ for given $\lambda \in \mathbb{R}$.

(ii) Explain why, if $k \in L^1 \cap C$ and $k(t) = 0$ for $t \leqslant 0$ and for $t \geqslant R$, then $\mathcal{L}k$ can be extended to an analytical function $\mathcal{L}k: \mathbb{C} \to \mathbb{C}$. If k is once continuously differentiable, show that $\mathcal{L}k(re^{i\theta}) \to 0$ as $r \to \infty$ uniformly for $-\pi/2 < \theta < \pi/2$.

(ii)′ (Optional) Show that, if $k \in L^1 \cap C$ and $k(t) = 0$ for $t \leqslant 0$, then $\mathcal{L}k(z) = \int_{-\infty}^{\infty} k(t)e^{-zt}\,dt$ defines a continuous function $\mathcal{L}k: \{z: \operatorname{Re} z \geqslant 0\} \to \mathbb{C}$ which is analytic on $\{z: \operatorname{Re} z > 0\}$ and such that $\mathcal{L}k(re^{i\theta}) \to 0$ as $r \to \infty$ uniformly for $-\pi/2 \leqslant \theta \leqslant \pi/2$. (If you have done parts (i) and (iii) in Question 60.19 you need only to translate the result obtained there into new language. If not you may wish to look at Question 60.19 for hints.)

(iii) Without overdoing the rigour, convince yourself that Theorem 77.3 can be restated as follows. Let $\gamma: [0,2\pi] \to \mathbb{C}$ be a contour C with $\gamma(0) = \gamma(2\pi)$. Then the number of zeros of h within C is equal to the number of times $h(\gamma(t))$ winds round zero as t runs from 0 to 2π. If $h(z) \neq a$ for $z \in C$, show that the number of solutions of $h(z) = a$ within C is equal to the number of times $h(\gamma(t))$ winds round a as t runs from 0 to 2π.

(iv) Suppose the transfer functions k_1 and k_2 in Question 77.5 are well behaved (and, in particular, stable). Explain why you would expect the feedback system of that question to be unstable if $\mathcal{L}k_1(z)\mathcal{L}k_2(z) = \alpha^{-1}$ has a solution with $\operatorname{Re} z > 0$, and stable if $\mathcal{L}k_1(z)\mathcal{L}k_2(z) = \alpha^{-1}$ has no solution with $\operatorname{Re} z \geqslant 0$.

(v) Show that, if we connect our black boxes in series so that the input to our series is the input to box 1, the output from box 1 is the input to box 2 and the output from box 2 is the output from the series system, then (if k_1 and k_2 are well behaved) the transfer function K of the series system will be well behaved with $\mathcal{L}K(z) = \mathcal{L}k_1(z)\mathcal{L}k_2(z)$.

(vi) Let R be real, positive and very large. Let $\gamma: [0,2\pi] \to \mathbb{C}$ be given by $\gamma(t) = i(R - 2Rt/\pi)$ for $0 \leqslant t \leqslant \pi$, $\gamma(t) = R\exp(i(t + \pi/2))$ for $\pi \leqslant t \leqslant 2\pi$. Sketch $\gamma(t)$ and $\mathcal{L}\gamma(t))$ for t running from 0 to π and from π to 2π. Explain why (if everything is well behaved) knowledge of the behaviour of $\mathcal{L}K(\gamma(t))$ as t runs from 0 to π tells us whether $\mathcal{L}K(z) = \alpha^{-1}$ has a solution with $\operatorname{Re} z \geqslant 0$.

(vii) Explain how, using (i), (iv), (v) and (vi), it is possible to determine whether the feedback system of Question 77.5 is stable by a sequence of measurements on the series system of part (v) which (provided the constituent boxes are stable) is itself always stable.

79

MANY DIMENSIONS

79.1 (Minkowski's theorem)

(i) State and prove Parseval's formula (see Theorem 34.2) for many dimensions.

(ii) Suppose $g: \mathbb{R}^m \to \mathbb{R}$ is a continuous function such that $g(\mathbf{x}) = 0$ for all $\|\mathbf{x}\| > K$ where K is some constant. Define $f: \mathbb{T}^m \to \mathbb{R}$ by $f(\mathbf{x}) = \sum_{\mathbf{n}\in\mathbb{Z}^m} g(\mathbf{x} + 2\pi i\mathbf{n})$, for $0 \leqslant x_j < 2\pi$ $[1 \leqslant j \leqslant m]$ (compare the proof of Poisson's formula in Question 60.4). Show that f is continuous, that

$$\left(\frac{1}{2\pi}\right)^m \hat{g}(\mathbf{r}) = \left(\frac{1}{2\pi}\right)^m \int_{\mathbb{T}^m} f(\mathbf{x}) \exp(-i\mathbf{r}.\mathbf{x})\, dV(\mathbf{x}),$$

$$\text{and } \left(\frac{1}{2\pi}\right)^m \int_{\mathbb{T}^m} f(\mathbf{x})^2\, dV(\mathbf{x}) = \sum_{\mathbf{n}\in\mathbb{Z}^m} \left(\frac{1}{2\pi}\right)^m \int_{\mathbb{R}^m} g(\mathbf{x})g(\mathbf{x} + 2\pi\mathbf{n})\, dV(\mathbf{x}).$$

(iii) Using Parseval's formula, deduce that

$$\sum_{\mathbf{n}\in\mathbb{Z}^m} \left(\frac{1}{2\pi}\right)^m \int_{\mathbb{R}^m} g(\mathbf{x})g(\mathbf{x} + 2\pi i\mathbf{n})\, dV(\mathbf{x}) \geqslant |\hat{f}(\mathbf{0})|^2 = \left(\frac{1}{2\pi}\right)^m |\hat{g}(\mathbf{0})|^2$$
$$= \left(\left(\frac{1}{2\pi}\right)^m \int_{\mathbb{T}^m} g(\mathbf{x})\, dV(\mathbf{x})\right)^2.$$

In the particular case when $g(\mathbf{x})g(\mathbf{x} + 2\pi\mathbf{n}) = 0$ for all $\mathbf{x} \in \mathbb{R}^m$ and all $\mathbf{n} \neq \mathbf{0}$, deduce that

$$\left(\frac{1}{2\pi}\right)^m \int_{\mathbb{R}^m} g(\mathbf{x})^2\, dV(\mathbf{x}) \geqslant \left(\left(\frac{1}{2\pi}\right)^m \int_{\mathbb{R}^m} g(\mathbf{x})\, dV(\mathbf{x})\right)^2.$$

(iv) Let Ω be an open set such that $\Omega \subseteq \{\mathbf{x}: \|\mathbf{x}\| \leqslant K\}$. Explain why we can find $g_N: \mathbb{R}^m \to \mathbb{R}$ a continuous function such that $g_N(\mathbf{x}) = 0$ if $\mathbf{x} \notin \Omega$, $1 \geqslant g_N(\mathbf{x}) \geqslant 0$ for all $\mathbf{x}$ and $1 > g_N(\mathbf{x})$ implies $\|\mathbf{x} - \mathbf{y}\| \leqslant N^{-1}$ for some $y \notin \Omega$. Deduce that, if $\Omega \cap (\Omega + 2\pi\mathbf{n}) = \varnothing$ for all $\mathbf{n} \neq 0$, then

$$\text{volume}(\Omega) \leqslant (2\pi)^m.$$

(If you wish to add additional hypotheses on Ω this will not matter since we only use this result for simple Ω. We write $A + \mathbf{y} = \{\mathbf{a} + \mathbf{y} : \mathbf{a} \in A\}$, $\lambda A = \{\lambda\mathbf{a}: \mathbf{a} \in A\}$.)

(v) Think about the result of (iv) until you see why it is the equivalent for $\mathbb{R}^n$ of the statement 'you cannot put a quart in a pint pot'.

(vi) Let Ω be an open set such that $\Omega \subseteq \{\mathbf{x}: \|\mathbf{x}\| \leqslant K\}$. Suppose further that Ω is symmetric (i.e. $\mathbf{x} \in \Omega$ implies $-\mathbf{x} \in \Omega$) and convex (i,e, $\mathbf{x}, \mathbf{y} \in \Omega$, $1 \geqslant \lambda \geqslant 0$ implies $\lambda\mathbf{x} + (1 - \lambda)\mathbf{y} \in \Omega$). Show that, if $\Omega \cap 2\pi\mathbb{Z}^m = \{\mathbf{0}\}$, then $2^{-1}\Omega \cap (2^{-1}\Omega + 2\pi\mathbf{n}) = \varnothing$ for all $\mathbf{n} \neq \mathbf{0}$ and conclude, using (iv), that volume$(\Omega) \leqslant 4^m\pi^m$.

(vii) Let $T: \mathbb{R}^m \to \mathbb{R}^m$ be a non singular linear transformation. Let Λ be the lattice $\{T\mathbf{n}: \mathbf{n} \in \mathbb{Z}^m\}$. Show that, if Ω is a bounded symmetric convex open set with $\Lambda \cap \Omega = \{\mathbf{0}\}$, then volume$(\Omega) \leqslant 2^m \det T$. (To the casual observer this result may appear yet another pretty geometrical nick-nack. In fact Minowski made it the foundation of a profoundly new method in number theory, 'The geometry of numbers'. In the next few questions we give three simple examples of its use.)

79.2 (i) Suppose that a, b, c, d are real numbers with $ad - b = 1$. Show, using Minowski's theorem (Question 79.1), that, if $\eta > 0$ and $\varepsilon > 0$, there exists $(n, m) \in \mathbb{Z}^2$ with $(n, m) \neq (0, 0)$ such that

$$|an + bm| < (1 + \varepsilon)\eta, \ |cn + dm| < (1 + \varepsilon)\eta^{-1}.$$

(ii) By carefully considering what happens as $\varepsilon \to 0$, or otherwise, show that there exists $(n, m) \in \mathbb{Z}^2$ with $(n, m) \neq (0, 0)$ such that

$$|an + bm| \leqslant \eta, \ |cn + dm| \leqslant \eta^{-1}.$$

(iii) By considering the case when $b = 1$, $c = -1$, $d = 0$, show that, if a is real, then either a is rational or we can find coprime n_j, $m_j \in \mathbb{Z}$ with $n_j \to \infty$ such that $|a + m_j/n_j| \leqslant n_j^{-2}$.

79.3 (Another proof of Dirichlet's lemma) Use the method of Question 79.2 to show that, if $\lambda_1, \lambda_2, \ldots, \lambda_n \in \mathbb{T}$, $\varepsilon > 0$ and N is an integer with $N > \pi\varepsilon^{-1}$, then we can find an integer m with $0 < m \leqslant N^n$ and $|m\lambda_j| < \varepsilon$ for $1 \leqslant j \leqslant n$. (*Remark*. The proofs in Questions 8.2 and 79.3 are elegant, but give no way (apart from examining the $|k\lambda_j|$ for all integer values of k in a very large range) of finding the m whose existence they guarantee. The problem of finding m rapidly was open for 140 years and has only been solved very recently. It is one of a family of problems which succumb to the famous LLL algorithm of A. Lenstra, H. Lenstra and L. Lovász.)

79.4 The object of this question is to show that, if p is a prime with $p \equiv 1 \bmod 4$, then p is the sum of two squares of integers. We write $p = 4k + 1$. There are some algebraic preliminaries.

(i) Let $\mathbb{Z}_p$ be the field of integers modulo p. By using a factorisation argument, or otherwise, show that the equation $x^n - 1 = 0$ can have at most n roots in $\mathbb{Z}_p$.

(ii) Let $\mathbb{Z}_p^* = \mathbb{Z}_p \backslash \{0\}$. Show that $\mathbb{Z}_p^*$ is an Abelian group under multiplication. Show that, if an Abelian group G contains elements of order m_1 and m_2, then is contains an element of order the highest common factor of m_1 and m_2. Deduce that there exist an integer $N \geqslant 1$ and an element a of $\mathbb{Z}_p^*$ such that a has order N and $x^N = 1$ for all $x \in \mathbb{Z}_p^*$.

(iii) Combine (i) and (ii) (together with Lagrange's theorem) to show that $N = 4k$. By considering a^{2k}, or otherwise, show that there exists a $u \in \mathbb{Z}$ with $u^2 \equiv -1 \bmod p$.

(iv) Let

$$\Lambda = \{(n, m) \in \mathbb{Z}^2 : n \equiv mu \bmod p\}$$
$$\Omega = \{(x, y) \in \mathbb{R}^2 : x^2 + y^2 \leqslant 3p/2\}.$$

Show, using Minowski's theorem (Question 79.1), that there exists $(r, s) \in \Lambda \cap \Omega$ with $(r, s) \neq (0, 0)$.

(v) Show, using the fact that $(r, s) \in \Lambda$, that $r^2 + s^2 \equiv 0 \bmod p$ and deduce, using the fact that $(r, s) \in \Omega \backslash \{(0, 0)\}$, that $r^2 + s^2 = p$.

79.5 Our final example of the use of Minowski's theorem is a proof of Lagrange's theorem that every positive integer is the sum of four squares of integers. This question makes a preliminary simplification.

(i) Show that 7 cannot be written as the sum of three squares.

(ii) Verify the remarkable equality due to Euler:

$$\begin{aligned}(x_1^2 + x_2^2 + x_2^3 + x_4^2)(y_1^2 + y_2^2 + y_3^2 + y_4^2) \\ &= (x_1y_1 + x_2y_2 + x_3y_3 + x_4y_4)^2 + (x_1y_2 + x^2y_1 + x_3y_4 - x_4y_3)^2 \\ &\quad + (x_1y_3 - x_3y_1 + x_4y_2 - x_2y_4)^2 \\ &\quad + (x_1y_4 - x_4y_1 + x_2y_3 - x_3y_2)^2.\end{aligned}$$

(This equality has a natural interpretation in terms of quaternions where it states that $\|x\|\|y\| = \|xy\|$. See, e.g. E. G. Rees, *Notes on Geometry*.)

(iii) Show that, if n, m are sums of four squares, so is nm. Deduce that it suffices to prove Lagrange's theorem for primes alone.

(iv) Show that it suffices to prove Lagrange's theorem for odd primes.

(v) Express 143 as the sum of four squares.

79.6 The proof of Lagrange's theorem for an odd prime $p = 2k + 1$ echoes the ideas of Question 79.4.

(i) By factorising $x^2 - t^2$, or otherwise, show that each element of $\mathbb{Z}_p$ has at most two square roots. Deduce that $\{t^2 : t \in \mathbb{Z}_p\}$ has $k+1$ elements. Conclude that $\{t^2 : t \in \mathbb{Z}_p\} \cap \{-s^2 - 1 : s \in \mathbb{Z}_p\} \neq \varnothing$ and so we can find $u, v \in \mathbb{Z}$ with $u^2 + v^2 \equiv -1 \bmod p$.

(ii) Let

$$\Lambda = \{(a, b, c, d) \in \mathbb{Z}^4 : c \equiv au + bv, d \equiv av - bu \bmod p\},$$

$$\Omega = \{(x, y, z, w) \in \mathbb{R}^4 : x^2 + y^2 + z^2 + w^2 < 1.9p\}.$$

Show that there exists $(\alpha, \beta, \gamma, \delta) \in \Lambda \cap \Omega$ with $(\alpha, \beta, \gamma, \delta) \neq (0, 0, 0, 0)$. (The volume of an n-dimensional sphere was obtained in Question 48.3.)

(iii) Show, using the fact that $(\alpha, \beta, \gamma, \delta) \in \Lambda$, that $\alpha^2 + \beta^2 + \gamma^2 + \delta^2 \equiv 0 \bmod p$ and deduce, using the fact that $(\alpha, \beta, \gamma, \delta) \in \Omega \backslash \{(0, 0, 0, 0)\}$, that $\alpha^2 + \beta^2 + \gamma^2 + \delta^2 = p$. (For alternative proofs of the results of this question and Question 79.4 together with a discussion of related problems see Chapter 20 of Hardy and Wright's *The Theory of Numbers*.)

79.7$^{(-)}$ (An alternative treatment of Question 79.5(ii))

(i) Show that, if $x_1, x_2, x_3, x_4 \in \mathbb{R}$ and $z_1 = x_1 + ix_2$, $z_2 = x_3 + ix_4$, then

$$\det \begin{pmatrix} z_1 & z_2 \\ -z_2^* & z_1^* \end{pmatrix} = x_1^2 + x_2^2 + x_3^2 + x_4^2.$$

(ii) If now $x_1, x_2, x_3, x_4, y_1, y_2, y_3, y_4 \in \mathbb{Z}$ and $w_1 = y_1 + iy_2$, $w_2 = y_3 + iy_4$, show, by computing

$$\det\left(\begin{pmatrix} z_1 & z_2 \\ -z_2^* & z_1^* \end{pmatrix}\begin{pmatrix} w_1 & w_2 \\ -w_2^* & w_1^* \end{pmatrix}\right)$$

in two different ways, that $(x_1^2 + x_2^2 + x_3^2 + x_4^2)(y_1^2 + y_2^2 + y_3^2 + y_4^2)$ is the sum of the squares of four integers.

(iii) By considering matrices of the form $\begin{pmatrix} x & y \\ -y & x \end{pmatrix}$ show, similarly, that the product of two numbers which are each of the sum of the squares of two integers is itself the sum of the squares of two integers.

(iv) Can we deduce from Question 79.4 that, if n is a positive integer with $n \equiv 1 \bmod 4$, then n is the sum of the squares of two integers? Give reasons.

79.8 State Weierstrass's theorem in two dimensions and prove it using a suitable version of the method of Chapter 59. Now read the chapter entitled 'Cross-purposes, unconscious assumptions, howlers, misprints, etc.' in Littlewood's *A Mathematician's Miscellany*.

79.9 (i) Suppose $K > 0$. Let $E_K: \mathbb{R}^m \times \mathbb{R} \to \mathbb{R}$ be defined by

$$E_K(\mathbf{x}, t) = \prod_{r=1}^{m} (2^{-1}(\pi K t)^{-1/2} \exp(-x_r^2/4Kt)) \quad [t > 0]$$

$$E_K(\mathbf{x}, t) = 0. \qquad [t \leqslant 0]$$

Verify that, if $G: \mathbb{R}^m \to \mathbb{C}$ is continuous and bounded, then

$$\phi(\mathbf{x}, t) = (G * E_K(\ ,t))(\mathbf{x})$$

defines an infinitely differentiable function on $\mathbb{R}^m \times \mathbb{R}^+$ with

$$\frac{\partial \phi}{\partial t} = K\nabla^2\phi; \quad \phi(\mathbf{x}, t) \to G(\mathbf{x}) \quad \text{as } t \to 0+.$$

(ii) Now let us return to the insects of Question 55.4(vi). Find the appropriate $\psi(\mathbf{x}, t)$, first if the insects are confined between close parallel plates (call this the two dimensional case), and then if the insects are in a very large open space (the three-dimensional case).

(iii) Suppose $g(s) = 0$ for $s < 0$, $g(s) = c_2$ for $s \geqslant 0$ $[c_2 > 0]$. Show that in the two-dimensional case

$$\psi(\mathbf{x}, t) = \frac{A_2 c_2}{K} \int_{\eta^2}^{\infty} \frac{e^{-s/4}}{s} \, ds,$$

where $\eta = \|\mathbf{x}\|/(Kt)^{1/2}$ and A_2 is a constant to be determined. Show that the male is galvanised with activity after a time $t_2 = \alpha_2 K^{-1} l^2$ where $\alpha_2 = \alpha_2(c_2/(c_1 K))$ depends only on the value of $c_2/(c_1 K)$.

(iv) Show that, if $g(s) = 0$ for $s < 0$, $g(s) = c_3$ for $s \geqslant 0$ $[c_3 > 0]$, then in the three-dimensional case

$$\psi(\mathbf{x}, t) = \frac{A_3 c_3}{K} \frac{1}{\|\mathbf{x}\|} \int_{\eta}^{\infty} e^{-s^2/2} \, ds$$

where $\eta = \|\mathbf{x}\|/(2Kt)^{1/2}$ and A_3 is a constant to be determined. Show that now the male may never be galvanised into activity.

(v) Note in particular that, in the one- and two-dimensional cases, $\psi(\mathbf{x}, t) \to \infty$ as $t \to \infty$ but that, in the three-dimensional case, $\psi(\mathbf{x}, t)$ remains bounded.

(vi) Find ψ in the three-dimensional case for a simple model of the female sitting in a tree above a large flat lawn. (*Hint.* Look at Figures 57.1 and 57.2.) What about the case when the female is inside a cubical room?

79.10 In the next three questions we shall use Kronecker's theorem (Theorem 79.1) to prove the theorem of Kolmogorov given as Theorem 19.1. The proof is long but the the individual steps are easy. Let us say that $x_1, x_2, \ldots, x_n \in \mathbb{T}$ are independent if the only solution of $\sum_{j=1}^{n} m_j x_j = 0$ with $m_1, m_2, \ldots, m_n, \in \mathbb{Z}$ is the trivial solution $m_1 = m_2 = \cdots = m_n = 0$. In this question we establish some properties of such independent points.

(i) If $x_1, x_2, \ldots, x_n$ are independent and M is a non-zero integer, show that $Mx_1, Mx_2, \ldots, Mx_n$ are independent.

(ii) Suppose that $x_1, x_2, \ldots, x_n$ are independent and $t \in \mathbb{T}$. Suppose further that we can find integers $p_1, p_2, \ldots, p_n$ not all zero and integers $q_1, q_2, \ldots, q_n$ not all zero such that $\sum_{j=1}^{n} p_j(x_j - t) = 0$ and $\sum_{j=1}^{n} q_j(x_j - t) = 0$. Show that $\sum_{j=1}^{n} p_j \neq 0$, $\sum_{j=1}^{n} q_j \neq 0$ and that $p_k \sum_{j=1}^{n} q_j = q_k \sum_{j=1}^{n} p_j$ for all $1 \leqslant k \leqslant n$.

(iii) Suppose that $x_1, x_2, \ldots, x_n$ are independent and $t \in \mathbb{T}$. Suppose that $p(x_k - t) \neq 0$ for each integer $p \neq 0$ and each k with $1 \leqslant k \leqslant n$. Use (ii) to show that we can find a u with $2 \leqslant u \leqslant n$ such that the collection of points $x_i - t$ with $i \neq u$, $1 \leqslant i \leqslant n$ is independent.

(iv) Suppose that $x_1, x_2, \ldots, x_m$ are independent and that $y_0 \in \mathbb{T}, \delta_0 > 0$ are given. Show that we can find, inductively, $y_n \in \mathbb{T}$, $\delta_n > 0$ such that

(a) $|y_n - y_{n-1}| < \delta_{n-1}/4$,

(b) $0 < \delta_n < \delta^{n-1}/4$,

(c) if $|y - y_n| < \delta_n$ then

$$py \notin \left\{ \sum_{j=1}^{m} p_j x_j : \sum_{j=1}^{m} |p_j| \leqslant n, p_1, p_2, \ldots, p_m \in \mathbb{Z} \right\}.$$

for all $p \in \mathbb{Z}$, $0 < |p| \leqslant n$. Show that y_n tends to a limit x_{m+1}, say, such that $|y_0 - x_{m+1}| < \delta_0$ and $x_1, x_2, \ldots, x_{m+1}$ are independent.

(v) Conclude that, given $w_1, w_2, \ldots, w_n \in \mathbb{T}$ and $\delta > 0$, we can find $x_1, x_2, \ldots, x_n$ independent with $|w_j - x_j| < \delta$ for $1 \leqslant j \leqslant n$.

79.11 This question contains analytic preliminaries.

(i) Suppose $g_n: \mathbb{T} \to \mathbb{Z}$ is a sequence of continuous functions such that $\sup_{n \leqslant 1} g_n(x) \leqslant M_1$ for each $x \in \mathbb{T}$. Show that, if $M_2 < M_1$, we can find an N such that $\sup_{N \geqslant n \geqslant 1} g_n(x) \geqslant M_2$.

(If you cannot think up your own proof, observe that, if the result is false, we can find a sequence $x_n \in \mathbb{T}$ such that $g_r(x_n) < M_2$ for each $1 \leqslant r \leqslant n$. Now extract a convergent subsequence $x_{n(k)} \to x$. Note that $g_{N(0)}(x) \geqslant (M_1 + M_2)/2$ for some $N(0)$, and consider the behaviour of $g_{N(0)}(x_{n(k)})$.)

(ii) If $1 > \varepsilon > 0$, let us write $\Delta_\varepsilon(t) = \varepsilon^{-2}(\varepsilon - t)$ for $|t| \leqslant \varepsilon, \Delta_\varepsilon(t) = 0$ otherwise. If $x_1, x_2, \ldots, x_n \subset \mathbb{T}$ and set $f_\varepsilon(t) = 2\pi n^{-1} \sum_{j=1}^{n} \Delta_{\varepsilon(t - xj)}$ show that

(a) f_ε is continuous, $f_\varepsilon(t) \geqslant 0$ for all $t \in \mathbb{T}$ and $(1/2)\pi \int_{\mathbb{T}} f_\varepsilon(t)\, dt = 1$,

(b) $\hat{f}_\varepsilon(m) \to n^{-1} \sum_{j=1}^{n} \exp(-imx_j)$ as $\varepsilon \to 0+$ for each fixed m.

Deduce from (b) that, if Q is fixed,

(c) $S_q(f_\varepsilon, t) \to n^{-1} \sum_{j=1}^{n} \sum_{m=-q}^{q} \exp(im(x_j - t))$ uniformly for all $t \in \mathbb{T}$ and $0 \leqslant q \leqslant Q$ as $\varepsilon \to 0+$.

Deduce that, if we write

$$F_q(t) = \frac{1}{n}\sum_{j=1}^{n} \frac{\sin((q+1/2)(x_j - t))}{\sin(x_j - t)/2}$$

for $t \neq x_1, x_2, \ldots, x_n$ and define $F_q(x_j)$ appropriately $[1 \leqslant j \leqslant n]$, then

(c)′ $S_q(f_\varepsilon,t) \to F_q(t)$ uniformly for all $t \in \mathbb{T}$ and $0 \leqslant q \leqslant Q$ as $\varepsilon \to 0+$.

(See chapter 18 if you need a hint.)

(iv) By using Fejér's theorem (see Lemma 18.7 if you need a hint), or otherwise, show that, if $x_1, x_2, \ldots, x_n$ and Q are given, we can find a trigonometric polynomial G such that

(a) $G(t) \geqslant 0$ for all $t \in \mathbb{T}$ and $1/2\pi \int_{\mathbb{T}} G(t)dt \leqslant 1$,
(b) $|S_q(G,t)| \geqslant F_q(t) - 1$ for all $t \in \mathbb{T}$ and $0 \leqslant q \leqslant Q$.

(v) Suppose that, given $A > 0$, we can find an $n \geqslant 1$ and $x_1, x_2 \ldots, x_n$ such that $\sup_{q \geqslant 1} F_q(t) > A + 2$ for all $t \in \mathbb{T}$. Show, using (i) and (iv), that we can find a trigonometric polynomial G such that

(a) $G(t) \geqslant 0$ for all $t \in \mathbb{T}$ and $(1/2\pi)\int_{\mathbb{T}} G(t)\,dt \leqslant 1$,
(b) $\sup_q |S_q(G,t)| \geqslant A$ for all $t \in \mathbb{T}$.

Thus Theorem 19.1 will follow, provided we can prove the hypothesis stated in the first sentence of this part of the question.

79.12 (Kolmogorov's theorem) Let $n \geqslant 10$. By Question 79.10(v) we can find $x_1, x_2, \ldots, x_n$ independent with $|x_j - 2\pi j/n| < 10^{-3}n^{-1}$. Define F_q as in Question 79.11. We wish to study $F_q(t)$ for each $t \in \mathbb{T}$.

(i) Let $1 \leqslant j(0) \leqslant n$ be such that $|x_{j(0)} - t| \leqslant |x_j - t|$ for all $1 \leqslant j \leqslant n$. Show, using Question 79.10(iii), that *either*

(a) $x_1 - t, x_2 - t, \ldots, x_n - t$ are independent, *or*
(b) we can find a $j(1) \leqslant j(1) \leqslant n$ and a strictly positive integer M such that $M(x_{j(1)} - t) = 0$ and the $x_j - t$ with $j \neq j(1)$ are independent, *or*
(c) we can find a $j(2)$ with $j(2) \neq j(0)$ and $1 \leqslant j(2) \leqslant n$ such that the $x_j - t$ with $j \neq j(2)$ are independent.

(ii) Let us start with case (a). Show, using Kronecker's theorem (Theorem 79.1), that we can find an integer $r \geqslant 1$, such that, writing $k(j) = 1$ if $\sin(x_j - t)/2 > 0$, $k(j) = -1$ otherwise,

$$|r(x_j - t) - (k(j)\pi/2 - (x_j - t)/2)| \leqslant 10^{-3} \quad \text{for } 1 \leqslant j \leqslant n.$$

Recalling that $\sin x \geqslant 2x/\pi$ for $\pi/2 \geqslant x \geqslant 0$ (see e.g. Question 32.4), show that

$$F_r(t) \geqslant A_1 \log n$$

for constant $A_1 > 0$ ($A_1 = 10^{-3}$ will certainly work but the value of A_1 is of no importance). Show also that we can find an integer $u \geqslant 1$ such that

$$F_u(t) \leqslant -A_1 \log n.$$

(iii) Now look at case (b). By considering $F_{rM}(t)$, or otherwise, show that we can still find an integer $u \geqslant 1$ such that

$$|F_u(t)| \geqslant A_2 \log n$$

for some constant $A_2 > 0$.

(iv) Finally consider case (c). Using the fact that the term associated with $j(2)$ cannot be very large, show that, again we can find an integer $u \geqslant 1$ such that

$$|F_u(t)| \geqslant A_3 \log n$$

for some constant $A_3 > 0$.

(v) Using Question 79.11(v), deduce Kolmogorov's theorem.

80
SUMS OF RANDOM VECTORS

80.1 Let R be a rectangle cut up into smaller rectangles $R(1), R(2), \ldots, R(k)$ each of which has sides parallel to the sides of R. Then, if at least one pair of sides of each $R(j)$ has integer length, it follows that at least one pair of sides of R has integer length.

(i)$^{(+)}$ Prove this without using Fourier analysis. (There are several proofs but if you can find any one you may pat yourself on the back. For a general context see Chapter 2 of Bollobás, *Graph Theory*.)

(ii) Prove the result by looking at the Fourier transform

$$\hat{\chi}_{R(j)}(1, 1) = \iint_{R(j)} \exp(-2\pi i(x + y))\, dx\, dy.$$

(Orient R along the x and y axes.)

81

A CHI SQUARED TEST

81.1 The first 2000 decimal digits of the constant e have the following frequency table

0	1	2	3	4	5	6	7	8	9
196	190	208	202	201	197	204	198	202	202

Is this evidence that the digits of e are more uniform that we would expect?

81.2 We work in $\mathbb{R}^n$ with the usual inner product $(\mathbf{x}, \mathbf{y}) = \sum_{i=1}^n x_i y_i$ and norm $\|\mathbf{x}\| = (\mathbf{x}, \mathbf{x})^{1/2}$. If $A = (a_{ij})$ is a matrix we write A^{T} for the transpose given by $A^{\mathrm{T}} = (a_{ji})$.

(i) Show that $4(\mathbf{x}, \mathbf{y}) = \|\mathbf{x} + \mathbf{y}\|^2 - \|\mathbf{x} - \mathbf{y}\|^2$.

(ii) Suppose $\omega: \mathbb{R}^n \to \mathbb{R}^n$ is a linear map which preserves length (i.e. $\|\omega\mathbf{x}\| = \|\mathbf{x}\|$ for all $\mathbf{x} \in \mathbb{R}^n$). Show that ω preserves inner products (i.e. $(\omega\mathbf{x}, \omega\mathbf{y}) = (\mathbf{x}, \mathbf{y})$ for all $\mathbf{x}, \mathbf{y} \in \mathbb{R}^n$

(iii) Conclude that, if ω is represented by a matrix W, then, if ω is length preserving, $\mathbf{x}^{\mathrm{T}} W^{\mathrm{T}} W \mathbf{y} = \mathbf{x}^{\mathrm{T}} \mathbf{y}$ and, by varying $\mathbf{x}$ and $\mathbf{y}$, deduce that $W^{\mathrm{T}} W = I$.

(iv) Show that, if ω is represented by a matrix W, then ω is length preserving if and only if $W^{\mathrm{T}} W = I$. Such a matrix is called orthogonal.

(v) Show that, if W is orthogonal, then $(\det W)^2 = 1$ and so $\det W = \pm 1$. Show, by example, that both cases can occur. If $\det W = 1$, we say that ω is a rotation, if $\det W = -1$, we say that ω is a reflection.

(vi) Show that an $n \times n$ matrix W is orthogonal if and only if the columns $\boldsymbol{\omega}_1, \boldsymbol{\omega}_2, \ldots, \boldsymbol{\omega}_n$ of W form an orthonormal system (i.e. $(\boldsymbol{\omega}_i, \boldsymbol{\omega}_j) = 1$ if $i = j$, $(\boldsymbol{\omega}_i, \boldsymbol{\omega}_j) = 0$ otherwise).

(vii) In Question 39.2 (ii) to (vi) we showed that, if A is an $n \times n$ real symmetric matrix (i.e. $A^{\mathrm{T}} = A$), then there exists $\mathbf{u}_1, \mathbf{u}_2, \ldots, \mathbf{u}_n$ orthonormal vectors with $A\mathbf{u}_j = \lambda_j \mathbf{u}_j$ where λ_j is real. By taking W either to have columns $\mathbf{u}_1, \mathbf{u}_2, \ldots, \mathbf{u}_n$ or to have columns $-\mathbf{u}_1, \mathbf{u}_2, \ldots, \mathbf{u}_n$, show that, if A is an $n \times n$ real symmetric matrix, we can find a matrix W representing a rotation such that $W^{\mathrm{T}} A W = D$ where D is a real diagonal matrix.

(viii) If $\mathbf{x}, \mathbf{y} \in \mathbb{C}^n$, let us write $(\mathbf{x}, \mathbf{y}) = \sum_{j=1}^{n} x_j y_j^*$ and take $\|\mathbf{x}\|^2 = (\mathbf{x}, \mathbf{x})$, $\|\mathbf{x}\| \geqslant 0$. Show that, if A is a complex $n \times n$ matrix, then $\|A\mathbf{x}\| = \|\mathbf{x}\|$ for all $\mathbf{x} \in \mathbb{C}^n$ if and only if A is invertible with $A^{-1} = A^*$. (Here A^* represents the conjugate matrix. If $A = (a_{ij})$, $A^* = (b_{ij})$, then $b_{ij} = a_{ji}^*$.) If you get stuck, Lemma 34.3 may be helpful.

81.3 (i) Let $X_1, X_2, \ldots, X_m$ be real valued random variables with well-behaved continuous joint probability density $f: \mathbb{R}^m \to \mathbb{R}$. Show that $X_1, X_2, \ldots, X_m$ are independent if and only if we can find $g_k: \mathbb{R} \to \mathbb{R}$ a continuous probability density $[1 \leqslant k \leqslant m]$ such that

$$\hat{f}(\lambda_1, \lambda_2, \ldots, \lambda_m) = \prod_{k=1}^{m} \hat{g}_k(\lambda_k) \quad \text{for all } \boldsymbol{\lambda} \in \mathbb{R}^m.$$

If the formula above holds, what is the probability density of X_k?

(ii) Let $X_1, X_2, \ldots, X_n$ be normally distributed, independent random variables such that $\mathbb{E}X_k = 0$ $[1 \leqslant k \leqslant n]$. Let $Y_j = \sum_{k=1}^{n} a_{jk} X_k$ $[1 \leqslant j \leqslant m]$ for some $a_{jk} \in \mathbb{R}$ $[1 \leqslant j \leqslant m,\ 1 \leqslant k \leqslant n]$. Show that $Y_1, Y_2, \ldots, Y_m$ are independent if and only if $\mathbb{E}(Y_r Y_s) = 0$ for $1 \leqslant r < s \leqslant m$ (we say $Y_1, Y_2, \ldots, Y_m$ are uncorrelated).

(iii) Give an example of two real valued random variables Y_1, Y_2 such that $\mathbb{E}Y_1 = \mathbb{E}Y_2 = \mathbb{E}Y_1Y_2 = 0$ yet Y_1 and Y_2 are not independent.

(iv) Suppose that $X_1, X_2, \ldots, X_n$ are normally distributed, independent random variables such that $\mathbb{E}X_k = 0$ $[1 \leqslant k \leqslant n]$, that A is an $n \times n$ real matrix with entries a_{jk} and that $Y_k = \sum_{j=1}^{n} a_{jk} X_j$. Show that $Y_1, Y_2, \ldots, Y_n$ are independent if and only if A is orthogonal (see Question 81.2).

(v) By using Question 81.2(vi), or otherwise, show that there is an $n \times n$ real orthogonal matrix A whose first row is

$$\mathbf{w}_1 = (n^{-1/2}, n^{-1/2}, \ldots, n^{-1/2}).$$

(vi) If A is as in part (v) and $X_1, X_2, \ldots, X_n, Y_1, Y_2, \ldots, Y_n$ as in part (iv), show that $Y_1 = n^{1/2}\bar{X}$ and

$$Y_2^2 + Y_3^2 + \cdots + Y_n^2 = \sum_{i=1}^{n} X_i^2 - n\bar{X}^2 = ((n-1)/n)\hat{\sigma}^2.$$

(We write $\bar{X} = n^{-1}\sum_{i=1}^{n} X_i$, $\hat{\sigma}^2 = (n-1)^{-1}\sum_{i=1}^{n}(X_j - \bar{X})^2$.)

(vii) Continuing (vi), deduce that $\bar{X}$ and $\hat{\sigma}^2$ are independent random variables. Show that $\bar{X}/\hat{\sigma}$ is a random variable with the same probability distribution as

$$W = \frac{W_1}{\sqrt{((W_2^2 + W_3^2 + \cdots + W_n^2)/(n-1))}},$$

where $W_1, W_2, \ldots, W_n$ are independent normally distributed random variables each with mean 0 and variance 1. (The distribution of W is called Student's t distribution with $n-1$ degrees of freedom.)

(viii) Let $Z_1, Z_2, \ldots, Z_n$ be the result of independent trials so that $Z_1, Z_2, \ldots, Z_n$ are independent, identically distributed, real random variables with mean μ and variance σ^2. Suppose that we do not know σ^2 or μ, but that we have some reason to hope that $\mu = 0$. The following ideas are suggested for deciding whether the hypothesis that $\mu = 0$ is to be accepted or not. (We write $\bar{Z} = n^{-1}\sum_{j=1}^{n} Z_j$, $\hat{\sigma}^2 = (n-1)^{-1}\sum_{j=1}^{n}(Z_j - \bar{Z})^2$.)

(a) Accept if $|\bar{Z}| < \alpha$.
(b) Accept if $\hat{\sigma} < \alpha$.
(c) Accept if $|\bar{Z}|/\hat{\sigma} < \alpha$.
(d) Accept if $|\bar{Z}|^2/\hat{\sigma}^2 < \alpha$.
(e) Accept if $|\bar{Z}|/(n^{-1}\sum_{j=1}^{n}|Z_j - \bar{Z}|) < \alpha$.

State, with reasons, which of these proposals are absurd. (You are not asked to choose among the others.)

(ix) If proposal (c) is adopted and if the Z_j are normally distributed, show how knowledge of Student's t distribution allows us to find the probability of rightly accepting the hypothesis $\mu = 0$ if the hypothesis is true. (Student was the nom de plume of W. S. Gosset who worked for the brewing firm of Guinness. Although the firm was sufficiently interested in what would now be called quality control to send him for a year of postgraduate study under Pearson in 1906, they insisted that his published results should appear under a pseudonym.)

81.4$^{(-)}$ As part of a drive to improve road safety, traffic lights are installed at the road junction where most accidents occurred last year. This year there were no accidents at the junction. Discuss briefly the extent to which this demonstrates the efficacy of traffic lights in reducing accidents.

82

HALDANE ON FRAUD

82.1 The author of a scientific paper claims to have done the following experiment 3600 times. The subject wrote down a number, then a die was thrown and the number shown on the die compared with the prediction. He claims that the results were as shown.

Number of die	1	2	3	4	5	6	Total
Number of times thrown	604	425	694	664	495	718	3600
Number of correct predictions	95	84	181	171	114	203	848

(a) If the experimental data are not fraudulent do they provide good evidence that the future may be predicted?

(b) Looking at the table, do you find evidence that the data are fraudulent? If so, explain why.

(The example just given is fictitious. For an interesting real life (and, almost inevitably, more complicated) case see the fascinating discussion of the Soal–Goldney experiment in Volume 56 of the *Proceedings of the Society for Psychical Research*.)

86

WILL A RANDOM WALK RETURN?

86.1 (i) Show that, if $\alpha > 0$, the integral

$$\int_1^\infty \int_1^\infty \cdots \int_1^\infty \frac{1}{(x_1^2 + x_2^2 + \cdots + x_n^2)^\alpha} \, dx_1 \, dx_2 \ldots dx_n$$

converges if and only if $\alpha > n/2$.

(ii) Show that, if $\alpha(1), \alpha(2), \ldots, \alpha(n) > 0$, the integral

$$\int_1^\infty \int_1^\infty \cdots \int_1^\infty \frac{1}{x_1^{\alpha(1)} + x_2^{\alpha(2)} + \cdots + x_n^{\alpha(n)}} \, dx_1 \, dx_2 \cdots dx_n$$

converges if and only if $\alpha(1)^{-1} + \alpha(2)^{-1} + \cdots + \alpha(n)^{-1} < 1$.

(iii) Investigate the convergence of

$$\int_0^1 \int_0^1 \cdots \int_0^1 \frac{1}{(x_1^2 + x_2^2 + \cdots + x_n^2)^\alpha} \, dx_1 \, dx_2 \ldots dx_n$$

and

$$\int_0^1 \int_0^1 \cdots \int_0^1 \frac{1}{x_1^{\alpha(1)} + x_2^{\alpha(2)} + \cdots + x_n^{\alpha(n)}} \, dx_1 \, dx_2 \ldots dx_n.$$

86.2 Dubrovsky sits down to a night of gambling with his fellow officers. Each time he stakes u roubles there is a probability p that he will win and receive back $2u$ roubles (including his stake), and a probability $1 - p$ that he will lose his stake. At the beginning of the night he has 8000 roubles. If ever he has 256 000 roubles he will marry the beautiful Natasha and retire to his estates in the country. Otherwise he will commit suicide. He decides to follow one of two courses of action:

(a) to stake 1000 roubles each time until the issue is decided,
(b) to stake everything each time until the issue is decided.

Advise him (i) if $p = 1/4$ and (ii) if $p = 3/4$. What are the chances of a conventional happy ending in each case if he follows your advice?

(More generally, gamble boldly if the odds are against you, gamble cautiously if the odds are for you. Casinos restrict the boldness of their customers by limiting the amount they can stake on each turn. Although the going rapidly gets tougher, the introduction to *How to Gamble If You Must* by Dubins and Savage is most illuminating.)

86.3 (i) Consider two particles performing independent random walks on the lattice $\mathbb{Z}^m$ according to the rules laid down in Theorem 86.1. If they both start at $\mathbf{0}$ at time $t = 0$, what is

(a) Pr(both particles at the same point for some $t > 0$),
(b) Pr(both particles at $\mathbf{0}$ for some $t > 0$)?

(Pólya considers two friends strolling at random through a town laid out on a square grid. We then ask for the probability that they meet at some time on their promenade and for the probability that they meet outside the hotel they started from.)

(ii) Three very rich, very mean, players with a great deal of time on their hands play three independent fair games of heads and tails with three other equally rich, mean and unoccupied players. The games are synchronised with one toss every minute. Naturally the games can only stop when none of the six players has an advantage. Show that, with probability 1, any pair of games will be simultaneously in a drawn state infinitely often but the probability that all three will be in a drawn state at some time after they have started is less than 1.

87

WILL A BROWNIAN MOTION RETURN?

87.1 (An alternative proof of Pólya's theorem) Consider a system which can be in one of a collection of states A and suppose that, if it is in state $i \in A$ after m seconds, the probability it will be in state $j \in A$ after $m+1$ is p_{ij} (independent of m and its previous history). Let $p_{ij}^{(n)}$ be the probability that it is in state j after n seconds, given that it was in state i initially, and let $f_{ij}^{(n)}$ be the probability that it reaches state j for the *first* time after n seconds.

(i) Explain why $p_{ij}^{(n)} = f_{ij}^{(n)} + \sum_{r=1}^{n-1} f_{ij}^{(r)} p_{jj}^{(n-r)}$.

(ii) Deduce that, if we write $P(x) = \sum_{n=1}^{\infty} p_{jj}^{(n)} x^n$, $F(x) = \sum_{n=1}^{\infty} f_{jj}^{(n)} x^n$ for x real and $|x| < 1$, then $P(x) = F(x) + P(x)F(x)$.

(iii) By considering what happens as $x \to 1-$, show that $\sum_{n=1}^{\infty} f_{jj}^{(n)} = 1$ if and only if $\sum_{n=1}^{\infty} p_{jj}^{(n)} = \infty$.

(iv) Deduce that, if the system starts at j, it returns to j with probability 1 if and only if $\sum_{n=1}^{\infty} p_{jj}^{(n)} = \infty$.

(v) Specialise to the random walk on $\mathbb{Z}$ described in Lemma 86.3 with $p + q = 1$. Explain why

$$Pr(\text{particle started at 0 returns to 0 at time } 2n+1) = 0 \text{ and}$$

$$Pr(\text{particle started at 0 returns to 0 at time } 2n) = \binom{2n}{n} p^n q^n.$$

By using Stirling's formula (Question 18.16(i)) and (iv), show that

$$Pr(\text{Particle started at 0 returns to 0 at some later time}) = 1$$

if and only if $p = q = 1/2$.

(vi) Prove Theorem 86.1.

(Neither of our proofs represents the final word. What happens between $n = 2$ and $n = 3$? What happens if our grids $\mathbb{Z}^n$ have vertices removed? For some answers you may wish to read *Random Walks and Electric Networks*, by Doyle and Snell.)

89

WILL A BROWNIAN MOTION TANGLE?

89.1 (i) Show that, if A is a real $n \times n$ matrix, then, if A is invertible, so is A^{T} (the transpose of A) and that then $(A^{\mathrm{T}})^{-1} = (A^{-1})^{\mathrm{T}}$.

(ii) For the rest of the question A will be a real invertible $n \times n$ matrix. If $f: \mathbb{R}^n \to \mathbb{C}$ is in $L^1 \cap C$, let us write $S_A f(\mathbf{x}) = f(A\mathbf{x})$. Show, by using the Jacobian change of variable formula, or otherwise, that $(S_A f)\hat{\ } = |\det A|^{-1} S_{(A^{-1})^{\mathrm{T}}}(\hat{f})$, i.e. that

$$(S_A f)\hat{\ }(\zeta) = |\det A|^{-1} \hat{f}((A^{-1})^T \zeta) \quad \text{for all } \zeta \in \mathbb{R}^n.$$

(iii) If $\mathbf{a} \in \mathbb{R}^n$, let us write $T_{\mathbf{a}} f(\mathbf{x}) = f(\mathbf{x} - \mathbf{a})$. Show that $(T_{\mathbf{a}} f)\hat{\ }(\zeta) = (\exp - i\mathbf{a} \cdot \zeta)\hat{f}^*(\zeta)$.

(iv) Verify that (ii) and (iii) correspond to known results if $n = 1$.

(v) Show that, if the map $\mathbf{x} \mapsto A\mathbf{x}$ is a rotation (see Question 81.2 if necessary), then $(S_A f)\hat{\ } = S_A \hat{f}$. Conclude that, if f is radially symmetric, then $\hat{f}$ is.

94

WHY DO WE COMPUTE?

94.1 (Phased array radars) We work in the (x, y) plane since the extension to three dimensions is easy. Consider an array of $N+1$ transmitters broadcasting on a frequency ω. Let the kth transmitter be at $(0, kl)$ and let the signal at (x, y) due to this transmitter be

$$(N+1)^{-1} r_k^{-2} \exp\left(i(\omega t - \lambda^{-1} r_k - \phi_k)\right),$$

where $r_k^2 = x^2 + (y - kl)^2$, $r_k > 0$. We can control the phase ϕ_k. The total signal at (x, y) is

$$S(x, y) = \sum_{k=0}^{N} (N+1)^{-1} r_k^{-2} \exp\left(i(\omega t - \lambda^{-1} r_k - \phi_k)\right).$$

(i) Show that, if $x = R\cos\theta$, $y = R\sin\theta$ where R is very large compared with $a = N\lambda$, then, to a good approximation,

$$S(R\cos\theta, R\sin\theta) = R^{-2} \exp\left(i(\omega t - \lambda^{-1} R)\right) P(u)$$

where $P(u) = (N+1)^{-1} \sum_{k=0}^{N} \exp\left(i(ku - \phi_k)\right)$ and $u = \lambda^{-1} l \sin\theta$.

(ii) If we choose $\phi_0 = \phi_1 = \phi_2 = \ldots = \phi_N = 0$, we get a forward pointing beam

$$S(R\cos\theta, R\sin\theta) = R^{-2} \exp(i(\omega t - \lambda^{-1} R)) P_N(u).$$

Compute and graph $P_N(u)$ showing that $|P_N(u)|$ is a maximum when $u = 0$.

(iii) Show that, given any $-\pi/2 < \theta_0 < \pi/2$, we can choose the phases ϕ_0, $\phi_1, \ldots, \phi_N$ so that

$$S(R\cos\theta, R\sin\theta) = R^{-2} \exp(i(\omega t - \lambda^{-1} R)) P_N(u - u_0)$$

where $u_0 = \lambda^{-1} l \sin\theta_0$. We have thus 'electronically steered our beam to point in the direction θ_0'. Why is it desirable to have $\lambda^{-1} l \leq 1/2$?

(iv) We would like our beam to be 'tight'. Specifically let us say that we would like the interval $[-\delta, \delta]$, such that

$$|P_N(u)| \geq \tfrac{1}{2}|P_N(0)| \text{ for } |u| \leq \delta \text{ and } |P_N(\delta)| = |P_N(-\delta)| = \tfrac{1}{2}|P_N(0)|$$

to be as small as possible. Show that allowing $N \to \infty$ does not substantially decrease δ, but that increasing l without changing N does decrease δ. Why is it extremely desirable to have Nl fairly large? (Thus, if we want a single tight steerable beam, it would appear that we should have $l < \lambda/2$, i.e. the distance between transmitters less than half a wavelength, and $a = N\lambda$ the width of the array as large as possible and, in any case, fairly large compared with the wavelength l. More sophisticated analysis supports these conclusions. If we want more flexibility in structuring the beam we must decrease λ and increase N.)

(v) Indicate how, if $\lambda \leq l/4$ and N is reasonably large, you can produce two independently steerable beams from your array. (Ballistic missile tracking radars may have a 100×100 array of transmitters.)

94.2 Now consider the problem of tracking a radio source (perhaps an object reflecting a radar beam of the type described above) with an array of $N+1$ receivers. Let the source be at (x, y), let the receiver be at $(0, kl)$ and let the signal received at this transmitter be

$$Ar_k^{-2} \exp(i(\omega t - \lambda^{-1} r_k - \psi)),$$

where $r_k^2 = x^2 + (y - kl)^2$, $r_k > 0$. Here $A > 0$ and ψ is real and, normally, we would not expect to know their values. We observe the total of the signals modified by phase changes as

$$S(x, y) = \sum_{k=0}^{N} Ar_k^{-2} \exp(i(\omega t - \lambda^{-1} r_k - \psi - \phi_k)).$$

Here the phase changes $\phi_0, \phi_1, \ldots, \phi_N$ are under our control.

(i) Let $(x, y) = (R\cos\theta, R\sin\theta)$ with R very large compared with $a = N\lambda$ and $-\pi/2 < \theta < \pi/2$. Describe how you would determine θ by systematically changing $\phi_0, \phi_1, \ldots, \phi_N$. Explain why you cannot determine θ unambiguously unless $\lambda/2 \geq l$. Will it be possible to use your method in practice if Nl is small compared with λ?

(ii) Let us write

$$T(\phi) = \sum_{k=0}^{N} Ar_k^{-2} \exp\left(i(\omega t - \lambda^{-1} r_k - \psi - k\lambda^{-1} l \sin\phi)\right).$$

Suppose that $\lambda/2 > l$, that N is large and Nl is large compared with λ. Explain what $T(\phi)$ corresponds to. Show that $T(\phi)$ has a maximum when $\phi = \theta$ and find the zeros of T.

(iii) Now suppose you are listening to $T(\phi)$ through headphones. Is it easier to locate the maximum of T or its zeros? How would you use knowledge of the zeros of T to find θ? (In the Second World War people operating radio detection equipment knew that they should listen for silence. The design of

ESPRIT, one of the modern programs for operating phase array radars, uses the same idea.)

(iv) Suppose now that there are M sources which we are endeavouring to find. Let the jth source be at $(R\cos\theta_j, R\sin\theta_j)$ and the signal received at the kth receiver be

$$A_j r_k^{-2} \exp\left(i(\omega t - \lambda^{-1} r_{kj} - \psi_j)\right)$$

where $r_{kj}^2 = (R\cos\theta_j - kl)^2 + (R\sin\theta_j)^2$. We are thus faced with interpreting

$$T(\phi) = \sum_{j=1}^{M} T_j(\phi)$$

with $T_j(\phi) = \sum_{k=0}^{N} A_j r_{kj}^{-2} \exp\left(i(\omega t - \lambda^{-1} r_{kj} - \psi_j - k\lambda^{-1}\sin\theta)\right).$

Observe that we do not know A_j or ψ_j.

Suppose that $\theta_1, \theta_2, \ldots, \theta_M$ are close to 0. Discuss the problem of finding $\theta_1, \theta_2, \ldots, \theta_M$ and indicate why, in practice, even if M is small (say 2 or 3), we can only hope to succeed if $\min_{r\neq s}|\theta_r - \theta_s|$ is large compared with $\lambda(Nl)^{-1}$. (Thus the discrimination of our system is inversely proportional to the wave length and directly proportional to the array length Nl. This is another aspect of the uncertainty principle of Question 60.15.)

(v) At the beginning of the Second World War all BBC medium range wireless transmitters were synchronised into groups with common frequencies. Explain why this allowed broadcasting to continue without allowing enemy aircraft to use the transmitters to locate themselves. Explain why it is none the less possible to navigate using a group of lighthouses emitting light of the same frequency.

(vi) The shortest wavelength avalable to British and German radar at the beginning of the Second World War was 50 cm. In February 1940, Randall and Boot produced a valve giving 10 cm wavelength. 'Including the two half pennies used to seal the end plates, the prototype . . . cost an estimated £200: had it cost £2 million it would have been a bargain.' Why?

(vii) The Hubble space telescope has a mirror 2.4 metres across. According to its makers, it could distinguish the left and right headlights of a car at a distance of 4000 kilometres. Use back-of-an-envelope calculations to discuss whether such performance is possible. (Unfortunately, although the mirrors for the Hubble telescope were ground to the highest precision ever attained, they were ground precisely to the wrong shape. The predicted performance did not materialise.)

(viii) In MERLIN a collection of steerable radio telescopes many tens of kilometres apart (and thus very many radio wavelengths apart) are interconnected to give 'the performance of a radio telescope 200 kilometres across'. Discuss (The reader is also directed to the discussion of a bee's eye in the marvellous *Feynman Lectures on Physics*.)

94.3 (CAT and the Radon transform) 'All the business of war', said the Duke of Wellington, 'and, indeed all the business of life, is to endeavour to find out what you don't know by what you do'.

In a CAT scanner, beams of X-rays are shone through a human skull, say, and we seek to map the interior of the skull by looking at the shadows that it casts.

(i) Consider the one-dimensional problem in which a ray of light is shone along the x axis in the direction x increasing. Suppose the light has intensity $V(x)$ at x and that it passes through translucent material of density (attenuating power) $f(x)$ at x. Convince yourself that, to first order in δx, it is plausible that

$$V(x + \delta x) \approx (1 - f(x)\delta x)V(x).$$

Assuming that this is in fact the case, find a differential equation for V and show that, if f is a well-behaved function, then

$$\frac{V(X)}{V(Y)} = \exp\left(-\int_X^Y f(x)\,dx\right).$$

This 'exponential law of absorption' has been verified by experiment. Observe that, assuming $f(x) = 0$ for large x, this means that we can find

$$\int_{-\infty}^{\infty} f(x)\,dx$$

by measuring the attenuation of the ray of light due to its passage through the region in which f is non-zero.

(ii) Now consider the two-dimensional problem in which we have a well-behaved, but unknown, function $f\colon \mathbb{R}^2 \to \mathbb{R}$ which is zero outside a bounded region. By measuring the attenuation of light rays (or X-rays) we can find the line integral

$$\int_l f ds$$

of f along any given straight line l. Can we reconstruct f? This is the 'tomographic problem' (the mysterious initials 'CAT' given above stand for 'computer assisted tomography'). It has higher-dimensional analogues and generalisations but this two-dimensional version corresponds to 'slice tomography' in which a picture of a three-dimensional object is built up slice by slice out of two-dimensional cross sections.

Our first step is to establish some notation. Write $\mathbf{u}_\theta$ for the unit vector making an angle θ to the x axis and $\mathbf{v}_\theta$ for the perpendicular unit vector making an angle $\theta + \pi/2$ to the x axis. We know the line integral

$$(\mathcal{R}f)(r, \theta) = \int_{-\infty}^{\infty} f(r\mathbf{u}_\theta + t\mathbf{v}_\theta)\,dt$$

for each real θ and r and we wish to find f. $\mathcal{R}f$ is called the Radon transform of f. (The name comes from the eminent mathematician Radon who gave the first extensive treatment of the problem in a paper in 1917. At the time the results were purely theoretical and his paper was more or less ignored.)

Show that $(\mathcal{R}f)(-r, \theta+\pi) = (\mathcal{R}f)(r, \theta)$. Compute $(\mathcal{R}f)(r, \theta)$ in the cases

(a) $f(x, y) = (2\pi)^{-1/2} \exp -(x^2+y^2)/2$,
(b) $f(x, y) = 1$ if $|x|, |y| \leqslant 1$, $f(x, y) = 0$ otherwise.

(iii) The Radon transform has a remarkable connection with the Fourier transform. Write

$$\mathcal{F}_1(\mathcal{R}f)(s, \theta) = \int_{-\infty}^{\infty} (\mathcal{R}f)(r, \theta) \exp(-irs)\, dr$$

for the one-dimensional Fourier transform of $\mathcal{R}f$ with respect to the first variable. Show that, if $\hat{f}$ is the usual two-dimensional Fourier transform then (provided f is well behaved)

$$\hat{f}(s\mathbf{u}_\theta) = \mathcal{F}_1(\mathcal{R}f)(s, \theta).$$

Conclude that f is uniquely determined by its Radon transform and that f can be recovered from $(\mathcal{R}f)$ ('we can invert the Radon transform') by using one- and two-dimensional Fourier transforms. Carry out this recovery in the case of the Radon transform you computed in (ii)(a) above.

The fact that the inversion can be done using Fourier transforms does not mean that it must be done that way. (Remember that, in the real world, we only have a finite number of measurements none of which are perfect.) However a Fourier transform technique (different from but related to the one above) is frequently used. CAT represents a happy collaboration of advanced engineering technique, modern computer power and ingenious mathematics in the service of medicine recognised by the award of a Nobel prize to two of its pioneers, Cormack and Hounsfield, in 1979. Interesting and important mathematical problems still remain to be solved in this area.

94.4 (A simple version of Herglotz's theorem) A sequence $(a_n)_{n\in\mathbb{Z}}$ is said to be positive definite if, given any $N \geqslant 1$ and any $z_r \in \mathbb{C}$ $[|r| \leqslant N]$, we have

$$\sum_{|n|,|m|\leqslant N} a_{n-m} z_n z_m^* \geqslant 0.$$

(i) If $f\colon \mathbb{T} \to \mathbb{C}$ is continuous, show that

$$\sum_{|n|,|m|\leqslant N} \hat{f}(n-m) z_n z_m^* = \frac{1}{2\pi} \int_{\mathbb{T}} f(t) \left| \sum_{n=-N}^{N} z_n e^{-int} \right|^2 dt.$$

Conclude that, if $f(t)$ is real and $f(t) \geqslant 0$ for all $t \in \mathbb{T}$, then the sequence $(\hat{f}(n))_{n\in\mathbb{Z}}$ is positive definite.

(ii) If $f: \mathbb{T} \to \mathbb{C}$ is continuous show that

$$\sigma_{2N}(f, t) = (2N+1)^{-1} \sum_{|n|,|m| \leqslant N} \hat{f}(n-m) e^{int} (e^{imt})^*.$$

Conclude that, if the sequence $(\hat{f}(n))_{n \in \mathbb{Z}}$ is positive definite, then $f(t)$ is real and $f(t) \geqslant 0$ for all $t \in \mathbb{T}$.

95

THE DIAMETER OF STARS

95.1$^{(-)}$ Let R be a rectangle in $\mathbb{R}^2$ and let $\chi_R(x) = 1$ if $x \in \mathbb{R}$, $\chi_R(x) = 0$ otherwise. Show that knowledge of the zeros of $\hat{\chi}_R$ enables us to find the length of the sides of R. Does it enable us to find the orientation (i.e. the angle the longest side makes with the x axis)? Does it enable us to find the position of the centre of gravity of R?

95.2 (i) Explain why, if $z \in \mathbb{C}$, we can find $J_n(z)$ such that

$$\exp(iz\sin\theta) = \sum_{n=-\infty}^{\infty} J_n(z)\exp(in\theta).$$

Show that $J_n(z) = \frac{1}{2\pi}\int_{\mathbb{T}} \exp(-i(n\theta - z\sin\theta))\,d\theta$. (We call J_n the nth Bessel function.)

(ii) Prove in, particular, that

$$\int_0^{2\pi} \exp(-ia\cos\theta)\,d\theta = 2\pi J_0(a) \quad \text{for all } a \in \mathbb{C}.$$

(iii) Show that $J_n(z) = (-1)^n J_{-n}(z)$.

(iv) By differentiating under the integral, or otherwise, show that J_n is analytic and

$$z^2 J_n''(z) + zJ_n'(z) + (z^2 - n^2)J_n(z) = 0.$$

(v) Since $J_n\colon \mathbb{C} \to \mathbb{C}$ is analytic, it has a Taylor series of infinite radius. By using (iv), or otherwise, obtain it as

$$J_n(z) = \sum_{k=0}^{\infty} \frac{(-1)^k (z/2)^{n+2k}}{k!(n+k)!}.$$

(vi) Show that $d(z^n J_n(z))/dz = z^n J_{n-1}(z)$ and deduce that, if x is real, $\int_0^x y^n J_{n-1}(y)\,dy = x^n J_n(x)$.

95.3 (i) How many terms in the sum

$$J_0(x) = \sum_{k=0}^{\infty} \frac{(-1)^k (x/2)^{2k}}{(k!)^2}$$

(obtained in Question 95.2(v)) are required to obtain $J_0(x)$ to six-figure accuracy for x real and $0 \leqslant x \leqslant 1$? How many terms are required to compute $J_0(10)$ by this means? Is it reasonable to try to compute $J_0(10)$ in this way?

(ii) Is it sensible to try and compute $J_0(x)$ for small real positive x by using the differential equation (obtained in Question 95.2(iv))

$$x^2 J_0''(x) + xJ_0'(x) + x^2 J_0(x) = 0?$$

Why? Is it reasonable to compute $J_0(x)$ for $x \geqslant 1$ using the differential equation? How would you obtain $J_0(1)$ and $J_0'(1)$?

(iii) If you have access to a computer and some rudimentary skill in using it for mathematical purposes, calculate and plot $J_0(x)$ for real x with $0 \leqslant x \leqslant 40$. Find the zeros $\alpha_1, \alpha_2, \alpha_3, \ldots$ of J_0 in the range given and compute $\alpha_m - \alpha_{m-1}$ for $m = 1, 2, \ldots$. Do you agree that an optimist might hope that $J_0(x) \approx f(x) \cos(x + \theta)$ for large x?

(iv) A real optimist would conjecture that f takes the simple form $f(x) = Ax^{-\beta}$. Plot $\log|J_0((\alpha_m + \alpha_{m-1})/2)|$ against $\log|(\alpha_m + \alpha_{m-1})/2|$. Do you agree that an optimist might hope that $\beta = 1/2$?

(v) Let us guess that $J_0(x) \approx Ax^{-1/2} \cos(x + \theta)$. Using the numerical evidence available, estimate A and θ. (You may well be able to guess the exact value of θ.) Plot $g(x) = Ax^{-1/2} \cos(x + \theta)$ and $J_0(x)$ using the same graph.

(vi) By generating and examining numerical examples, see whether it is reasonable to guess that $J_n(x) \approx A_n x^{-1/2} \cos(x + \theta_n)$. Have you any guesses about A_n and θ_n? (You are warned that, for fixed x, the quality of the approximation deteriorates rapidly with n, so you will need to be both optimistic and lucky.)

(If you have no skill in numerical analysis or no computer, do the best you can with tables.)

95.4 (The Bessel functions have real zeros.) Such is the power of suggestion that I doubt if one reader in ten paused during the paragraph preceding Lemma 95.1 to ask if she had actually seen a proof that J_1 has zeros. The fact that J_n has an infinity of real zeros is an immediate consequence of several different general theorems, but I know of no totally elementary proof. The proof below is suggested by the discussion of Question 95.3. Throughout, n is a fixed integer and x is real and positive.

(i) Explain why the zeros of J_n, if they exist, must be isolated (i.e. if $J_n(x_1) = 0$, then we can find a $\delta > 0$ such that $J_n(x) \neq 0$ for $0 < |x - x_1| < \delta$).

(ii) Writing $y(x) = x^{1/2} J_n(x)$, show that

$$y'' + \left(1 - \frac{(4n^2 - 1)}{4x^2}\right) y = 0.$$

Deduce that, given $\varepsilon > 0$, we can find x_0, depending on ε, such that

$$|y''(x) + y(x)| \leqslant \varepsilon |y(x)| \quad \text{for all } x \geqslant x_0.$$

In what follows we shall take ε fixed and small ($\varepsilon < 10^{-2}$, say) and consider $x \geqslant x_0$. How do you expect $y(x)$ and $y'(x)$ to behave?

(iii) How do you expect $y(x)^2 + y'(x)^2$ to behave? By considering the derivative of $\log(y(x)^2 + y'(x)^2)$, or otherwise, show that, if $x \geqslant x_1 \geqslant x_0$, then

$$y(x)^2 + y'(x)^2 \geqslant (y(x_1)^2 + y'(x_1)^2) \exp(-\varepsilon(x - x_1)).$$

(iv) How do you expect

$$f(x) = \frac{e^{ix}(y(x) + iy'(x))}{(y(x)^2 + y'(x)^2)^{1/2}}$$

to behave? Show that, if $x \geqslant x_1 \geqslant x_0$, then

$$|f(x) - f(x_1)| \leqslant 2\varepsilon |x - x_1|.$$

(v) Conclude that, if x_1 is large enough, then y, and so J_n, has a zero in the interval $[x_1, x_1 + 2\pi]$. Deduce that J_n has an infinity of zeros.

(vi) By tightening up the argument above, show that, if the zeros of J_n are $\alpha_1, \alpha_2, \ldots$ with $0 \leqslant \alpha_1 < \alpha_2 < \cdots$, then $\alpha_m - \alpha_{m-1} \to \pi$ as $m \to \infty$. (Although the results of this question strengthen our confidence in the conjectures of Question 95.3, they do not prove them. However, the conjectures are true and we shall prove them in Question 95.6.)

95.5 (i) In this question n is a fixed integer. Let $a, b \in \mathbb{C}$ with $a, b \neq 0$ and write $u(z) = J_n(az)$, $v(z) = J_n(bz)$. Show that

$$z(zu')' + (a^2z^2 - n^2)u = 0$$
$$z(zv')' + (b^2z^2 - n^2)v = 0$$

and deduce that

$$v(zu')' - u(zv')' + (a^2 - b^2)zuv = 0.$$

(ii) Suppose now that a is a zero of J_n and $b \neq a$. Show that

$$\int_0^1 xJ_n(ax)J_n(bx)\,dx = \frac{uJ'_n(u)J_n(b)}{b^2 - a^2}.$$

Deduce that, if a and b are distinct zeros of J_n, then

$$\int_0^1 xJ_n(ax)J_n(bx)\,dx = 0.$$

(iii) By allowing $b \to a$ in the first formula of (ii), show that, if a is a zero of J_n,

$$\int_0^1 x(J_n(ax))^2\,dx = \frac{(J'_n(a))^2}{2}.$$

(iv) By using 95.2(v), or otherwise, show that, if $a \in \mathbb{C}$, $x \in \mathbb{R}$, then $J_n(ax)^* = J_n(a^*x)$. Using the final formula in (ii), deduce that all the zeros of J_n are real.

(v) Now use (iii) to show that all the non zero roots of $J_n(z) = 0$ are real and simple.

(vi) Can $z = 0$ be a repeated root of $J_n(z) = 0$?

(vii) Show that $2nJ_n(z) = z(J_{n-1}(z) + J_{n+1}(z))$ and that $d(z^n J_n(z))/dz = z^n J_{n-1}(z)$ for all $n \geqslant 1$.

(viii) By considering the restriction of the formulae in (vii) to the positive real axis (i.e. $d(x^n J_n(x))/dx = x^n J_{n-1}(x)$ for $x > 0$), show that exactly one zero of J_n lies between each pair of zeros of J_{n-1} on the positive real axis.

95.6 In this question we prove the conjecture of Question 95.3 using Watson's lemma (Question 75.23). We shall also need some properties of the factorial function discussed in Questions 75.13, 75.14 and 75.17.

(i) By using results about the Beta function obtained in Question 75.17 and the duplication formula of Question 75.14(vii), show that, if n is an integer with $n \geqslant 0$,

$$\begin{aligned}\int_{-1}^1 (1-t^2)^{n-1/2} e^{iwt}\,dt &= \sum_{k=0}^{\infty} \frac{(iw)^k}{k!} \int_{-1}^1 (1-t^2)^{n-1/2} t^k\,dt \\ &= \sum_{r=0}^{\infty} \frac{(-1)^r w^{2r}}{(2r)!} \frac{(n-1/2)!(r-1/2)!}{(n+r)!} \\ &= (-1/2)!(n-1/2)! \sum_{r=0}^{\infty} \frac{(-1)^r w^{2r}}{r!(n+r)!2^{2r}}\end{aligned}$$

and deduce from Question 95.2(v) that, if $w \in \mathbb{C}$,

$$J_n(w) = c_n^{-1}\left(\frac{w}{2}\right)^n \int_{-1}^1 (1-t^2)^{n-1/2} e^{iwt}\,dt$$

with $c_n = (-1/2)!(n-1/2)! = \pi\prod_{r=1}^n (r-1/2)$.

(ii) (The manipulation of (i) is very much in the 'rabbit out of a hat' style and it would take us just too far out of our way to establish the formula for J_n in a transparent manner. However, the remainder of the question follows the standard 'contour pushing' way in which Watson's lemma is often applied.) Let x be real and strictly positive. Show, by considering integration round a suitable rectangular contour and proceeding to a limit, that

$$\int_{-1}^{1}(1-t^2)^{n-1/2}e^{ixt}\,dt=\int_{L(1)}(1-z^2)^{n-1/2}e^{ixz}\,dz-\int_{L(2)}(1-z^2)^{n-1/2}e^{ixz}\,dz,$$

where $L(1)$ and $L(2)$ are the paths described by $-1+is$ and $1+is$ as s runs from 0 to ∞. By rewriting the two path integrals as ordinary integrals and changing variables if necessary, show that

$$J_n(x)=\frac{x^{-1/2}}{\sqrt{2c_n}}\,(a_n(x)+a_n(x)^*)$$

with $a_n(x)=e^{-ix}\int_0^\infty e^{-u}\,u^{n-1/2}(i+u/2x)^{n-1/2}du$.

Now apply Watson's lemma to $a_n(x)$ to obtain

$$x^{1-\varepsilon}\,(J_n(x)-(2/\pi)^{1/2}x^{-1/2}\cos(x-(2n+1)\pi/4))\to 0$$

as $x\to\infty$ for all $\varepsilon>0$. (Thus we have not merely proved the conjecture of Question 95.3, but also found A_n and θ_n. Moreover, by using further terms of the Watson expansion, we could obtain even closer results (for sufficiently large x).)

97

FOURIER ANALYSIS ON THE ROOTS OF UNITY

97.1$^{(-)}$ Let r and N be integers with $N \geqslant 1$. Let G and $C(G)$ be as in Chapter 97. Check that if r and N have no common factor and we write $\omega = \exp(2\pi ir/N)$, $e_m(\omega^j) = \omega^{mj}$, $\hat{f}(e_m) = N^{-1} \sum_{j=0}^{N-1} f(\omega^j)\omega^{-mj}$, then it remains true that $f = \sum_{j=0}^{N-1} \hat{f}(e_j)e_j$ for all $f \in C(G)$. Does the result remain true if r and N are permitted to have a common factor? Give reasons. (If r and N have no common factor we call ω a primitive Nth root of unity.)

97.2 Gauss was very interested in the properties of the Gauss sum $S = \sum_{r=0}^{N-1} \omega^{r^2}$ where ω is a primitive Nth root of unity. In what follows, N will be odd.

(i) Show that $|S|^2 = \sum_{r=0}^{N-1} \sum_{s=0}^{N-1} \omega^{(r+s)(r-s)}$.

(ii) Show that, given any pair of integers u, v with $0 \leqslant u \leqslant N-1$, $0 \leqslant v \leqslant N-1$, we can find one and only one pair of integers r, s with $0 \leqslant r \leqslant N-1$, $0 \leqslant s \leqslant N-1$ and $r+s \equiv u \bmod N$, $r-s \equiv v \bmod N$. Conclude that

$$|S|^2 = \sum_{u=0}^{N-1} \sum_{v=0}^{N-1} \omega^{uv} = \sum_{u=1}^{N} \left(\sum_{v=0}^{N-1} (\omega^u)^v \right) = N.$$

In fact Gauss knew that S took one of the four values $N^{1/2}$, $-N^{1/2}$, $iN^{1/2}$ or $-iN^{1/2}$. But, as he wrote to his friend the astronomer Olbers, 'The determination of the sign of the root was what always troubled me. This lack clouded everything else that I found. For four years scarcely a week has gone by in which I have not made some vain attempt or other to untie this knot. A few days ago I finally suceeded, not through my own painstaking search but, if I may say so, through the grace of God. Like a flash of lightning the puzzle resolved itself.'

In the next two questions we attack this problem, not by Gauss's method, but by one due to Schur.

97.3$^{(-)}$ (We revise some linear algebra for the next question.)

(i) If A is an $n \times n$ (complex) matrix with (i, j)th entry a_{ij}, we define the trace of A by $\operatorname{tr} A = \sum_{i=1}^{n} a_{ii}$. Show that, if A and B are $n \times n$ matrices and $\lambda \in \mathbb{C}$, then $\operatorname{tr}(A+B) = \operatorname{tr} A + \operatorname{tr} B$, $\operatorname{tr} I = 1$, $\operatorname{tr}(\lambda A) = \lambda \operatorname{tr} A$ and $\operatorname{tr} AB = \operatorname{tr} BA$. If $n = 2$ and $A = \begin{pmatrix} 0 & 1 \\ 1 & 0 \end{pmatrix}$ is it true that $\operatorname{tr} A^2 = (\operatorname{tr} A)^2$?

(ii) Show that, if A is an $n \times n$ matrix and B an invertible $n \times n$ matrix, then $\operatorname{tr}(BAB^{-1}) = \operatorname{tr} A$. Conclude that, if V is an n-dimensional vector space over $\mathbb{C}$ and $\alpha: V \to V$ is a linear map which has matrix representation A_1 with respect to one basis of V and A_2 with respect to another, then $\operatorname{tr} A_1 = \operatorname{tr} A_2$. Thus we may define $\operatorname{tr} \alpha$ unambiguously by $\operatorname{tr} \alpha = \operatorname{tr} A_1$.

97.4 We have used only the simplest linear algebra in this book. In this question we shall allow ourselves the use of ideas which appear in the first part of most second courses in linear algebra. The ideas are due to Schur and will give us a glimpse of several avenues which we have left unexplored.

(i) Suppose N is a positive odd integer. With the notation of Question 97.1, let $\omega = \exp(2\pi ir/N)$ be a primitive Nth root of unity and define $F: C(G) \to C(G)$ by $(Ff)(\omega^k) = N^{-1/2} \sum_{j=0}^{N-1} f(\omega^j)\omega^{jk}$. Show that F is a linear map and $F(Ff) = \check{f}$ where $\check{f}(\omega^j) = f(\omega^{-j})$.

(ii) Let $U = \{f \in C(G): f = \check{f}\}$ and $V = \{f \in C(G): f = -\check{f}\}$. By using the formula $f = 2^{-1}((f + \check{f}) + (f - \check{f}))$, or otherwise, show that U, V are subspaces of $C(G)$ and $C(G) = U \oplus V$. By choosing suitable bases, show that $\dim U = (N+1)/2$ and $\dim V = (N-1)/2$. Show also that $F(U) \subseteq U$, $F(V) \subseteq V$. Writing F_U for the restriction of F to U and F_V for the restriction of F to V, show that $F_U^2 = I_U$, $F_V^2 = -I_V$ where I_U is the identity on U and I_V the identity on V.

(iii) Write $U_1 = \{f \in U: Ff = f\}$, $U_2 = \{f \in U: Ff = -f\}$, $V_1 = \{f \in V: Ff = if\}$, $V_2 = \{f \in V: Ff = -if\}$. By using the formula

$$f = (2i)^{-1}((Ff + if) - (Ff - if)),$$

or otherwise, show that $V = V_1 \oplus V_2$. Show similarly that $U = U_1 \oplus U_2$ and deduce that $C(G) = U_1 \oplus U_2 \oplus V_1 \oplus V_2$.

(iv) Choose a basis $e_1, \ldots, e_{n(1)}$ for U_1, a basis $e_{n(1)+1}, \ldots, e_{n(1)+n(2)}$ for U_2, a basis $e_{n(1)+n(2)+1}, \ldots, e_{n(1)+n(2)+n(3)}$ for V_1 and a basis $e_{n(1)+n(2)+n(3)+1}, \ldots, e_{n(1)+n(2)+n(3)+n(4)}$ for V_2. Show that $e_1, e_2, \ldots, e_{n(1)+n(2)+n(3)+n(4)}$ is a basis for $C(G)$. Write down the matrix of F with respect to this basis and conclude that

(a) $\operatorname{tr} F = (n(1) - n(2)) + i(n(3) - n(4))$.
(b) $\det F = (-1)^{n(2)+n(4)} i^{n(3)+n(4)}$.
(c) $n(1) + n(2) = (N+1)/2$, $n(3) + n(4) = (N-1)/2$.

(v) Now return to the obvious basis for $C(G)$ given by $v_0, v_1, \ldots, v_{N-1}$

where $v_j(\omega^k)=1$ if $j=k$, $v_j(\omega^k)=0$, otherwise $[0\leqslant j,\ k\leqslant N-1]$. Write down the matrix of F with respect to this basis and conclude, possibly using Question 6.2(ii), that

(a)′ $\operatorname{tr} F = N^{-1/2}\sum_{j=0}^{N}\omega^{j^2}$.

(b)′ $\det F = N^{-N/2}\prod_{0\leqslant j<k\leqslant N-1}(\omega^k-\omega^j)$.

(vi) Using the two formulae for the trace, show that the Gauss sum S is given by $S=\varepsilon N^{-1/2}$ where $\varepsilon=(n(1)-n(2))+i(n(3)-n(4))$. Using Question 97.2, observe that $|\varepsilon|=1$ and so

(d) Either $n(1)=n(2)$, $|n(3)-n(4)|=1$ or $|n(1)-n(2)|=1$, $n(3)=n(4)$.

(vii) Using the two formulae for the determinant, show that

$$(-1)^{n(2)+n(4)}i^{n(3)+n(4)}$$
$$=N^{-N/2}\exp\left(\sum_{0\leqslant j<k<N-1}(j+k)\pi i/N\right)\prod_{0\leqslant j<k\leqslant N-1}(2i\sin\pi(k-j)/N)$$

Show by direct computation, or otherwise, that $\sum_{0\leqslant j<k<N-1}(j+k)$ is divisible by $2N$ and so

$$(-1)^{n(2)+n(4)}i^{n(3)+n(4)}=2i^{N(N-1)/2}N^{-N/2}\prod_{0\leqslant j<k\leqslant N-1}\sin(\pi(k-j)/N).$$

Hence, deduce that

(e) $2n(2)+n(3)+3n(4)\equiv N(N-1)/2 \bmod 4$.

(viii) Using the results (c), (d) and (e) (by, for example, considering the four cases $N=8m+1$, $8m+3$, $8m+5$, $8m+7$, or otherwise), show that, if $N\equiv 1 \bmod 4$, then $n(1)=n(2)+1$, $n(3)=n(4)$ and, if $N\equiv 3 \bmod 4$, then $n(3)=n(4)+1$ and $n(1)=n(2)$. Conclude, using (vi), that

$$\sum_{j=0}^{N-1}\omega^{j^2}=\begin{cases}N^{1/2} & \text{if } N\equiv 1 \bmod 4\\ iN^{1/2} & \text{if } N\equiv 3 \bmod 4\end{cases}.$$

(ix) We now drop the condition N odd. By experiment, or otherwise, determine $\sum_{j=0}^{N-1}\omega^{j^2}$ for $N\equiv 2 \bmod 4$, and then prove your result. The remaining case $N\equiv 0 \bmod 4$ is not trivial, but, although we shall not do this, it can be derived by standard number theoretic techniques from the cases $N\not\equiv 0 \bmod 4$. (Indeed, the full result can be obtained from the case N an odd prime.)

We shall continue this train of thought in Questions 103.4 and 103.5. The next question gives an evaluation of S for all N.

97.5 The idea of investigating the eigenspace decomposition of F used in Question 97.4 is a natural one though, of course, the particularly happy outcome could hardly be foreseen. Here is another natural evaluation of Gauss sums due to Dirichlet. It makes use of a circle of ideas that we have seen repeatedly in these exercises (e.g. Questions 60.4 and 79.1) which we could refer to generically as Poisson methods.

(i) Suppose $g: \mathbb{R} \to \mathbb{C}$ is once continuously differentiable. By applying Theorem 16.4 to an appropriately defined $f: \mathbb{T} \to \mathbb{C}$, show that

$$\sum_{k=-m}^{m} \int_r^{r+1} g(t) \exp(-2\pi ikt)\, dt \to \frac{g(r+1)+g(r)}{2}$$

as $m \to \infty$. Conclude that, if N is a strictly positive integer,

$$\sum_{k=-m}^{m} \int_0^N g(t) \exp(2\pi ikt)\, dt \to \sum_{r=0}^{N-1} g(r) + (g(N) - g(0))/2$$

as $m \to \infty$.

(ii) Deduce that, if we write

$$S_N = \sum_{r=0}^{N-1} \exp(2\pi ir^2/N),$$

we have

$$\sum_{k=-m}^{m} \int_0^N \exp(2\pi ikt + 2\pi it^2/N)\, dt \to S_N$$

as $m \to \infty$. By a suitable change of variables, deduce that

$$\sum_{k=-m}^{m} N \exp(-\pi iNk^2/2) \int_{k/2}^{1+k/2} \exp(2\pi iNy^2)\, dy \to S_N$$

as $m \to \infty$. Now set $m = 2M$ and, by considering the cases k even and k odd separately, deduce that

$$N \int_{-M}^{M+1} \exp(2\pi iNy^2)\, dy + Ni^{-N} \int_{-M+1/2}^{M+1/2} \exp(2\pi iNy^2)\, dy \to S_N$$

as $M \to \infty$.

(iii) Although $\int_{-\infty}^{\infty} \exp(2\pi iy^2)\, dy$ is not absolutely integrable (why?), there are many ways of showing that $\lim_{R,S\to\infty} \int_{-S}^{R} \exp(2\pi iy^2)\, dy$ exists. For example, the reader could show that

$$\int_1^R \exp(2\pi iy^2)\, dy = \frac{1}{2} \int_1^{R^2} u^{-1/2} \exp(2\pi iu)\, du,$$

and then integrate the right hand side by parts to obtain an expression involving an absolutely convergent integral. Using this method, or any other, show that

$$\int_{-S}^{R} \exp(2\pi i y^2)\, dy \to A, \text{ say,}$$

as $R, S \to \infty$.

(iv) Conclude that

$$S_N = N^{1/2} A(1 + i^{-N}).$$

Calculate A by setting $N = 1$ and deduce that

$$\begin{aligned} S_N &= (1+i)N^{1/2} && \text{if } N \equiv 0 \bmod 4,\\ S_N &= N^{1/2} && \text{if } N \equiv 1 \bmod 4,\\ S_N &= 0 && \text{if } N \equiv 2 \bmod 4,\\ S_N &= iN^{1/2} && \text{if } N \equiv 3 \bmod 4. \end{aligned}$$

99

HOW FAST CAN WE MULTIPLY

99.1 (How fast can we compute π?)

(i) By differentiation, or otherwise, show that

$$\tan^{-1}\alpha = \alpha - \frac{\alpha^3}{3} + \cdots + (-1)^n \frac{\alpha^{2n+1}}{2n+1} + R_n(\alpha)$$

with $R_n(\alpha) = (-1)^{n+1}\int_0^\alpha x^{2n+2}/(1+x^2)dx$.

(ii) By using the inequalities

$$0 \leqslant \frac{x^{2n+2}}{2} \leqslant \frac{x^{2n+2}}{1+x^2} \leqslant x^{2n+2} \quad \text{for } |x| \leqslant 1,$$

obtain upper and lower bounds on the number of terms you would need to compute $\tan^{-1}\alpha$ to within an error of ε when $|\alpha| \leqslant 1$. What happens if $|\alpha| > 1$?

(iii) How many terms are required to compute π to six-figure accuracy using the formula

$$\frac{\pi}{4} = \tan^{-1} 1 = 1 - \frac{1}{3} + \frac{1}{5} - \cdots ?$$

(iv) By considering $(5-i)^4(1+i)$, show that

$$\pi/4 = 4\tan^{-1} 1/5 - \tan^{-1}(1/239).$$

Prove similarly that

$$\pi/4 = 3\tan^{-1} 1/4 + \tan^{-1}(1/20) + \tan^{-1}(1/1985).$$

(v) How many terms are required to compute $\tan^{-1} 1/5$ to seven-figure accuracy using the formula $\tan^{-1}\alpha = \sum_{r=0}^{\infty}(-1)^r\alpha^{r+1}/(r+1)$? How many are needed for $\tan^{-1}(1/239)$? Compute π to six-figure accuracy using the first formula of (iv).

(vi) Give a rough estimate of the number of computations required to obtain π to n figure accuracy in the manner of (v).

(The method outlined above was the standard method used for the 200 years up to 1980. Calculating π by hand to 500 places in Victorian times took several years. The same calculation using a mechanical calculator took a couple of months in 1945. By 1980 a large computer could manage a million places in a few hours.)

99.2 (What Gauss knew) Gauss enjoyed calculating and was interested in formulae like those in Question 99.1(iv). One of his first mathematical investigations concerned the following ideas.

(i) Let $a \geqslant b > 0$ and take $a = a_0$, $b = b_0$. Set $a_{n+1} = (a_n + b_n)/2$, $b_{n+1} = \sqrt{(a_n b_n)}$. If the reader takes any a, b she pleases (for example $a = 100$, $b = 1$) and uses a hand calculator to compute a_n, b_n for $n = 1, 2, 3, \ldots, 10$, she will surely see why this process caught Gauss's fancy.

(ii) Show that a_n is a decreasing sequence bounded below by b and deduce that a_n converges to α say. Show, similarly, that $b_n \to \beta$ for some β. Show that $\alpha = (\alpha + \beta)/2$ and so $\alpha = \beta$. We write $m(a, b) = \alpha$.

(iii) Show that

$$\frac{a_{n+1} - b_{n+1}}{a_{n+1} + b_{n+1}} = \left(\frac{\sqrt{a_n} - \sqrt{b_n}}{\sqrt{a_n} + \sqrt{b_n}}\right)^2 \leqslant \left(\frac{a_n - b_n}{a_n + b_n}\right)^2.$$

Conclude that, if $3b \geqslant 2a$, then $|a_{n+1} - m(a, b)| \leqslant m(a, b)2^{-n}$.

(iv) Show that

$$\frac{a_{n+1}}{b_{n+1}} \leqslant \frac{a_n}{\sqrt{(a_n b_n)}} = \left(\frac{a_n}{b_n}\right)^{1/2}.$$

Conclude that $3b_N \geqslant 2a_N$ if $N \geqslant \log_2 \log_2 a/b - \log_2 \log_2 3/2$.

(v) In particular show that, if we write $h(A) = A/m(A, 1)$, then (taking $a = A$, $b = 1$) we will need to take n proportional to $\log k$ in order for A/a_n to approximate $h(A)$ to k decimal places.

(vi) We have shown that a_n converges like a bat out of hell but it is quite another matter to find what it converges to. Here Gauss plays a master stroke. Consider

$$I(a, b) = \int_{-\infty}^{\infty} \frac{dx}{\sqrt{((x^2 + a^2)(x^2 + b^2))}}.$$

By carefully making the change of variables $t = (x - ab/x)/2$ show that

$$\int_0^{\infty} \frac{dx}{\sqrt{((x^2 + a^2)(x^2 + b^2))}} = \frac{1}{2}\int_{-\infty}^{\infty} \frac{dt}{\sqrt{((t^2 + (a+b)/2)^2)(t^2 + ab))}}$$

and deduce that $I(a_n, b_n) = I(a_{n+1}, b_{n+1})$.

(vii) Show that, given any $\varepsilon > 0$, we can find an $R(\varepsilon) > 0$ such that

$$\int_{|x| \geqslant R(\varepsilon)} dx/\sqrt{((x^2 + a_n^2)(x^2 + b_n^2))} < \varepsilon$$

for all n. Show also that

$$1/\sqrt{((x^2+a_n^2)(x^2+b_n^2))} \to 1/(x^2+m(a,b)^2),$$

uniformly on $[-R(\varepsilon), R(\varepsilon)]$ and deduce that we have $I(a_n, b_n) \to \int_{-\infty}^{\infty}(x^2 + m(a,b)^2)^{-1}\,dx$. Hence, conclude that $\pi/m(a,b) = I(a,b)$ and $a_n \to \pi/I(a,b)$.

99.3 (What Gauss missed) The results of Question 99.2 belong to the beginning of elliptic function theory. This theory formed a major theme of nineteenth-century mathematics (in the exercises for Chapter 38 we saw some glimpses of it). Mathematical fashions change and, by 1960, it was possible for undergraduates to pass three years without seeing an elliptic integral. In the mid 1970s Brent and Salamin discovered an unexpected link between this neglected theory and the problem of computing e^x and π. This question and the previous one are based on a paper of Newman (see bibliography) which simplifies their work.

(i) Show that, if $h(A)$ is defined as in Question 99.2, then

$$h(A) = \frac{2}{\pi}\int_0^\infty \frac{dx}{\sqrt{((1+x^2)(1+A^{-2}x^2))}}.$$

(ii) By considering the change of variable $t = A/x$, show that

$$\int_0^{\sqrt{A}} \frac{dx}{\sqrt{((1+x^2)(1+A^{-2}x^2))}} = \int_{\sqrt{A}}^\infty \frac{dx}{\sqrt{((1+x^2)(1+A^{-2}x^2))}}$$

and so

$$h(A) = \frac{4}{\pi}\int_0^{\sqrt{A}} \frac{dx}{\sqrt{((1+x^2)(1+A^{-2}x^2))}}.$$

(iii) Show that, if $0 \leqslant \alpha \leqslant 1$, $1 \geqslant (1+\alpha)^{-1/2} \geqslant 1 - \alpha/2$ and $1 - \alpha/2 + 3\alpha^2/8 \geqslant (1+\alpha)^{-1/2}$. Hence show that

$$\left|\frac{1}{\sqrt{((1+x^2)(1+A^{-2}x^2))}} - \frac{1-x^2/(2A^2)}{\sqrt{(1+x^2)}}\right| \leqslant \frac{x^4}{A^4} \quad \text{for } 0 \leqslant x \leqslant A,$$

$$\left|\frac{x^2}{\sqrt{(1+x^2)}} - x\right| \leqslant \frac{1}{x} \quad \text{for } x > 1.$$

Deduce that there is a constant K_1 such that, if A is large,

$$\left|h(A) - \frac{4}{\pi}\int_0^{\sqrt{A}}\left(\frac{1}{\sqrt{(1+x^2)}} - \frac{x}{2A^2}\right)dx\right| \leqslant \frac{K_1}{A^2}$$

and so $|h(A) - 4\pi^{-1}(\log(\sqrt{A} + \sqrt{(A+1)}) - 1/(4A))| \leqslant K_1A^{-2}$.

(iv) By using Taylor expansions with remainder estimates, or otherwise, show that there is a constant K_2 such that, for A large,

$$|\log(\sqrt{A} + \sqrt{(A+1)}) - 1/(4A) - \log(2\sqrt{A})| \leqslant K_2A^{-2}.$$

Hence conclude that $|h(A) - 2\pi^{-1}\log 4A| \leqslant K_3A^{-2}$ where $K_3 = K_1 + K_2$.

(v) Deduce that there exists a K_4 such that for A large

$$|A(h(A+1) - h(A)) - 2\pi^{-1}| \leqslant K_4A^{-1}.$$

(vi) Now suppose we wish to know π to about k decimal places. To do this it clearly suffices to know $2\pi^{-1}$ to about k decimal places. If k is large, then, if we take $A = 10^k$, we see from (v) that it suffices to know $h(A)$ and $h(A+1)$ to about $2k$ decimal places. Question 99.2(v) tells us that $h(A)$ and $h(A+1)$ can be computed to this degree of accuracy using a number of separate computations proportional to log k. Thus π can be computed to k decimal places using less than $K_5 \log k$ operations where K_5 is a constant independent of k.

(vii) Compute π to six decimal places using the method above and a hand calculator.

(viii)$^{(+)}$ Show that, if $3 \leqslant x \leqslant 9$, we can compute $\log x$ to k decimal places using a number of separate operations which grow like $\log k$. Why does this give us a quick method of computing $\exp y$ for suitable y? (*Hint*. Recall Newton's method.)

(In the discussion above, we considered multiplication and addition as single operations, however many places of decimals are involved. This is not the case in practice, nor is it the point of view we took in Chapter 99. The reader may wish to consider what difference the more realistic point of view would make.)

102

A GOOD CODE?

102.1 (Euler's totient function)

(i) By using the same kind of argument as in Lemma 102.1, or otherwise, show that, if n, m are strictly positive integers with no common factor, then $\phi(nm) = \phi(n)\phi(m)$.

(ii) Show that, if p is a prime and m a strictly positive integer, then $\phi(p^m) = p^m - p^{m-1}$.

(iii) Conclude that, if n is a strictly positive integer divisible by the distinct primes $p_1, p_2, \ldots, p_k$ and no others, then $\phi(n) = n\prod_{j=1}^{k}(1 - p_j^{-1})$.

103

A LITTLE MORE GROUP THEORY

103.1$^{(-)}$ Let G be a finite group. Show that if $a, g, h \in G$ satisfy $ag = ha$, then $e(gh^{-1}) = 1$ for all $e \in \hat{G}$. Conclude, using Theorem 103.8, that, if G is any finite non commutative group, $\hat{G}$ has fewer elements than G.

103.2 (i) Write $S^1 = \{\lambda \in \mathbb{C}: |\lambda| = 1\}$. Is it true that, if λ, $\mu \in S^1$ and $\lambda^2 = \mu^2$, then $\lambda = \mu$?

(ii) Suppose $e: \mathbb{R} \to S^1$ is a continuous homomorphism (i.e. e is a continuous function and $e(x + y) = e(x)e(y)$ for all $x, y \in \mathbb{R}$). Show that there exists a $\delta > 0$ such that $|e(x) - 1| < 1$ for all $|x| < \delta$. Choose some η with $0 < \eta < \delta$ and let θ be the unique solution of $e(\eta) = \exp i\theta$ with $|\theta| < \pi/2$. Show by induction, or otherwise, that $e(2^{-n}\eta) = \exp(i2^{-n}\theta)$ for all integers $n \geqslant 0$. Hence show that $e(r2^{-n}\eta) = \exp(ir2^{-n}\theta)$ for all integers r, n with $n \geqslant 0$. Conclude that setting $\lambda = \eta^{-1}\theta$ we have $e(x) = \exp(i\lambda x)$ for all $x \in \mathbb{R}$, and so the continuous homomorphisms $e: \mathbb{R} \to S^1$ are precisely the functions e_λ given by

$$e_\lambda(x) = \exp(i\lambda x) \text{ for all } x \in \mathbb{R}\ [\lambda \in \mathbb{R}].$$

(iii) Find the continuous homomorphisms $e: \mathbb{T} \to S^1$.

(iv) If $e: \mathbb{T}^n \to S^1$ is a continuous homomorphism, what can you say about the map $\tilde{e}: \mathbb{T} \to S^1$ given by $\tilde{e}(x) = e(x, 0, \ldots, 0)$? Hence, or otherwise, show that the continuous homomorphisms $e: \mathbb{T}^n \to S^1$ are precisely the functions $e_{\boldsymbol{m}}: \mathbb{T}^n \to S^1$ given by $e_{\boldsymbol{m}} = \exp(i\boldsymbol{m} \cdot \boldsymbol{x})$ for all $\boldsymbol{x} \in \mathbb{R}^n$ $[\boldsymbol{m} \in \mathbb{Z}^n]$.

(v) Find the continuous homomorphisms $e: \mathbb{R}^n \to S^1$.

(vi) Find the continuous homomorphisms from $\mathbb{T}^n$ to $\mathbb{T}$.

(vii) Find the continuous homomorphisms from S^1 to $\mathbb{R}$.

(The reader may well ask if there exist homomorphisms $\varepsilon: \mathbb{R} \to S^1$ which are not continuous. This question cannot be answered without introducing ideas from the foundations of mathematics. The standard tools of classical mathematics show that, if such non continuous homomorphisms exist, they are very odd indeed but do not, and cannot, show that these objects do not exist.)

103.3 (An inductive proof of Kronecker's theorem) The following proof may appear rather roundabout but the underlying idea has been exploited by Varopoulos to great effect. We shall say that $\gamma_1, \gamma_2, \ldots, \gamma_m$ are independent if they satisfy the hypothesis of Theorem 79.1, and we shall write $P(m)$ for the statement that the theorem holds for any set of m independent points.

(i) Let G be a subgroup of $\mathbb{T}$. Show, in the style of Question 103.2, that either G is dense in $\mathbb{T}$ or $G = \{2\pi r/N: 0 \leqslant r \leqslant N-1\}$ for some integer $N \geqslant 1$.

(ii) Let $\gamma_1 \in \mathbb{R}$. By considering $G = \{2\pi n\gamma_1: n \in \mathbb{Z}\}$ show that, either, given $0 \leqslant a_1 < b_1 \leqslant 1$, we can find an n with $a_1 \leqslant \langle n\gamma_1\rangle \leqslant b_1$, or $N\gamma_1 \in \mathbb{Z}$ for some $N \in \mathbb{Z}$. Conclude that $P(1)$ is true.

(iii) Now suppose $P(m)$ is true and $\gamma_1, \gamma_2, \ldots, \gamma_{m+1}$ are independent points. Let us say that $x \in G$ if and only if we can find a sequence $n(j)$ of integers (not necessarily distinct) such that $2\pi n(j)\gamma_k \to 0$ as $j \to \infty$ for each $1 \leqslant k \leqslant m$, and $2\pi n(j)\gamma \to x$. Show that G is a subgroup of $\mathbb{T}$ and deduce that either G is dense in $\mathbb{T}$ or $G = \{2\pi r/N: 0 \leqslant r \leqslant N-1\}$ for some $N \geqslant 1$.

(iv) If G is dense in $\mathbb{T}$, we proceed as follows. Suppose $x_1, x_2, \ldots, x_{m+1} \in \mathbb{T}$. Use the inductive hypothesis to find a sequence $p(j)$ of integers such that $2\pi p(j)\gamma_k \to x_k$ as $j \to \infty$ for $1 \leqslant k \leqslant m$. Observe, giving reasons, that the sequence $2\pi p(j)\gamma_{m+1}$ must have a convergent subsequence and deduce that we can find a $y \in \mathbb{T}$ and a sequence $q(j)$ such that $2\pi q(j)\gamma_k \to x_k$ as $j \to \infty$ for $1 \leqslant k \leqslant m$ and $2\pi q(j)\gamma_{m+1} \to y$. Now use the fact that G is dense to show the existence of a sequence $r(j)$ with $2\pi r(j)\gamma_k \to 1$ for $1 \leqslant k \leqslant m$ and $2\pi r(j)\gamma_{m+1} \to x_{m+1} - y$. By looking at the sequence $q(j) + r(j)$, show that the conclusion of Theorem 79.1 hold for $\gamma_1, \gamma_2, \ldots, \gamma_{m+1}$.

(v) If $G = \{2\pi r/N: 0 \leqslant r \leqslant N-1\}$, observe that $\alpha_1 = N^{-1}\gamma_1$, $\alpha_2 = N^{-1}\gamma_2, \ldots, \alpha_m = N^{-1}\gamma_m$ are independent and that, if $2\pi r(j)\alpha_k \to 0$ for $1 \leqslant k \leqslant m$, then $2\pi r(j)\gamma_{m+1} \to 0$. Show from this that, if $2\pi r(j)\alpha_k \to x_k$ for $1 \leqslant k \leqslant m$, then $2\pi r(j)\gamma_{m+1}$ must converge to a unique point $\chi(x_1, x_2, \ldots, x_m)$. Use the inductive hypothesis to show that $\chi(x_1, x_2, \ldots, x_m)$ is defined for each $x_1, x_2, \ldots, x_m \in \mathbb{T}$.

(vi) Now show that $\chi: \mathbb{T}^m \to \mathbb{T}$ is a homomorphism. Observe, using vectorial notation that, if $\|\mathbf{x}(j) - \mathbf{x}\| \to 0$ as $j \to \infty$, we can find integers $r(j)$ such that $\|2\pi r(j)\boldsymbol{\alpha} - \mathbf{x}(j)\| \leqslant 2^{-j}$ and $\|2\pi r(j)\gamma_{m+1} - \chi(\mathbf{x}(j))\| \leqslant 2^{-j}$. Conclude that $\chi(\mathbf{x}(j)) \to \chi(\mathbf{x})$ and so χ is continuous.

(vii) Using Question 103.2 (vi) (or (iii)), deduce that

$$\chi(\mathbf{x}) = \mathbf{n} \cdot \mathbf{x} = \textstyle\sum_{k=1}^{m} n_k x_k$$

for some integers $n_1, n_2, \ldots, n_m$. Observe that $2\pi\alpha_k \to 2\pi\alpha_k$ as $j \to \infty$ (for trivial reasons). Conclude that

$$2\pi\gamma_{m+1} = \chi(2\pi\alpha_1, 2\pi\alpha_2, \ldots, 2\pi\alpha_m) = 2\pi \sum_{k=1}^{m} n_k\alpha_k$$

and so $\sum_{k=1}^{m} n_k\gamma_k - N\gamma_{m+1} = 0$ contradicting our hypothesis.

(viii) Bring together the threads of argument to obtain Kronecker's theorem.

103.4 Let p be an odd prime and let G be the group of units modulo p. Number theorists, from the eighteenth century onwards, have been deeply interested in the congruence $x^2 \equiv r \bmod p$. Following Legendre, we write

$$\left(\frac{r}{p}\right) = 1 \text{ if there is a solution and } r \not\equiv 0 \bmod p,$$

$$\left(\frac{r}{p}\right) = -1 \text{ if there is no solution,}$$

$$\left(\frac{r}{p}\right) = 0 \text{ if } r \equiv 0 \bmod p.$$

(i) Show that, if we define $e: G \to \mathbb{C}$ by $e([r]) = \left(\frac{r}{p}\right)$, then our definition is unambiguous and $e \in \hat{G}$.

(ii) Show that the set $\{[1]^2, [2]^2, \ldots, [p-1]^2\}$ has less than $p-1$ elements. Deduce, using elementary group theory, or otherwise, that $Q = \{[r] \in G: e([r]) = 1\}$ and $Q' = ([r] \in G: e([r]) = -1\}$ both have $(p-1)/2$ elements. Show further that, if $[r] \neq 0$, the equation $[x]^2 = [r]$ either has exactly two solutions or no solutions.

103.5 Let p be an odd prime as in the previous question, but, this time, consider the group G of pth roots of unity. We adopt the notation of Chapter 97.

(i) Show that we can define a function $\chi \in C(G)$ by taking $\chi(\omega^r) = \left(\frac{r}{p}\right)$.

(ii) Show that, if we write

$$Q = \left\{\omega^r: \left(\frac{r}{p}\right) = 1\right\},\ Q' = \left\{\omega^r: \left(\frac{r}{p}\right) = -1\right\},$$

then

$$p\hat{\chi}(e_s) = \sum_{\omega^r \in Q'} \omega^{rs} - \sum_{\omega^r \in Q} \omega^{rs} = 1 + 2 \sum_{\omega^r \in Q} \omega^{rs}.$$

Conclude, using the previous question, that

$$\hat{\chi}(e_s) = p^{-1} \sum_{j=0}^{p-1} \omega^{sj^2}.$$

(iii) Using the methods of previous question show that, if $\omega^s \in Q$, then ω^{sj^2} runs through each value of Q twice as j runs from 1 to $p-1$. What happens if $\omega^s \in Q'$? Deduce that $\hat{\chi}(e_s) = \left(\frac{s}{p}\right)\hat{\chi}(e_1)$ for all s, i.e. that $\hat{\chi}(e_s) = \hat{\chi}(e_1)\chi(\omega^s)$ for all s.

(iv) Let us write $G_N(s) = \sum_{j=0}^{N-1} \exp(2\pi i s j^2/N)$ whenever $s \in \mathbb{Z}$, $N \in \mathbb{Z}$, $N \geqslant 1$. Rewrite (iii) to show that, if p is an odd prime,

$$\left(\frac{r}{p}\right) = \frac{G_p(r)}{G_p(1)}$$

and deduce that, if p and q are odd primes, then

$$\left(\frac{q}{p}\right)\left(\frac{p}{q}\right) = \frac{G_p(q)G_q(p)}{G_p(1)G_q(1)}$$

(v) Show that, if j runs from 0 to $p-1$ and k runs from 0 to $q-1$, then, modulo pq, $kq + jp$ takes each value between 0 and $pq-1$ exactly once. Deduce that

$$G_p(q)G_q(p) = G_{pq}(1)$$

and so

$$\left(\frac{q}{p}\right)\left(\frac{p}{q}\right) = \frac{G_{pq}(1)}{G_p(1)G_q(1)}$$

(vi) In Question 97.4 we evaluated the Gauss sum $G_N(1)$ for all odd N. Use this evaluation to show that, if p, q are odd primes,

$$\left(\frac{q}{p}\right)\left(\frac{p}{q}\right) = -1 \qquad \text{if } p, q \equiv 3 \bmod 4,$$

$$\left(\frac{q}{p}\right)\left(\frac{p}{q}\right) = 1 \qquad \text{otherwise.}$$

This is the famous law of quadratic reciprocity conjectured by Euler in 1783. Both Euler and Legendre sought hard to find a proof but the first complete proof of this 'gem of arithmetic' was found by Gauss at the age of nineteen.

103.6 The definition of the Legendre symbol in Question 103.4 does give a way of computing it (just evaluate $[1]^2, [2]^2, \ldots, [p-1]^2$), but not a very efficient one. In this question we given another method for computing it.

(i) (Wilson's theorem) By observing that $x^2 - 1 \equiv (x-1)(x+1) \bmod p$, show that the only self inverse elements in the group of units modulo p are 1 and -1. Conclude, by pairing inverse elements, that

$$(p-1)! \equiv 1(-1) \equiv -1 \bmod p.$$

(ii) Use Fermat's little theorem (Theorem 101.3) to show that, if $x^2 \equiv -1 \bmod p$ for some x, then $(-1)^{(p-1)/2} \equiv 1 \bmod p$ and so $p \equiv 1 \bmod 4$.

(iii) Observe that $(p+1)/2 + r \equiv -((p-1)/2 - r)$ and use (i) to show that

$$-1 \equiv (p-1)! \equiv (-1)^{(p-1)/2}((p-1)/2)!)^2.$$

Conclude that, if $p \equiv 1 \bmod 4$, then the equation $x^2 \equiv -1 \bmod p$ has a solution.

(iv) Combine (ii) and (iii) to show that

$$\left(\frac{-1}{p}\right) = (-1)^{(p-1)/2}.$$

(v) In the proof of Wilson's theorem we paired x and y in the product $(p-1)! = \prod_{r=1}^{p-1} r$ if $xy = 1$ and $x \neq y$. By considering a similar pairing for the more general equation $xy = a$, show that

$$(p-1)! \equiv -\left(\frac{a}{p}\right) a^{(p-1)/2} \bmod p$$

whenever $a \equiv 0 \bmod p$. Conclude that

$$\left(\frac{a}{p}\right) \equiv a^{(p-1)/2} \bmod p$$

whenver $a \not\equiv 0 \bmod p$.

(vi) Calculate $(\frac{5}{11})$ and $(\frac{4}{13})$.

(Electronic computers are well adapted to the computation of $a^{(p-1)/2} \bmod p$. The reader should have no difficulty in showing that such a computation requires less than $A \log p$ steps for some constant A. However, human beings are less happy with such calculations. In the next two questions we develop a method which, while no faster for computers, is much faster for us.)

103.7 We introduce a generalisation of the Legendre symbol called the Jacobi symbol defined, for all comprime integers a, m such that m is odd and positive, by

$$\left(\frac{a}{m}\right) = \prod_{i=1}^{k} \left(\frac{a}{p_i}\right)$$

where $m = p_1 p_2 \dots p_k$ and the p_i are prime. (We take $\left(\frac{a}{1}\right) = 1$.)

(i) Perform the trivial verification that the Jacobi symbol is a well-defined generalisation of the Legendre symbol.

(ii) By induction on the number k of factors of m, or otherwise, show that, if m is odd and positive, then $\left(\frac{-1}{m}\right) = 1$ if $m \equiv 1 \bmod 4$, $\left(\frac{-1}{m}\right) = -1$ if $m \equiv 3 \bmod 4$.

(iii) Use the fact that $\left(\frac{n_1}{p}\right)\left(\frac{n_2}{p}\right) = \left(\frac{n_1 n_2}{p}\right)$ for p an odd prime (see Question 103.4 (i)) together with the law of quadratic reciprocity, to show that, if m and n are odd comprime positive integers, then

$$\left(\frac{n}{m}\right) = -\left(\frac{m}{n}\right) \quad \text{if } m \equiv n \equiv 3 \bmod 4,$$

$$\left(\frac{n}{m}\right) = \left(\frac{m}{n}\right) \quad \text{otherwise.}$$

(iv) Let m be an odd positive integer. Use earlier parts of this question to justify all the steps in the following computation

$$\left(\frac{2}{m}\right) = \left(\frac{m+2}{m}\right) = \left(\frac{m}{m+2}\right) = \left(\frac{-2}{m+2}\right) = \left(\frac{-1}{m+2}\right)\left(\frac{2}{m+2}\right).$$

By a careful induction on m, or otherwise, show that

$$\left(\frac{2}{m}\right) = 1 \quad \text{if } m \equiv \pm 1 \bmod 8,$$

$$\left(\frac{2}{m}\right) = -1 \quad \text{if } m \equiv \pm 3 \bmod 8,$$

(v) Check that the results of this question may be stated as follows. If m and n are odd positive integers, then

$$\left(\frac{-1}{m}\right) = (-1)^{(m-1)/2}$$

$$\left(\frac{2}{m}\right) = (-1)^{(m^2-1)/8}$$

$$\left(\frac{n}{m}\right) = (-1)^{(m-1)(n-1)/4}\left(\frac{m}{n}\right).$$

103.8 (i) Sketch an algorithm (computer program) to compute $\left(\frac{n}{m}\right)$ (where m and n are comprime integers and m is an odd positive integer) based on the following observations.

(A) $\left(\frac{0}{m}\right) = 0$, $\left(\frac{1}{m}\right) = 1$.

(B) If $n < 0$, then $\left(\frac{n}{m}\right) = \left(\frac{-1}{m}\right)\left(\frac{|n|}{m}\right)$.

(C) We can find $-m/2 < r \leqslant m/2$ an integer with $n \equiv r \bmod m$. Automatically

$$\left(\frac{n}{m}\right) = \left(\frac{r}{m}\right).$$

(D) If $n = 2s$, then $\left(\frac{n}{m}\right) = \left(\frac{2}{m}\right)\left(\frac{s}{m}\right)$.

(E) $\left(\frac{m}{n}\right) = (-1)^{(m-1)(n-1)/4}\left(\frac{n}{m}\right)$

whenever n is an odd positive integer.

(ii) By examining the behaviour of your algorithm when m and n are not coprime, show that it may be modified so that, given m an odd positive integer and n any integer, it will either show that m and n are not comprime or, if they are comprime, calculate the Jacobi symbol.

(iii) Using your calculator, apply your algorithm to a few randomly chosen eight-figure numbers.

(iv) Show that there exists a constant A such that, if $|n|$, $|m| \leqslant 2^k$, your algorithm uses less than Ak steps to deliver its answer. (Notice that since the Legendre symbol is a special case of the Jacobi symbol we have, as promised at the beginning of Question 103.6, another very rapid method for computing the Legendre symbol. But, as Solovay and Strassen noticed, still more can be done by combining the ideas behind the two methods.)

103.9 (Solovay and Strassen's probabilistic test for primes) In this question we are interested in whether a given odd number $m \geqslant 3$ is prime.

(i) Show, using previous results, that, if m is prime, then for all $1 \leqslant n \leqslant m-1$,

(a) n and m are comprime,

(b) $n^{(m-1)/2} \equiv \left(\frac{n}{m}\right) \bmod m$.

(ii) Suppose, on the other hand, that there is a prime p such that p^2 divides m. Set $n = (1 + m/p)$ and show, by using the binomial theorem or otherwise, that $n^p \equiv 1 \bmod m$. Conclude that $n^{m-1} \not\equiv 1 \bmod m$ and so condition (b) of (i) cannot hold.

(iii) The remaining possibility is that $m = p_1 p_2 \ldots p_r$ where $p_1, p_2, \ldots, p_r$ are distinct primes. Explain why we can find an a such that $\left(\frac{a}{p_1}\right) = -1$ and why the Chinese remainder theorem allows us to find an n such that $n \equiv a \bmod p_1$ and $n \equiv 1 \bmod p_i$ for $2 \leqslant i \leqslant r$. Show that n does not satisfy condition (b).

(iv) Let G be the group of units modulo m. Write

$$H = \{[n] \in G\colon n \text{ satisfies condition (b)}\}.$$

Show that H is a subgroup of G. Using (ii) and (iii), deduce that, if m is composite, $|H| \leqslant |G|/2$. (Here $|A|$ is the number of elements in A.)

(v) Suppose n is chosen at random from the integers $1, 2, \ldots, m-1$. Show that

$$Pr((a) \text{ and } (b) \text{ satisfied}) = 1 \quad \text{if } m \text{ is prime},$$

$$Pr((a) \text{ and } (b) \text{ satisfied}) \leqslant 1/2 \quad \text{if } m \text{ is composite}.$$

(vi) Suppose we pick an odd positive integer m at random and then test whether (a) and (b) are satisfied for M randomly and independently chosen integers $n_1, n_2, \ldots, n_M \in \{1, 2, \ldots, m-1\}$. Let B be the event m is prime, B^c the event m is not prime, and A_i the event that (a) and (b) are satisfied for n_i. We write $Pr(B^c)$ for the probability we assign to the event that m is not prime before we carry out the test, and $Pr(B^c|A_1 \cap A_2 \cap \ldots \cap A_M)$ the probability that we assign to the event that m is not prime after we know that $n_1, n_2, \ldots, n_M$ satisfy (a) and (b). Show that

$$\begin{aligned}
&Pr(B^c|A_1 \cap A_2 \cap \ldots \cap A_M)\\
&= \frac{Pr(A_1 \cap A_2 \cap \ldots \cap A_M|B^c)Pr(B^c)}{Pr(A_1 \cap A_2 \cap \ldots \cap A_M|B)Pr(B) + Pr(A_1 \cap A_2 \cap \ldots \cap A_M|B^c)Pr(B^c)}\\
&\leqslant \frac{Pr(A_1 \cap A_2 \cap \ldots \cap A_M|B^c)}{Pr(A_1 \cap A_2 \cap \ldots \cap A_M|B)Pr(B)}\\
&\leqslant 2^{-M}Pr(B)^{-1}.
\end{aligned}$$

(Memory jogger: recall Bayes' theorem.)

(vii) Suppose m is chosen at random from the odd integers between 3 and 10^{110} (or, if you wish to be more realistic, from the integers between 10^{100} and 10^{110}). Use Tchebychev's theorem (Question 43.2) or the prime number theorem (Question 43.4) to estimate $Pr(B)$. How large should M be taken (very roughly) to give $Pr(B^c|A_1 \cap A_2 \cap \ldots \cap A_M) \leqslant 10^{-8}$? Using Question 103.8 (iv) estimate how many operations you need to carry out the tests on $n_1, n_2, \ldots, n_M$.

(viii) The first m chosen at random will usually fail one of the tests. Very roughly how many randomly chosen m of the type considered in (vii) must we expect to try before we get one which satisfies M tests? Give a very rough estimate of the total number of operations we must expect to perform before our first success.

(ix) Discuss how the effort involved in finding 'engineer's primes m' using the ideas above increases as the required size of m increases.

103.10$^{(-)}$ (The Mellin transform) If, as has been suggested in this chapter, we should look at Fourier analysis in the context of group theory, then the notion of isomorphism should play a role. One of the most important isomorphisms is the map log: $(\mathbb{R}^+, \times) \to (\mathbb{R}, +)$, where $(\mathbb{R}, +)$ is the additive group of the real numbers and $(\mathbb{R}^+, \times)$ the multiplicative group of the strictly positive real numbers.

(i) Verify that log: $(\mathbb{R}^+, \times) \to (\mathbb{R}, +)$ is an isomorphism and that the map log and its inverse are continuous.

(ii) Show that the continuous maps $e\colon \mathbb{R}^+ \to \mathbb{C}$ such that $e(xy) = e(x)e(y)$ and $|e(x)| = 1$ for all $x, y \in \mathbb{R}^+$ are precisely the maps $e_\lambda(x) = x^{i\lambda}$ $[x > 0]$ where λ is real.

(iii) Let us say $f \in \mathcal{M}$ if $f\colon \mathbb{R}^+ \to \mathbb{C}$ is continuous and $\int_{\mathbb{R}^+} |f(t)t^{-1}|\, dt < \infty$. Show that, if $f_a(t) = f(at)$ for all $t \in \mathbb{R}^+$ $[a > 0]$, then, if $f \in \mathcal{M}$, it follows that $f_a \in \mathcal{M}$ and $\int_{\mathbb{R}^+} f_a(t)t^{-1}\, dt = \int_{\mathbb{R}^+} f(t)t^{-1}\, dt$.

(iv) If $f \in \mathcal{M}$, let us write

$$Mf(\lambda) = \int_{\mathbb{R}^+} f(t)t^{-i\lambda-1}\, dt \quad [\lambda \in \mathbb{R}].$$

Show that, if f is differentiable and $f, f' \in \mathcal{M}$, then

$$f(t) = \lim_{R\to\infty} \frac{1}{2\pi} \int_{-R}^{R} Mf(\lambda)t^{i\lambda}\, d\lambda \quad [t \in \mathbb{R}^+].$$

If $f, g \in \mathcal{M}$, show that

$$f*g(t) = \int_{\mathbb{R}^+} f(ts^{-1})g(s)s^{-1}\, ds \quad [t \in \mathbb{R}^+]$$

defines a function $f*g \in \mathcal{M}$ and that

$$M(f*g)(\lambda) = (Mf)(\lambda)(Mg)(\lambda) \quad \text{for all } \lambda \in \mathbb{R}.$$

(Number theorists are sometimes interested in the multiplicative properties of the positive real numbers rather than the additive properties of the reals. However, they usually use a wider definition of the Mellin transform and different notational conventions. We shall set out the standard conventions in Question 108.1.)

104

FOURIER ANALYSIS ON FINITE ABELIAN GROUPS

104.1 (This question and the one that follows are simple but assume that the reader has taken an elementary course in group theory.)

(i) Let $C(r)$ be the cyclic group with r elements. Show that if r and s are comprime, then $C(rs)$ is isomorphic to $C(r) \times C(s)$.

(ii) Show that, if $r \geqslant 2$, then $C(r^2)$ is not isomorphic to $C(r) \times C(r)$.

(iii) Show, by using Theorem 104.8, or otherwise, that every finite Abelian group is isomorphic to the product of cyclic groups whose orders are powers of primes $p_1^{n(1)}, p_2^{n(2)}, \ldots, p_k^{n(k)}$, say. Show that, apart from order, the integers $p_1^{n(1)}, p_2^{n(2)}, \ldots, p_k^{n(k)}$ are unique.

104.2 (This question uses the notation but not the results of the previous question.)

(i) Let $G(n)$ be the group of units modulo n. Show that, if p is a prime, then $G(p)$ is isomorphic to $C(p-1)$. (This is essentially proved in Question 79.11 (i), (ii) and the first sentence of (iii).)

(ii) Suppose that n and m are comprime. Show that $G(nm)$ is isomorphic to $G(n) \times G(m)$.

(iii) In Chapters 106 to 109 we shall prove a result (Dirichlet's theorem) which states, as a special case, that, if d is an integer with $d \geqslant 1$, then the sequence $d+1, 2d+1, \ldots$ contains at least one odd prime. Use this result together with Theorem 104.8 and the previous parts of this question to show that, given any finite Abelian group G, we can find an N such that $G(N)$ contains a subgroup isomorphic to G.

(Thus the structure of the groups of units modulo n is essentially as complicated as that of general finite Abelian groups and there would have been no advantage in restricting Chapter 104 to studying only groups of units.)

(iv) (This is independent of earlier questions on the Legendre symbol.) If n is odd and $r \in G(n)$ set

$$\chi_q = \left(\frac{r}{n}\right) = 1 \qquad \text{if } r = x^2 \text{ for some } x \in G(n),$$

$$\chi_q = \left(\frac{r}{n}\right) = -1 \quad \text{otherwise.}$$

Show that χ_q is an element of $\widehat{G(n)}$ of order 2.

(v) Let n be an odd number. Is it true that there always exists an element χ of $\widehat{G(n)}$ of order 2 with $\chi \neq \chi_q$ if

(a) n is a prime with $n \geqslant 5$?
(b) n is a sufficiently large prime?
(c) n is a composite (i.e. nonprime) number?
(d) n is a sufficiently large composite number?

(Hint. Recall Question 102.1. Dirichlet's Theorem also shows that if d is an integer with $d \geqslant 1$, then the sequence $d-1, 2d-1, \ldots$ contains at least one prime. A very special case of this was proved in Question 54.1 (v).)

105

A FORMULA OF EULER

105.1 Let X and Y be independent, identically distributed, random variables taking positive integral values according to the rule

$$Pr(X = n) = Pr(Y = n) = \zeta(s)^{-1}n^{-s}$$

where $\zeta(s) = \sum_{n=1}^{\infty} n^{-s}$. Let p_j be the jth prime.

(i) Show that $Pr(p_j \text{ divides } X) = p_j^{-s}$.

(ii) Show that the events (p_1 divides X), (p_2 divides X), ..., (p_m divides X) are independent.

(iii) Show that

$$Pr(\text{none of } p_1, p_2, \ldots, p_m \text{ divide both of } X \text{ and } Y) = \prod_{j=1}^{m}(1 - p_j^{-2s}).$$

(iv) Conclude that the probability that X and Y have no common factor is $\zeta(2s)^{-1}$. (Unless you know sufficient probability theory do not attempt to make this step rigorous.) What happens to $\zeta(2s)^{-1}$ as $s \to 1+$?

105.2 (Alternative proof of Theorem 105.7) If N is a strictly positive integer, let us write $B(N)$ for the set of ordered pairs (n, m) with n, m integers such that $1 \leqslant n, m \leqslant N$ and n, m, N have no common factor. Write $|B(N)|$ for the number of elements in $B(N)$.

(i) Show that, if $N = \prod_{j=1}^{k} p_j^{n(j)}$ with $p_1, p_2, \ldots, p_k$ distinct primes and $n(1), n(2), \ldots, n(k) \geqslant 1$ integers, then

$$N^{-2}|B(N)| = \prod_{j=1}^{k}(1 - p_j^{-2}).$$

(ii) Show that, if we have a sequence $N(1), N(2), \ldots$ of positive integers such that, for each $k \geqslant 1$, all but finitely many of the $N(r)$ are divisible by the first k primes, then

$$N(r)^{-2}|B(N(r))| \to 6/\pi^2.$$

(iii) Show that we can find a sequence $N(r)$ with the property stated in the hypothesis of (ii) such that, in addition, $N(r+1)/N(r) \to 1$. Deduce Theorem 105.7.

105.3 (The inclusion exclusion formula)

(i) Let Ω be a set. If $A \subseteq \Omega$, let us write $\chi_A(x) = 1$ if $x \in A$, $\chi_A(x) = 0$ if $x \notin A$. Show that, if $A, B \subseteq \Omega$, then $\chi_{A \cap B}(x) = \chi_A(x)\chi_B(x)$ and $\chi_{A \cap B} = \chi_A(x) + \chi_B(x) - \chi_{A \cap B}(x)$ for all $x \in \Omega$.

(ii) Find a similar formula for $\chi_{A \triangle B}(x)$ (recall that

$$A \triangle B = (A \cup B) \setminus (A \cap B))$$

and use it to show that $A \triangle (B \triangle C) = (A \triangle B) \triangle C$ for all $A, B, C \subseteq \Omega$.

(iii) Let us write $|A|$ for the number of elements of A. Explain why, if A is finite, $\sum_{x \in \Omega} \chi_A(x) = |A|$. Suppose that Ω is finite and $A(1), A(2), \ldots, A(n)$ are subsets of Ω. By expanding $\prod_{i=1}^{n}(1 - \chi_{A(i)}(x))$ and then summing over Ω show that

$$\left|\left\{x \in \Omega : x \notin \bigcup_{i=1}^{n} A(i)\right\}\right| = |\Omega| - \sum_{i=1}^{n} |A(i)| + \sum_{i \neq j} |A(i) \cap A(j)| - \cdots$$

and deduce that

$$\left|\bigcup_{i=1}^{n} A(i)\right| = \sum_{i=1}^{n} |A(i)| - \sum_{i \neq j} |A(i) \cap A(j)| + \sum_{i,j,k \text{ distinct}} |A(i) \cap A(j) \cap A(k)| - \cdots.$$

(iv) Now let X be a random variable taking values in Ω. Explain why $\mathbb{E}(\chi_A(X)) = Pr(X \in A)$ and show by an argument paralleling (iii) that

$$Pr\left(X \in \bigcup_{i=1}^{n} A(i)\right) = \sum_{i=1}^{n} Pr(X \in A(i)) - \sum_{i \neq j} Pr(X \in A(i) \cap A(j)) + \sum_{i,j,k \text{ distinct}} Pr(X \in A(i) \cap A(j) \cap A(k)) - \cdots.$$

(The results of (iii) and (iv) are called inclusion exclusion formulae.)

(v) Now let us return to the wrong envelope problem of Question 3.8. Suppose that I have n letter to send and that, not merely do I place the letters in my addressed envelopes at random (one letter to one envelope), but that, owing to the chaotic state of my desk, the chance of any envelope actually being posted is p (independent of what happens to the other envelopes). Show that the chance that none of my correspondents actually get their correct letter tends to e^{-p} as $n \to \infty$. Is the convergence uniform in p?

107

PRIMES IN SOME ARITHMETICAL PROGRESSIONS

107.1 (Euler's γ)

(i) Show that

$$\frac{1}{n} \geqslant \int_n^{n+1} \frac{1}{x}\,dx \geqslant \frac{1}{n+1} \quad \text{for all } n \geqslant 1.$$

Hence show that $g(n) = \sum_{r=1}^{n} r^{-1} - \log n$ is a decreasing function of n with $1 \geqslant g(n) \geqslant 0$ for all $n \geqslant 1$. Deduce that $1 + 1/2 + \ldots + 1/n - \log n$ tends to a limit as $n \to \infty$. This limit is called Euler's constant and written γ.

(ii) Deduce that

$$f(n) = 1 + \frac{1}{3} + \frac{1}{5} + \ldots + \frac{1}{2n+1} - \frac{1}{2}\log n \to \frac{1}{2}\gamma + \log 2 \quad \text{as } n \to \infty$$

and so

$$1 - \frac{1}{2} + \frac{1}{3} - \frac{1}{4} + \ldots + \frac{1}{2n+1} - \frac{1}{2n+2} \to \log 2 \quad \text{as } n \to \infty.$$

Deduce that $\sum_{r=1}^{N} (-1)^{r+1}/r \to \log 2$ as $N \to \infty$ so we may write $1 - \frac{1}{2} + \frac{1}{3} - \frac{1}{4} + \ldots = \log 2$.

(iii) Show, however, that $1 + \frac{1}{3} - \frac{1}{2} + \frac{1}{5} + \frac{1}{7} - \frac{1}{4} + \ldots = (3/2)\log 2$. More generally, show that the rearranged sum

$$1 + \frac{1}{3} + \ldots + \frac{1}{(2p-1)} - \frac{1}{2} - \frac{1}{4} - \ldots - \frac{1}{(2q)} + \frac{1}{(2p+1)} + \ldots$$

in which p positives alternate with q negatives has sum $\log 2 + \frac{1}{2}\log(p/q)$.

(Euler's γ is a mysterious number clearly linked to the numbers $\zeta(k) = \sum_{n=1}^{\infty} n^{-k}$ with $k > 1$ (see e.g. Question 43.5). It is not known after three centuries whether γ is rational or not. G. H. Hardy is said to have offered to give his chair to any student of his who could prove γ irrational. (On the other hand Cassels recalls Mordell suggesting the problem of deciding whether γ was transcendental as a possible subject for a PhD!))

107.2 (i) If you have access to an electronic computer, calculate $\sum_{r=1}^{10^6} r^{-1}$. Discuss briefly the accuracy of your answer.

(ii) Use the Euler–Maclaurin formula together with the error estimate in Question 18.11 (viii) to show that

$$\left|\sum_{r=1}^{n} r^{-1} - \left(K_M + \log n + 2n^{-1} + \sum_{k=1}^{M-1} (B_{2k}/2k)n^{-2k}\right)\right| \leqslant |B_{2M}/2M|n^{-2M}$$

where K_M does not depend on n.

(iii) By allowing $n \to \infty$, show that $K_M = \gamma$ (Euler's constant, see Question 107.1). Deduce that

$$\left|\sum_{r=1}^{n} r^{-1} - \left(\gamma + \log n + 2n^{-1} + \sum_{k=1}^{M-1} (B_{2k}/2k)n^{-2k}\right)\right| \leqslant |B_{2M}/2M|n^{-2M}.$$

(iv) By choosing M and n appropriately, use a hand calculator to obtain γ to a guaranteed accuracy of 10^{-8}.

(v) Now use a hand calculator to obtain $\sum_{r=1}^{10^6} r^{-1}$ to a guaranteed accuracy of 10^{-7}.

(Euler used the Euler–Maclaurin formula to compute γ to sixteen decimal places. One aspect of Euler's many-sided genius was a love of, and talent for, calculation. Condorcet relates that, when two of his pupils disagreed on a digit in the fifteenth place of a complicated computation, Euler redid the whole calculation correctly in his head.)

108
EXTENSION FROM REAL TO COMPLEX VARIABLE

108.1$^{(-)}$ (The Mellin transform) This question extends the ideas of Question 103.10 to match some of the ideas of this chapter. However, I only seek to establish notation and, if the reader has read the chapters on the Laplace transform, she should find the results trivial. (Note that the choice of signs in the definition is different from elsewhere in this book.)

Let us write $\mathbb{R}^+ = \{x \in \mathbb{R}: x > 0\}$. If α, β are real with $\alpha < \beta$, let us say that $f \in \mathcal{M}(\alpha, \beta)$ if $f: \mathbb{R}^+ \to \mathbb{C}$ is continuous (in practice, this condition can be weakened) and $\int_{\mathbb{R}^+} |f(x)x^{\sigma-1}|\, dx$ converges for all real σ with $\alpha < \sigma < \beta$.

(i) Show that, if $f \in \mathcal{M}(\alpha, \beta)$, then

$$Mf(z) = \int_{\mathbb{R}^+} f(x)x^{z-1}\, dx$$

defines an analytic function

$$Mf: \{z: \alpha < \operatorname{Re} z < \beta\} \to \mathbb{C}.$$

(ii) If $f, f' \in \mathcal{M}(\alpha, \beta)$, show that

$$f(x) = \lim_{R\to\infty} \frac{1}{2\pi} \int_{-R}^{R} (Mf)(\sigma + it)x^{-(\sigma+it)}\, dt$$

for all $x \in \mathbb{R}^+$ and all $\alpha < \sigma < \beta$.

(iii) Recall the definition of $z!$ in Question 75.13. Show that, if σ is real and strictly positive, then

$$e^{-x} = \lim_{R\to\infty} \frac{1}{2\pi} \int_{-R}^{R} x^{-(\sigma+it)}(\sigma + it - 1)!\, dt,$$

for all $x > 0$.

(iv) If $f, g \in \mathcal{M}(\alpha, \beta)$, show that

$$f * g(x) = \int_{\mathbb{R}^+} f(xy^{-1})g(y)y^{-1}\, dy$$

defines a function $f * g \in \mathcal{M}(\alpha, \beta)$ and

$$M(f * g)(z) = (Mf)(z)(Mg)(z)$$

for all z with $\alpha < \operatorname{Re} z < \beta$.

(v) Show that, if f is well behaved and we set $h_n(x) = x^n f^{(n)}(x)$, then

$$Mh_n(z) = (-1)^n z(z+1) \dots (z+n-1) Mf(z).$$

Specify precisely sufficient conditions on f which will make this true and the appropriate range of z.

108.2$^{(-)}$ (i) Let $a_n \in \mathbb{C}$ $[n \geqslant 0]$. Show that, if there exists a real number κ and a positive number A such that $|\sum_{n=1}^{N} a_n| \leqslant AN^{\kappa}$, then $\sum_{n=1}^{\infty} a_n n^{-s}$ converges to an analytic function on $\{s\colon \operatorname{Re} s > \kappa\}$.

(ii) Let $a_n \in \mathbb{C}[n \geqslant 0]$. Suppose $\kappa(1), \dots, \kappa(r), \kappa$ are real numbers with $\kappa < \kappa(1) < \cdots < \kappa(r)$ and $A, A_1, \dots, A_r$ are real numbers with $A > 0$ and $|\sum_{n=1}^{N} a_n - \sum_{j=1}^{r} A_j N^{\kappa(j)}| \leqslant AN^{\kappa}$. Show that $\sum_{n=1}^{\infty} a_n n^{-s}$ extends to an analytic function on $\{s\colon \operatorname{Re} s > \kappa\} \setminus \{\kappa(1), \dots, \kappa(r)\}$. What is the nature of the singularities at $\kappa(1), \dots, \kappa(r)$?

108.3 (i) Let us define $\zeta\colon \{z\colon \operatorname{Re} z > 1\} \to \mathbb{C}$ by $\zeta(z) = \sum_{n=1}^{\infty} n^{-z}$. Using the ideas of previous chapters, show that ζ is a well-defined analytic function and $\prod_{p<N}(1-p^{-z})^{-1} \to \zeta(z)$ as $N \to \infty$ uniformly in every region of the form $\{z\colon \operatorname{Re} z > 1 + \delta\}$ with $\delta > 0$.

(ii) Use Theorem 75.3 (iv) to show that

$$-\frac{\zeta'(z)}{\zeta(z)} = \sum_{p} \frac{\log p}{p^z - 1} \quad \text{for all } \operatorname{Re} z > 1.$$

(iii) Use the methods of Chapter 108 to show that there exists a unique analytic function $f\colon \{z\colon \operatorname{Re} z > 0\} \to \mathbb{C}$ with $f(z) = \zeta(z) - (z-1)^{-1}$ for $\operatorname{Re} z > 1$. (Thus we can extend our range of definition of ζ and say that $\zeta\colon \{z\colon \operatorname{Re} z > 0\} \to \mathbb{C}$ is a meromorphic function analytic except at 1 where it has a simple pole with residue 1. The function ζ is the famous Riemann zeta function.)

(iv) Observe that ζ'/ζ is meromorphic in $\{z\colon \operatorname{Re} z > 0\}$ with simple poles occurring at 1 and at the zeros of ζ. Show that the pole at 1 has residue -1, i.e. that $\zeta'(z)/\zeta(z) + 1/(z-1)$ has a removable singularity at 1, and so can be redefined to be analytic there.

108.4 Here is another proof of the result of the previous question.

(i) Show, by grouping pairs of terms, that $\sum_{n=1}^{\infty}(-1)^{n-1} n^{-z}$ converges to an analytic function $g(z)$ in the region $\{z\colon \operatorname{Re} z > 0\}$.

(ii) Show that we can find A and B such that $g(z) = A\zeta(z) + B2^{-z}\zeta(z)$ for $\operatorname{Re} z > 1$.

(iii) Conclude that there is a meromorphic extension of ζ to $\{z: \operatorname{Re} z > 0\}$ which is analytic except for a simple pole at 1 with residue 1.

(iv) Explain why $|\sum_{n=N+1}^{\infty}(-1)^{n+1}n^{-1/2}| \leqslant (N+1)^{-1/2}$ and, using a hand calculator, prove that $\sum_{n=1}^{\infty}(-1)^{n+1}n^{-1/2} \neq 0$. Hence show that $\zeta(1/2) \neq 0$.

108.5 In fact, we can show that ζ can be extended analytically to the entire complex plane by using the Euler–Maclaurin formula of Question 18.9.

(i) Show that the formula of Question 18.9 (iv) remains true for an M times continuously differentiable function $f: \mathbb{R} \to \mathbb{C}$. Hence show that, if $s \in \mathbb{C}$, $s \neq 1$,

$$\sum_{r=1}^{n} r^{-s} = \frac{n^{1-s}-1}{1-s} + \frac{1}{2}\left(1 + \frac{1}{n^s}\right)$$
$$+ \sum_{k=2}^{M} \frac{B_k}{k!}(-1)^{k-1}s(s+1)\ldots(s+k-2)\left(\frac{1}{n^{s+k-1}} - 1\right) + R_{M,n}(s)$$

where $R_{M,n}(s) = \int_1^n b_M(t)(-1)^M \dfrac{s(s+1)\ldots(s+M-1)}{t^{M+s}}\,dt$.

(ii) Suppose $\varepsilon > 0$. Show that

$$\frac{1}{2} + \sum_{k=2}^{M} \frac{B_k}{k!}(-1)^k s(s+1)\ldots(s+k-2)$$
$$+ \int_1^n b_M(t)(-1)^M \frac{s(s+1)\ldots(s+M-1)}{t^{M+s+1}}\,dt$$

tends to a limit $R_M(s)$, say, uniformly in $\operatorname{Re} s > 1 - M + \varepsilon$ as $n \to \infty$. Deduce that

$$\left(\sum_{r=1}^{n} r^{-s} - \frac{n^{1-s}-1}{1-s}\right)$$
$$- \left(\frac{n^{-s}}{2} + \sum_{k=2}^{M} \frac{B_k}{k!}(-1)^{k-1}s(s+1)\ldots(s+k-2)n^{-s-k+1}\right) \to R_M(s)$$

uniformly in $\operatorname{Re} s > 1 - M + \varepsilon$, $s \neq 1$, and that R_M is analytic on $\{s: \operatorname{Re} s > 1 - M, s \neq 1\}$.

(iii) By allowing $n \to \infty$, show that $R_M(s) = \zeta(s) - (s-1)^{-1}$ for $\operatorname{Re} s > 1$. Deduce that $R_{M(1)}(s) = R_{M(2)}(s)$ whenever $\operatorname{Re} s > 1 - \min(M(1), M(2))$, $s \neq 1$, that there is an analytic function $R: \mathbb{C}\backslash\{1\} \to \mathbb{C}$, say, such that $R(s) = \zeta(s) - (s-1)^{-1}$ for $\operatorname{Re} s > 1$ and $R(s) = R_M(s)$ for $\operatorname{Re} s > 1 - M$.

(iv) By estimating $|\int_n^{\infty} b_2(t)s(s+1)/t^{2+s}\,dt|$, or otherwise, show that there exists an $A > 0$ such that

$$\left|\left(\sum_{r=1}^{n} r^{-s} - \frac{n^{1-s}-1}{1-s}\right) - R(s)\right| \leqslant \frac{A}{n^{1/2}}$$

for all $0<|1-s|<1/2$ and all $n\geq 1$. Hence, deduce that, given $\varepsilon>0$, we can find an $N(\varepsilon)\geq 1$ such that

$$\left|\left(\sum_{r=1}^{N} r^{-s}-\frac{N^{1-s}-1}{1-s}\right)-R(s)\right|\leq\varepsilon$$

for all $0<|1-s|<1/2$ and all $N>N(\varepsilon)$. Now show that for any fixed $N>N(\varepsilon)$ there exists a $\delta(N)>0$ such that $|\sum_{r=1}^{N} r^{-1}-\log N-R(s)|\leq\varepsilon$ for all $|s-1|<\delta(N)$. Conclude that $R(s)\to\gamma$ as $s\to 1$ where γ is Euler's constant (see Question 107.1). Use the principle of removable singularities to extend R to an analytic function on the whole of $\mathbb{C}$. What is the value of $R(1)$?

(v) Observe that we have now produced a meromorphic expansion of ζ to $\mathbb{C}$ which is analytic except at 1 where it has a simple pole. For s near 1, $\zeta(s)$ behaves like $(s-1)^{-1}+\gamma$.

(vi) Explain briefly why the techniques of this question do not apply to $\sum_{n=1}^{\infty}\chi(n)n^{-s}$.

108.6 (Riemann's functional equation for the zeta function) This question requires the reader to know of the existence of the meromorphic function $z!$ with the property that $z!=\int_0^\infty e^{-t}t^z\,dt$ for $\operatorname{Re} z>-1$ (see Question 75.13). The reader must also know the definition of the theta function $\psi(t)=\sum_{n=-\infty}^{\infty}\exp(-n^2\pi t)$ [t real, $t>0$] together with Jacobi's identity $\psi(t)=t^{-1/2}\psi(t^{-1})$ (see Question 60.22). In return, we not only obtain a second proof that we can extend ζ to the whole complex place, but also a great deal of information about the behaviour of ζ.

(i) Let $\operatorname{Re} z>0$. Show that

$$\int_0^\infty t^{z-1}\sum_{n=1}^{N}\exp(-n^2\pi t)\,dt=\frac{(z-1)!}{\pi^z}\sum_{n=1}^{N}n^{-2z}.$$

(ii) We wish to allow $N\to\infty$. Show that, if we write $g(t)=(\psi(t)-1)/2$, we have $\sum_{n=1}^{N}\exp(-n^2\pi t)\to g(t)$ uniformly for $t>\delta$ and deduce that

$$\int_\delta^1 t^{z-1}\sum_{n=1}^{N}\exp(-n^2\pi t)\,dt\to\int_\delta^1 t^{z-1}g(t)\,dt$$

for all z with $\operatorname{Re} z>0$. Show that, if $t\geq 1$, then

$$\left|g(t)-\sum_{n=1}^{N}\exp(-n^2\pi t)\right|\leq 2\exp(-N^2\pi t)$$

(clearly we can do better than this, but we do not need to) and deduce that

$$\int_\delta^\infty t^{z-1}\sum_{n=1}^{N}\exp(-n^2\pi t)\,dt\to\int_\delta^\infty t^{z-1}g(t)\,dt$$

as $N\to\infty$ for all z with $\operatorname{Re} z>0$. Deduce also that $|g(t)|\leq 3\exp(-\pi t)$ for all $t\geq 1$.

(iii) There remains only the problem of the gap $(0, \delta)$. Use Jacobi's formula to show that

$$g(t) = -1/2 + t^{-1/2}/2 + t^{-1/2}g(t^{-1}) \qquad (*)$$

and deduce that $0 \leqslant \sum_1^N \exp(-n^2\pi t) \leqslant g(t) \leqslant 4t^{-1/2}$ for all $0 < t \leqslant 1$. Conclude that, if $\operatorname{Re} z > 1/2$, then $\int_0^\delta |t^{z-1}\sum_{n=1}^N \exp(-n^2\pi t)|\,dt \leqslant \eta(\delta)$ for some $\eta(\delta)$ indeper ˙ ιt of N and with $\eta(\delta) \to 0$ as $\delta \to 0+$. Hence show that

$$\frac{(z-1)!}{\pi^z}\sum_{n=1}^N n^{-2z} \to \int_0^\infty t^{z-1}g(t)\,dt$$

for all z with $\operatorname{Re} z > 1/2$.

(iv) Now write

$$\int_0^\infty t^{z-1}g(t)\,dt = \int_0^1 t^{z-1}g(t)\,dt + \int_1^\infty t^{z-1}g(t)\,dt.$$

By using $(*)$ (i.e. by using Jacobi's identity) followed by a simple change of variable, show that

$$\begin{aligned}\int_0^1 t^{z-1}g(t)\,dt &= \left(\frac{1}{2z-1} - \frac{1}{2z}\right) + \int_0^1 g(t^{-1})t^{z-3/2}\,dt \\ &= \left(\frac{1}{2z-1} - \frac{1}{2z}\right) + \int_1^\infty t^{-z-1/2}g(t)\,dt\end{aligned}$$

and so

$$\int_0^\infty t^{z-1}g(t)\,dt = \frac{1}{2z-1} - \frac{1}{2z} + F(z),$$

with

$$F(z) = \int_1^\infty (t^{z-1} + t^{-z-1/2})g(t)\,dt.$$

(v) Using the estimate $|g(t)| \leqslant 3\exp(-\pi t)$ for $t \geqslant 1$ obtained in (ii) and standard techniques (see e.g. the method of proof of Theorem 75.2), show that F is an analytic function on $\mathbb{C}$. Thus

$$\frac{(z-1)!}{\pi^z}\zeta(2z) = \frac{1}{2z-1} - \frac{1}{2z} + F(z) \quad \text{for } \operatorname{Re} z > 1/2$$

with F analytic on $\mathbb{C}$. We have extended ζ to a function meromorphic on the whole of $\mathbb{C}$.

(vi) We can do more. Observe that ζ (which we now extend once and for all to a meromorphic function on $\mathbb{C}$) satisfies

$$\frac{(z/2-1)!}{\pi^{z/2}}\zeta(z) = \frac{1}{z-1} - \frac{1}{z} + F(z/2). \qquad (**)$$

Verify by inspection that $F(z/2) = F((1-z)/2)$, and conclude that

$$(z/2-1)!\pi^{-z/2}\zeta(z) = (-z/2-1)!\pi^{-(1-z)/2}\zeta(1-z)$$

i.e. that $G(z) = (z/2-1)!\pi^{-z/2}\zeta(z)$ is symmetric about $z = 1/2$ (in the sense that $G(z) = G(1-z)$).

(vii) By looking at (**) and using Question 75.13, show that ζ has only one pole in $\mathbb{C}$, that this is at 1, is simple and has residue 1. By modifying the ideas of Lemma 108.7 (and so of Lemma 107.4), show that $\prod_{p<N}(1-p^{-z})^{-1} \to \zeta(z)$ as $N \to \infty$ for $\operatorname{Re} z > 1$ and that $\zeta(z) \neq 0$ for $\operatorname{Re} z > 1$. (Alternatively use Exercise 76.2.) Find all the solutions of $\zeta(z) = 0$ with $\operatorname{Re} z < 0$ (these are the so-called trivial zeros of the Riemann zeta function).

(viii) By multiplying both sides of (**) by $z(z-1)/2$ and setting $z = 1/2 + iy$ show that $\zeta(1/2 + iy) = 0$ with $y \in \mathbb{R}$ if and only if $G(y) = 0$ where

$$G(y) = 1/2 - (y^2 + 1/4)\int_1^\infty g(t)t^{-3/4}\cos((y/2)\log t)\,dt.$$

Observe that G is a real valued continuous function on $\mathbb{R}$ and explain why this means that, if, by sufficiently accurate numerical computations, we can show $G(y_1) > 0 > G(y_2)$ for some $y_1, y_2 \in \mathbb{R}$ it will follow that G has a zero and so $\zeta(1/2 + iy_3) = 0$ at some y_3 between y_1 and y_2. If you have access to an electronic computer and some grasp of numerical analysis you may plot a graph G for y between 0 and 35. (If you do not have access to such a machine accept the author's assurance that ζ does have zeros of the form $1/2 + iy$ with $y \in \mathbb{R}$.)

108.7 There are similarities between the behaviour of Dirichlet series $\sum a_n n^{-s}$ and power series $\sum a_n z^n$ but the behaviour of Dirichlet series is more subtle.

(i) (Revision) Suppose $a_n \in \mathbb{C}$. Show that there exists an R (which may take the value 0 or ∞) such that $\limsup_{n\to\infty}|a_n z^n| = \infty$ if $|z| > r$ but $\sum_{n=0}^\infty a_n z^n$ converges absolutely if $|z| < R$.

(ii) Suppose $a_n \in \mathbb{C}$. Show by comparison, or otherwise, that if $\sum a_n n^{-z(1)}$ converges absolutely then $\sum a_n n^{-z}$ converges absolutely for all $\operatorname{Re} z > \operatorname{Re} z(1)$. Deduce the existence of a σ_a (which may take the value $-\infty$ or ∞) such that $\sum a_n n^{-z}$ converges absolutely for $\operatorname{Re} z > \sigma_a$ and fails to converge absolutely for $\operatorname{Re} z < \sigma_a$. Show by means of examples that σ_a may take any real value and either of the values $\pm\infty$.

(iii) Suppose $b_n \in \mathbb{C}$ and $\sum b_n$ converges. Show, by writing $S(x) = \sum_{1\leq m\leq x} b_m$ and imitating the proof of Lemma 108.2, that $\sum b_n n^{-z}$ converges for $\operatorname{Re} z > 0$. Show further that $\sum b_n n^{-z}$ defines a function analytic on $\{z: \operatorname{Re} z > 0\}$.

(iv) Suppose $a_n \in \mathbb{C}$. Show that, if $\sum a_n n^{-z(1)}$ converges, then $\sum a_n n^{-z}$ converges for $\operatorname{Re} z > \operatorname{Re} z(1)$. Deduce the existence of a σ_c (which may take the value $-\infty$ or ∞) such that $\sum a_n n^{-z}$ converges when $\operatorname{Re} z > \sigma_c$ and diverges when $\operatorname{Re} z < \sigma_c$. Show that $\sum a_n n^{-z}$ is analytic on $\{z: \operatorname{Re} z > \sigma_c\}$. Show by means of examples that σ_c may take any real value and either of the values $\pm\infty$.

(v) Show by means of an example that we may have $\sigma_a - \sigma_c = 1$. (Such examples occur in Chapter 108, so there is no need to search very hard.) Show however that, if σ_c is finite, then σ_a is finite and $1 \geqslant \sigma_a - \sigma_c \geqslant 0$. (If you need a hint, note that in (i) you proved, in effect, that $R_a = R_c$.) What happens if $\sigma_c = -\infty$? What happens if $\sigma_a = \infty$?

(vi) In Questions 108.5 and 108.6 we showed that we could find a function ζ analytic on all of $\mathbb{C}$ except for a simple pole at 1 with residue 1 such that $\sum_{n=1}^{\infty} n^{-z} = \zeta(z)$ for $\operatorname{Re} z > 1$. By considering $\zeta(z) - 2\zeta(2z-1)$, or otherwise, find $a_n \in \mathbb{C}$ such that $\sum a_n n^{-z}$ has σ_c finite yet there is an analytic function $f: \mathbb{C} \to \mathbb{C}$ with $f(z) = \sum_{n=1}^{\infty} a_n n^{-z}$ for all $\operatorname{Re} z$. (Contrast this with the case of power series which must always have a singularity on the circle of convergence.)

109

PRIMES IN GENERAL ARITHMETICAL PROGRESSIONS

109.1 (i) By approximating a sum by an integral, show that $(s-1)\sum_{n=1}^{\infty} n^{-s} \to 1$ as $s \to 1+$ (i.e. as $s \to 1$ through real values greater than 1).

(ii) By refining the argument leading up to Theorem 105.4, show that

$$\left(\sum p^{-s}\right)\Big/\log(s-1)^{-1} \to 1 \quad \text{as } s \to 1+.$$

(iii) By refining the argument leading up to Theorem 109.8, show that, if a and d are coprime positive integers and

$$\Lambda = \{p\colon p \text{ prime and } p = a + nd \text{ for some integer } n\},$$

then

$$\left(\sum_{p\in\Lambda} p^{-s}\right)\Big/\log(s-1)^{-1} \to 1/\phi(d) \quad \text{as } s \to 1+.$$

(This is evidence for the rather vague statement that 'the primes are equally distributed amongst the arithmetic sequences $a + nd$ with a coprime to some fixed d'. The statement can be made precise and proved.)

Our final sequence of questions makes use of the results and notations of Questions 43.2, 43.3 and 108.3. In it we prove the famous prime number theorem which states that, if $\pi(x)$ is the number of primes less than x, then $(\pi(x)\log x)/x \to 1$ as $x \to \infty$. Our first two questions are intended to provide motivation.

109.2 Let $q_1, q_2, q_3, \ldots$ be a strictly increasing sequence of strictly positive integers. Let $\alpha(x)$ be the number of q_j less than x and let $\beta(x) = \sum_{q<x} \log q$ (where, for simplicity, we write $\sum_{q<x} \log q = \sum_{q_j<x} \log q_j$).

(i) Show that

$$\alpha(x)\log x \geqslant \beta(x) \geqslant (\alpha(x) - x^{1-\varepsilon})(1-\varepsilon)\log x$$

(look at Question 43.4 if you need a hint). Deduce that, if $(\alpha(x)\log x)/x \to 1$ as $x \to \infty$, then $x^{-1}\beta(x) \to 1$ as $x \to \infty$.

(ii) Suppose that $x^{\delta}|x^{-1}\beta(x) - 1| \to 0$ as $x \to \infty$ for some $\delta \geq 0$. By directly following the method of Lemma 108.2 (ii) or by quoting Exercise 108.2, show that there exists an analytic function F: $\{z: \operatorname{Re} z > 1 - \delta\} \to \mathbb{C}$ such that

$$\frac{1}{z}\sum_{q<N} q^{-z}\log q \to F(z) + (z-1)^{-1}$$

as $N \to \infty$ for all z with $\operatorname{Re} z > 1$.

109.3 Under certain circumstances we might hope to reverse the reasoning of Question 109.2. Let $q_1, q_2, q_3, \ldots$ be a strictly increasing sequence of strictly positive integers and let α, β be as before.

(i) Check that, without any further hypotheses, $\int_1^\infty \beta(x)x^{-z}\,dx$ is a well-defined analytic function on $\{z: \operatorname{Re} z > 1\}$ and that $\sum_{q<N} q^{-z}\log q \to z\int_1^\infty \beta(x)x^{-z-1}\,dx$ for $\operatorname{Re} z > 1$.

(ii) Suppose now that we know that there exists a $\delta_0 > 0$ and an analytic function F: $\{z: \operatorname{Re} z > 1 - \delta_0\} \to \mathbb{C}$ such that

$$\frac{1}{z}\sum_{q<N} q^{-z}\log q \to F(z) + (z-1)^{-1}$$

as $N \to \infty$ for all z with $\operatorname{Re} z > 1$. Show that

$$\int_1^\infty x(x^{-1}\beta(x) - 1)x^{-z-1}\,dx = F(z) \quad \text{for } \operatorname{Re} z > 1$$

and so

$$\int_0^\infty f(t)e^{-zt}\,dt = F(z) \quad \text{for } \operatorname{Re} z > 1$$

where $f(t) = e^t(e^{-t}\beta(e^t) - 1)$ for $t \geq 0$, $f(t) = 0$ for $t \leq 0$.

(iii) Show that $f(t)\exp(-at) \to 0$ as $t \to \infty$ for all $a > 1$. In imitation of the opening of Chapter 74, we might say that $f \in \mathscr{C}_a^*$ for $a > 1$. Thus, provided we can extend Theorem 74.5, Theorem 76.1 and Lemma 76.2, it will follow that, if F is reasonably well behaved, then $f \in \mathscr{C}_b^*$ for all $b > 1 - \delta_0$. (Convince yourself that the previous sentence is plausible, but do not go into the details.) Deduce that, if F is reasonably well behaved, then $x^{\delta}|x^{-1}\beta(x) - 1| \to 0$ as $x \to \infty$ for all δ with $\delta_0 > \delta \geq 0$.

(iv) Now drop all our previous hypotheses concerning β and assume only that $|x^{-1}\beta(x) - 1| \to 0$ as $x \to \infty$. Show, by using the inequality established in Question 109.2, that $(\alpha(x)\log x)/x \to 1$ as $x \to \infty$. (Questions 109.2 and 109.3 belong to the 'if we had some eggs, we could have some ham and eggs, if we had some ham' school of mathematics. At the present time it is not known whether the sequence of primes $p_1, p_2, \ldots$ obeys any of our optimistic hypotheses on $q_1, q_2, \ldots$. However, as we shall see, it is possible, using hard work and some very clever ideas, to obtain just enough information to give the prime number theorem.)

109.4 We now return to the sequence of primes $p_1, p_2, \ldots$. Let $\pi(x)$ be the number of primes less than x and $\theta(x) = \sum_{p<x} \log p$.

(i) By quoting Question 109.3, or direct calculation, show that, if we can prove that $|x^{-1}\theta(x) - 1| \to 0$ as $x \to \infty$, then it will follow that $(\pi(x)\log x)/x \to 1$ as $x \to \infty$. Show, using Question 43.4 (i), that $\int_1^\infty \theta(x)x^{-z}\,dx$ is a well-defined analytic function on $\{z: \operatorname{Re} z > 1\}$ and that $\sum_p p^{-z}\log p = \int_1^\infty \theta(x)x^{-z-1}\,dx$ for $\operatorname{Re} z > 1$. Our first problem is to extend the range of definition of the analytic function $\Phi(z) = \sum p^{-z}\log p$ just introduced and for this we make use of the zeta function. (Although this will involve some mild repetition, we shall not use any of the results obtained in the exercises to Chapter 108.)

(ii) Recall that $\zeta(z) = \sum_{n=1}^\infty n^{-z}$ for $\operatorname{Re} z > 1$. Show, using ideas from previous chapter, that $\prod_{p<N}(1 - p^z)^{-1} \to \zeta(z)$ uniformly on $\{z: \operatorname{Re} z > 1 + \delta\}$ as $N \to \infty$ for each $\delta > 0$. Use Theorem 75.3 (iv) to show that

$$\frac{\zeta'(z)}{\zeta(z)} = -\sum_p \frac{\log p}{p^z - 1} \quad \text{for all } \operatorname{Re} z > 1.$$

(iii) Show that

$$\Phi(z) = -\frac{\zeta'(z)}{\zeta(z)} - \sum_p \frac{\log p}{p^z(p^z - 1)} \quad \text{for all } \operatorname{Re} z > 1.$$

(iv) Show by simple arguments that $\sum_p \log p/(p^z(p^z - 1))$ converges to a well-defined analytic function for $\operatorname{Re} z > 1/2$. Deduce that Φ can be extended to a meromorphic function on $\{z: \operatorname{Re} z > 1/2\}$ analytic except possibly for poles at the zeros of $\zeta(z)$ and at 1.

(v) By considering the Laurent expansion of ζ near 1 and the Taylor expansion of ζ near its zeros, show that Φ has simple poles at the zeros of ζ and a simple pole with residue 1 at 1. Show that, if w is a zero of order n of ζ, then the residue of Φ at w is $-n$.

(vi) Conclude that

$$\sum_{p<N} p^{-z}\log p \to T(z) + (z - 1)^{-1}$$

as $N \to \infty$ for all z with $\operatorname{Re} z > 1$, where T is a meromorphic function on $\{z: \operatorname{Re} z > 1/2\}$ having simple poles at the zeros of ζ and nowhere else.

109.5 (This question constitutes a digression from the main sequence and will not be used later.)

(i) Extend the arguments of Question 109.4 (iii) and (iv) to show that

$$\Phi(z) = -\frac{\zeta'(z)}{\zeta(z)} - \Phi(2z) - G(z) \quad \text{for } \operatorname{Re} z > 1$$

where G is an analytic function on $\{z: \operatorname{Re} z > 1/3\}$. Conclude, using the results already obtained in Question 109.3, that Φ can be extended to a meromorphic function on $\{z: \operatorname{Re} z > 1/3\}$.

(ii) Recall from Question 109.4 (v) that Φ has a simple pole at 1 and from Question 108.3(iii) that $\zeta(1/2) \neq 0$. Conclude, giving your reasons, that Φ has a pole at 1/2.

(iii) By using Question 109.2 (ii) with $\delta = \eta + 1/2$ and the uniqueness of analytic continuation (Lemma 75.7), show that, if $\eta > 0$, then $x^{\eta+1/2}|x^{-1}\theta(x) - 1| \not\to 0$ as $x \to \infty$. Conclude that, if $A > 0$ and $\varepsilon > 0$, then

$$\left|\sum_{p<N} \log p - N\right| \geqslant AN^{1/2-\varepsilon}$$

for infinitely many integer values of N.

109.6 If we knew that ζ had no zeros (so equivalently that Φ had no poles) in $\{z: \operatorname{Re} z > 1 - \delta\}\backslash\{1\}$, we might hope to proceed somewhat along the lines of Exercise 109.2 and show that π must behave even better than the prime number theorem demands. In the brief historic paper which first linked the distribution of the prime numbers with the positions of the zeros of ζ Riemann went further. 'It is very likely [that ζ has no zeros in $\{z: \operatorname{Re} z > 1/2\}\backslash\{1\}$]. One would of course like to have a rigorous proof of this, but I have put aside the search for such a proof after some fleeting vain attempts.' (For the full paper and much interesting historical detail see Edward's *Riemann's Zeta Function*.)

In spite of over 100 years of investigation, not merely does Riemann's zeta hypothesis remain unsettled, but we know of no $\delta > 0$ such that ζ has no zeros in $\{z: \operatorname{Re} z > 1 - \delta\}\backslash\{1\}$. However Hadamard and de la Vallée Poussin did prove independently that ζ has no zeros on $\{z: \operatorname{Re} z = 1\}\backslash\{1\}$, and went on to prove the prime number theorem. Let us adopt the convention that, if g is analytic near z_0 and $g(z) = (z - z_0)^k h(z)$ near z_0 with $h(z_0) \neq 0$ and $k \geqslant 0$ an integer, then g has a zero of order k at z_0. (This is the standard definition except that the statement 'g has a zero of order 0 at z_0' means $g(z_0) \neq 0$.) Suppose that ζ has a zero of order μ at $1 + i\alpha$ where α is real and $\alpha > 0$.

(i) By observing that $\zeta(z^*)^* = \zeta(z)$ for $\operatorname{Re} z > 1$, or otherwise, show that ζ has a zero of order μ at $1 - i\alpha$.

(ii) Suppose now that ζ has a zero of order ν at $1 + 2i\alpha$. (Recall that we allow $\nu = 0$.) Using Question 109.4 (v), show that as $\varepsilon \to 0$ through positive real values

$$\varepsilon\Phi(1 + \varepsilon \pm i\alpha) \to -\mu$$

$$\varepsilon\Phi(1 + \varepsilon \pm 2i\alpha) \to -\nu$$

and

$$\varepsilon\Phi(1+\varepsilon)\to 1.$$

(iii) Show that, taking ε real and positive and setting $s=1+\varepsilon$, we have

$$0\leqslant \sum_p p^{-s}\log p\,(e^{(i\alpha\log p)/2}+e^{(i\alpha\log p)/2})^4$$
$$=\Phi(s+2i\alpha)+\Phi(s-2i\alpha)+4(\Phi(s+i\alpha)+\Phi(s-i\alpha))+6\Phi(s).$$

Multiplying by ε and allowing $\varepsilon\to 0+$, show that

$$0\leqslant -2\nu-8\mu+6$$

and explain why this implies $\mu=0$ i.e. that ζ has no zero at $1+i\alpha$.

(iv) Let T be as in Question 109.4 (vi). Show that T has no poles in $\{z\colon \operatorname{Re} z\geqslant 1\}$. Let $R>0$. Explain why T can have only a finite number of poles in $\{x+iy\colon |y|\leqslant 2R, x\geqslant 3/4\}$ and deduce that there exists a $\delta(R)>0$ such that T has no poles in $\{x+iy\colon |y|\leqslant 2R, x\geqslant 1-2\delta(R)\}$.

109.7 In Question 109.9 we shall see that the results obtained so far are sufficient to show that $\int_1^X(\theta(x)-x)x^{-2}\,dx$ tends to a finite limit as $X\to\infty$. Since this is the trickiest part of the proof, we delay it until we have shown that its conclusion will give the prime number theorem.

(i) Suppose $\beta\colon[1,\infty)\to\mathbb{R}$ is an increasing (so Riemann integrable) function. Show that, if $\lambda>1$, $y>1$ and $y^{-1}\beta(y)>\lambda$, then

$$\int_y^{\lambda y}\frac{\beta(x)-x}{x^2}\,dx\geqslant A(\lambda),$$

where $A(\lambda)$ does not depend on y and $A(\lambda)>0$. Conclude, using the general principle of convergence for integrals, that, if $\int_1^X(\beta(x)-x)x^{-2}\,dx$ tends to a limit as $X\to\infty$, then $\limsup_{x\to\infty}x^{-1}\beta(x)\leqslant 1$. Prove similarly that, under the same hypotheses, we have $\liminf_{x\to\infty}x^{-1}\beta(x)\geqslant 1$ and so $x^{-1}\beta(x)\to 1$ as $x\to\infty$.

(ii) By applying this result to θ, conclude that, if $\int_1^X(\theta(x)-x)x^{-2}\,dx$ tends to a limit as $X\to\infty$, then $x^{-1}\theta(x)\to 1$ and so $(\pi(x)\log x)/x\to 1$ as $x\to\infty$.

109.8 This question is mainly concerned with trivial notational matters. The exception is part (i) which brings in the Tchebychev estimates.

(i) Using Questions 43.2 and 43.4, show that there exists a constant K such that $|(\theta(x)-x)x^{-1}|\leqslant K$.

(ii) Let $f(t)=\theta(e^t)e^{-t}-1$ for $t\geqslant 0$, $f(t)=0$ for $t<0$. By making a suitable change of variable, show that

$$\mathscr{L}f(z)=\int_0^\infty f(t)e^{-tz}\,dt$$

is well defined for $\operatorname{Re} z > 0$ and

$$\mathscr{L}f(z) = \frac{\Phi(z-1)}{z} - \frac{1}{2},$$

again for $\operatorname{Re} z > 0$. Show that $\int_1^X ((\theta(x) - x)/x^2)\,dx$ converges as $X \to \infty$ if and only if $\int_0^T f(t)\,dt$ converges as $T \to \infty$. We shall write $F(z) = \Phi(z-1)$ for $\operatorname{Re} z > -1/2$.

(Part (ii) does nothing except to translate from the language of Mellin transforms to that of Laplace transforms and shift everything to a more convenient set of axes. Presumably a number theorist would do neither, but Question 109.9 is hard enough to make any possible simplification desirable.)

109.9 This question is devoted to proving the following lemma which bears the same relation to the results of Chapter 76 as a fine picklock does to a simple sledge hammer. Let Ω be an open set in $\mathbb{C}$ such that $\Omega \subseteq \{z: \operatorname{Re} z \geqslant 0\}$, let $F: \mathbb{R} \to \mathbb{C}$ be a function and $f: [0, \infty) \to \mathbb{R}$ a function which is Riemann integrable on each $[0, u]$ for $u > 0$. Suppose that

(A) there is a K with $|f(t)| \leqslant K$ for all $t \geqslant 0$,

(B) $F(z) = \mathscr{L}f(z) = \int_0^\infty f(t)e^{-tz}\,dz$ for $\operatorname{Re} z > 0$,

(C) for any $R > 0$ we can find a $\delta(R) > 0$ such that F is analytic on $\{x + iy: x > -2\delta(R), |y| \leqslant 2R\}$.

We shall conclude that $\int_0^T f(t)\,dt$ converges to a limit as $T \to \infty$.

(i) Use Questions 109.7, 109.6 and 109.4 (vi) to show that the lemma above yields the prime number theorem.

(ii) We now start on the proof. If $T > 0$, set $F_T(z) = \int_0^T f(t)e^{-tz}\,dt$. Explain why $F_T: \mathbb{C} \to \mathbb{C}$ is a well-defined analytic function.

(iii) If $R > 0$, let θ_R be the unique real number with $\pi > \theta_R > \pi/2$ such that $R \exp i\theta_R$ lies on $\{z: \operatorname{Re} z = -\delta(R)\}$. Let us define contours as follows.

(a) $C(1) = C(R, 1)$ is the semicircle $R \exp(i\theta)$ described as θ runs from $-\pi/2$ to $\pi/2$.

(b) $C(2) = C(R, 2)$ is the semicircle $R \exp(i\theta)$ described as θ runs from $\pi/2$ to $3\pi/2$.

(c) $C(3) = C(R, 3)$ consists of the arc $R \exp(i\theta)$ described as θ runs from $\pi/2$ to θ_R followed by the straight line from $R \exp(i\theta_R)$ to $R \exp(-i\theta_R)$ followed by the arc $R \exp(i\theta)$ described as θ runs from $2\pi - \theta_R$ to $3\pi/2$.

Sketch the contours carefully.

(iv) Let $\phi(z) = \phi_{T,R}(z) = e^{Tz}(1 + z^2R^{-2})$. Show that

$$F_T(0) - F(0) = \frac{1}{2\pi i}\int_{C(1)} (F_T(z) - F(z))\frac{\phi(z)}{z}\,dz + \frac{1}{2\pi i}\int_{C(3)} (F_T(z) - F(z))\frac{\phi(z)}{z}\,dz.$$

(This is really the Cauchy integral formula with a different choice of 'inserted pole'. We use $\phi(z)/z$ instead of $1/z$.) Since our theorem will follow if we can show $F_T(0) \to F(0)$ as $T \to \infty$, we try to show that both terms on the right hand side are small.

(v) Show that $|z^{-1}\phi(z)| \leqslant (\exp(T \operatorname{Re} z))(2R^{-2}|\operatorname{Re} z|)$ for all $|z| = R$.

(vi) Show that

$$|F_T(z) - F(z)| = \left|\int_T^\infty f(t)e^{-zt}\, dt\right| \leqslant K \frac{\exp(-T \operatorname{Re} z)}{\operatorname{Re} z}$$

for $z \in C(1)$, $\operatorname{Re} z \neq 0$ and conclude that

$$\left|\frac{1}{2\pi i}\int_{C(1)} (F_T(z) - F(z)) \frac{\phi(z)}{z}\, dz\right| \leqslant \frac{K}{R}.$$

(vii) Prove successively that

(a) $\displaystyle\int_{C(3)} F_T(z) \frac{\phi(z)}{z}\, dz = \int_{C(2)} F_T(z) \frac{\phi(z)}{z}\, dz,$

(b) $\displaystyle|F_T(z)| = \left|\int_0^T f(t)e^{-zt}\, dt\right| \leqslant K \frac{\exp(-T \operatorname{Re} z)}{|\operatorname{Re} z|}$ for $\operatorname{Re} z < 0$,

(c) $\displaystyle\left|\frac{1}{2\pi i}\int_{C(3)} F_T(z) \frac{\phi(z)}{z}\, dz\right| \leqslant \frac{K}{R}.$

(viii) Show that

(a) There exists an $A(R)$ depending only on R such that $|F(z)\phi_{T,R}(z)/z| \leqslant A(R)$ for all $z \in C(3)$.

(b) If $\eta > 0$ and R are fixed, then $F(z)\phi_{T,R}(z)/z \to 0$ uniformly as $T \to \infty$ for $z \in C(3)$, $\operatorname{Re} z < -\eta$.

Conclude that, if R is fixed, then

$$\int_{C(3)} F(z) \frac{\phi_{T,R}(z)}{z}\, dz \to 0 \quad \text{as } T \to \infty.$$

(ix) Deduce from (iv), (v), (vi) and (vii) that

$$|F_T(0) - F(0)| \leqslant 2KR^{-1} + \varepsilon(T, R)$$

where $\varepsilon(T, R) \to 0$ as $T \to \infty$ for any fixed R. By allowing T, $R \to \infty$ in a judiciously chosen manner, show that $F_T(0) \to F(0)$ as $T \to \infty$. We have proved the prime number theorem.

(The proof given above is the work of many hands: Tchebychev, Riemann, Hadamard, de la Vallée-Poussin, Hardy and Littlewood, Ingham and several others. Its present accessible form is due to D. J. Newman.)

The truth or falsity of the Riemann hypothesis is one of the two most famous insolved problems in mathematics. A concise and readable introduction to the surrounding mathematics is given in Patterson *An Introduction to the Theory of the Riemann Zeta-Function*. In recent years many long-standing and, apparently, just as intractable problems have been resolved and we live in a mathematical golden age 'when every morning brings a noble chance, and every chance brings out a noble knight'. It would not surprise me if the Riemann hypothesis was settled within the next twenty years.

APPENDIXES A, B, G

A.1$^{(-)}$ Identify the error in the following 'proof'.

$$1 = \sqrt{1} = \sqrt{(-1)^2} = \sqrt{(-1)}\sqrt{(-1)} = -1.$$

By writing $-1 = \exp(i\pi)$ and setting out the proof in the associated notation, show that the error is essentially the same as that warned against in the appendix.

B.1 By using the one-dimensional version after the manner of Lemma B.1, or otherwise, prove the general principle of convergence for $\mathbb{R}^n$. The sequence $\mathbf{x}_1$, $\mathbf{x}_2, \ldots$ in $\mathbb{R}^n$ converges to a limit $\mathbf{x}$ if and only if, given any $\varepsilon > 0$, we can find an $N(\varepsilon)$ such that $|\mathbf{x}_j - \mathbf{x}_k| < \varepsilon$ for all $j, k \geqslant N(\epsilon)$.

G.1 Let us consider one-dimensional Brownian motion on $[0, 1]$ started at 0. More specifically (compare Plausible Result 14.3), let us consider a family of random continuous functions $Z: [0, 1] \to \mathbb{R}$ such that $Z(0) = 0$ and such that, if $0 \leqslant t < t + \delta t \leqslant 1$, then

$$Pr\big((Z(t + \delta t) - Z(t))/\delta t^{1/2} \in [a, b]\big) = \frac{1}{\sqrt{(2\pi)}} \int_a^b \exp(-x^2/2)\, dx,$$

independent of previous history. Since Z is a continuous function, we could seek to express it by means of Fourier series. This can be done to great effect but, for reasons outlined later in Question G.6, requires more machinery than we have directly at our disposal.

However, we have seen that there are many useful ways of expanding continuous functions as infinite sums of simpler functions. The one that follows gives us a very convenient way of looking at one dimensional Brownian motion.

(i) Let us consider continuous functions $f: [0, 1] \to \mathbb{R}$ with $f(0) = 0$. Let us write

$$\Delta_0(x) = x \quad \text{for } 0 \leqslant x \leqslant 1$$

and, if r and m are integers with $1 \leqslant r \leqslant 2^{m-1}$, $1 \leqslant m$ let us write

$$\Delta_{r,m}(x) = 1 - 2^m|x - (2r-1)2^{-m}| \quad \text{for } |x - (2r-1)2^{-m}| \leqslant 2^{-m}$$

$$\Delta_{r,m}(x) = 0 \qquad\qquad \text{otherwise.}$$

Sketch $\Delta_{r,m}$ and show that

$$f_n(x) = f(1)\Delta_0(x) + \sum_{m=1}^{n}\sum_{r=1}^{2^{m-1}} a_{r,m}(f)\Delta_{r,m}(x),$$

with $a_{r,m}(f) = f((2r-1)2^{-m}) - 2^{-1}(f(2^{1-m}r) + f(2^{1-m}(r-1)))$, is the simplest piecewise linear function g such that $g(r2^{-n}) = f(r2^{-n})$ for $0 \leqslant r \leqslant 2^n$.

(ii) Show that $f_n \to f$ uniformly on $[0, 1]$. (You may wish to compare Question 5.3)

(iii) Now suppose $f(t) = Z(t)$. By writing

$$a_{r,m}(Z) = 2^{-1}(Z((2r-1)2^{-m}) - Z(2^{1-m}(r-1))) - 2^{-1}(Z(2^{1-m}r) - Z((2r-1)2^{-m})),$$

or otherwise, show that $a_{r,m}(Z)$ is a normally distributed random variable with mean 0 and variance 2^{-m-1}. By using Question 81.3 (ii), show that the random variables $a_{r,m}(Z)$ $[1 \leqslant r \leqslant 2^{m-1}, 1 \leqslant m]$ together with $Z(1)$ are all independent.

(iv) Conclude that

$$Z(t) = X_0 t + \sum_{m=1}^{\infty}\sum_{r=1}^{2^{m-1}} 2^{-(m+1)/2} X_{r,m}\Delta_{r,m}(t) \quad [0 \leqslant t \leqslant 1]$$

where X_0, $X_{r,m}$ $[1 \leqslant r \leqslant 2^{m-1}, 1 \leqslant m]$ are independent normal variables with mean 0 and variance 1.

G.2 (Construction of Brownian motion) In Question G.1 we assumed, without proof, the existence of a family of Brownian motions Z. By turning the argument of that question on its head, we can prove that such a family exists.

(i) If $R > 0$, show that

$\int_R^\infty \exp(-x^2/2)\,dx \leqslant \frac{1}{R}\int_R^\infty x\exp(-x^2/2)\,dx = R^{-1}\exp(-R^2/2)$.

(ii) Let X_0, $X_{r,m}$ $[1 \leqslant r \leqslant 2^{m-1}, 1 \leqslant m]$ be independent normal variables and set

$$g_m(t) = \sum_{r=1}^{2^{m-1}} 2^{-(m+1)/2} X_{r,m}\Delta_{r,m}(t) \quad [0 \leqslant t \leqslant 1].$$

Show that, if $\lambda > 0$,

$$Pr\left(\sup_{t\in[0,1]} |g_m(t)| > \lambda\right) = Pr\left(\max_{1\leqslant r\leqslant 2^{m-1}} |X_{r,m}| > 2^{(m+1)/2}\lambda\right)$$

$$\leqslant 2^{m-1}\,\Pr(|X_0| > 2^{(m+1)/2}\lambda)$$

and hence show that, if $A > 0$,

$$Pr(\sup|g_m(t)| > A2^{-m/4} \text{ for some } m \geq 1) \leq \sum_{m=1}^{\infty} Pr(\sup|g_m(t)| > A2^{-m/4})$$
$$\leq \varepsilon(A)$$

where $\varepsilon(A) \to 0$ as $A \to \infty$.

(iii) Conclude, by using the Weierstrass M test, or otherwise, that

$$Pr\left(\sum_{m=1}^{M} g_m \text{ converges uniformly on } [0, 1] \text{ as } M \to \infty\right) \geq 1 - \varepsilon(A)$$

and so, since A can be chosen arbitrarily, that $\sum_{m=1}^{M} g_m$ converges uniformly on [0, 1] with probability 1.

(iv) We may always exclude from consideration an event of zero probability (see the discussion preceding Plausible Lemma 88.5), so, from now on, we only consider the family of $X_{r,m}$ for which $\sum_{m=1}^{M} g_m$ converges uniformly, and we write

$$Z(t) = X_0 t + \sum_{m=1}^{\infty} g_m(t)$$
$$= X_0 t + \sum_{m=1}^{\infty} \sum_{r=1}^{2^{m-1}} 2^{-(m+1)/2} X_{r,m} \Delta_{r,m}(t).$$

Explain why Z is continuous.

(v) Explain why

$$Z(r2^{-n}) = X_0 r2^{-n} + \sum_{m=1}^{n} g_m(r2^{-n})$$

and, by using Question 81.3 (ii), or otherwise, show that the random variables $Y_1, Y_2, \ldots, Y_{2^n}$ defined by

$$Y_r = Z(r2^{-n}) - Z((r-1)2^{-n})$$

are independent, normally distributed random variables with mean 0 and variance $2^{-n/2}$. Conclude that, if $0 \leq r < s \leq 2^n$, then

$$Pr\left((Z(s2^{-n}) - Z(r2^{-n}))/((s-r)2^{-n})^{1/2} \in [a, b]\right)$$
$$= \frac{1}{\sqrt{(2\pi)}} \int_a^b \exp(-x^2/2)\, dx$$

independent of the values of $Z(0)$, $Z(2^{-n}), \ldots, Z((r-1)2^{-n})$. Allowing $n \to \infty$ and using the continuity of Z, it is extremely plausible that this last statement implies that, if $0 \leq t < t + \delta t \leq 1$, then

$$Pr\left((Z(t+\delta t) - Z(t))/\delta t^{1/2} \in [a, b]\right) = \frac{1}{\sqrt{(2\pi)}} \int_a^b \exp(-x^2/2)\, dx$$

independent of the previous history.

(Without an explicit statement of the underlying axioms there can be no watertight proof. However, the construction above is, in my opinion, essentially complete.)

(vi) Suppose that $Z_0, Z_1, \ldots$ are independent families of Brownian motions on [0, 1] started at 0, as specified in (v). Let us define $Z: [0, \infty) \to \mathbb{R}$ by

$$Z(t) = Z_n(t-n) + \sum_{r=0}^{n-1} Z_r(1) \quad [n \leqslant t < n+1].$$

Show that we have defined a family of random continuous functions $Z: [0, \infty] \to \mathbb{R}$ such that $Z(0) = 0$ and such that, if $0 \leqslant t < s$, then

$$Pr\big((Z(s) - Z(t))/(s-t)^{1/2} \in [a, b]\big) = \frac{1}{\sqrt{(2\pi)}} \int_a^b \exp(-x^2/2)\, dx.$$

Thus we have constructed Brownian motion on $[0, \infty]$ started at 0.

(vii) Suppose X, Y are independent families of Brownian motions on $[0, \infty]$ started at 0, as specified in (vi). Show that $(X, Y): [0, \infty) \to \mathbb{R}^2$ gives a family of two-dimensional Brownian motions started at time 0 at the origin (0, 0).

G.3 (Why are we not all called Smith?) Sir Thomas Browne writes that 'Generations pass while some trees stand and old families last not three oaks.' What is the probability that some family name will eventually become extinct? As a simplified model, let us suppose that family names are transmitted through the male line and that the probability that a single male has n male children is p_n $[0 \leqslant n \leqslant N]$. Explain why the discussion becomes trivial if $p_0 = 0$ or $p_0 = 1$. We shall assume $1 > p_0 > 0$.

(i) Show that

$$Pr\begin{pmatrix}\text{a name borne by } k \text{ males in the first}\\ \text{generation eventually becomes extinct}\end{pmatrix} = Pr\begin{pmatrix}\text{a name borne by 1 male in the first}\\ \text{generation eventually becomes extinct}\end{pmatrix}^k.$$

In particular observe that, if extinction is certain (with probability 1) for a name with only one male owner, it is certain however many males own it.

(ii) Let s be the probability of eventual extinction for a name with 1 male owner in the first generation. Show that

$$s = \sum_{r=0}^{N} p_r s^r.$$

(iii) If $f(s) = \sum_{r=0}^{N} p_r s^r$, show that $f: \mathbb{R} \to \mathbb{R}$ is convex, that $1 > f(0) > 0$ and that $f(1) = 1$. Deduce that the equation $f(s) = s$ has either one or two roots s with $0 \leqslant s \leqslant 1$. Show that, if $f'(1) > 1$, it has two roots and, if $f'(1) \leqslant 1$, it has one root.

(iv) Conclude that, if $\sum_{r=0}^{N} rp_r \leqslant 1$, then extinction is certain with probability 1. What is the probabilistic meaning of $\sum_{r=0}^{N} rp_r$?

If $\sum_{r=0}^{N} rp_r > 1$, the discussion above tells us that the equation $s = \sum_{r=0}^{N} p_r s^r$ has two roots in $[0, 1]$. One of these is 1, the other s_0, say. Thus the probability of extinction is either 1 or s_0. In two fascinating articles (*Journal of the London Mathematical Society*, Vol. 41 and *Bulletin of the London Mathematical Society*, Vol. 7) Kendall gives the history of this problem, showing how some of the early investigators were led to choose the root $s = 1$ rather than $s = s_0$ and conclude paradoxically that all surnames were doomed to eventual extinction. We discuss what actually happens in the next question.

G.4 (i) Continuing the argument and notation of Question G.3, let us set

$$f_n(t) = \sum_{k=0}^{\infty} Pr\ (k \text{ members of } n\text{th generation})t^k.$$

(Note that, for the model considered, f_n will be a polynomial. Note also that $f_1(t) = f(t)$.) Show that

$$f_{n+1}(t) = f_n(f(t)).$$

(Some readers will be able to give a one-line proof using generating functions. The rest need only compare coefficients of the two polynomials to obtain a two-line proof.)

(ii) Show by induction, inspection, or otherwise, that $f_n(f(t)) = f(f_n(t))$. (This is a simple general result on iteration and has nothing to do with polynomials.)

(iii) By setting $t = 0$, or otherwise, show that, if q_n is the probability a name becomes extinct at, or before, the nth generation, then $q_0 = 0$ and $q_{n+1} = f(q_n)$.

(iv) Let α be the least root of $f(s) = s$ in the interval $[0, 1]$. (Thus, if $\sum_{r=0}^{N} rp_r \leqslant 1$, we have $\alpha = 1$, if $\sum_{r=0}^{N} rp_r > 1$, then $\alpha = s_0$.) Show by induction, or otherwise, that $\alpha \geqslant q_{n+1} \geqslant q_n \geqslant 0$ for $n \geqslant 0$. Explain why this implies that q_n tends to a limit q as $n \to \infty$, and that the limit q must satisfy $\alpha \geqslant q \geqslant 0$ and $f(q) = q$ i.e. that $q = \alpha$. Conclude that extinction is certain with probability 1 if $\sum_{r=0}^{N} rp_r \leqslant 1$, but there is a positive probability $1 - s_0$ that a surname will never be extinguished if $\sum_{r=0}^{N} rp_r > 1$.

(v) It would thus appear that the reason that we have many different surnames is the population explosion, i.e. the fact that $\sum_{r=0}^{N} rp_r > 1$. However, even in isolated villages where the population numbers have been stable for a long time, i.e. $\sum_{r=0}^{N} rp_r = 1$, we still find several surnames. Use your pocket calculator to examine the behaviour of q_n in the two cases:

(a) $f(x) = 3/4 + x^2/4$,
(b) $f(x) = 1/2 + x^2/2$.

Can you account for the difference you observe as n gets large? (Two clear diagrams and one short sentence will suffice.)

(vi) Another way of looking at the problem is to consider e_n the expected number of males bearing the surname in the nth generation and $e = \sum_{n=0}^{\infty} e_n$ the expected total number of males of all generations bearing the surname. Show that $e_n = (\sum_{r=0}^{N} rp_r)^n$ and conclude that $e < \infty$ if $\sum_{r=0}^{N} rp_r < 1$ but that $e = \infty$ if $\sum_{r=0}^{N} rp_r \geqslant 1$. Thus, in particular, if $\sum_{r=0}^{N} rp_r = 1$, then extinction is certain with probability 1, but the expected number of bearers of a surname in infinite. Parts (v) and (iv) suggest that, for populations which merely replace themselves, i.e. $\sum_{r=0}^{N} rp_r = 1$, we will find a few surnames each held by a large number of people and this is in accordance with observation.

(The reader will not need to be told that the kind of question discussed here is relevant to a wide range of topics from chain reactions in atomic engineering to the survival of species in evolutionary biology. She will also have spotted various oversimplifications in the model presented here. For more information I refer you once again to Kendall's papers. It should perhaps be added that, in some sense, we *are* all called Smith. Mitochondrial DNA molecules are inherited from the female parent. We all have the same type of Mitochondrial DNA which geneticists believe (for references see Reader *Missing Links*) is inherited from a single African mother who lived 10 000 generations ago. All the other Mitochondrial DNA types (i.e. surnames) have died out. The lady is sometimes called the African Eve, but this is misleading. She is the only female of her generation who is connected without a break through the female line to the present generation, but, since most of our genetic heritage passes through both the male and female line, she is not the only female of her generation whose genes we have.)

(vii) If there was an African Adam, would he have had to meet the African Eve? Have you any other comments along these lines? (Upon reflection I find enough comments to form the basis of a long essay.)

(viii) Do the following first year Tripos question.

No-one in their right mind would wish to be a guest at the Virtual Reality Hotel. The rooms are numbered 0 to $(3^N - 3)/2$ where N is a very large integer. If $0 \leqslant i \leqslant (3^{N-1} - 3)/2$ and $j = 1, 2, 3$ there is a door between room i and room $(3i + j)$ through which (if it is unlocked) guests may pass in both directions. In addition, any room with number higher than $(3^{N-1} - 3)/2$ has an open window through which guests can (and should) escape into the street. So far as the guests are concerned, there are no other doors or windows.

Each door in the hotel is locked with probability 1/3 independently of the others. An arriving guest is placed in room 0 and can then wander freely (in so far as the locked doors allow). Show that the guest's chance of escape is about $(9 - \sqrt{27})/4$.

G.5 (Nowhere differentiability of Brownian motion) The following elegant argument is due to Kahane. We use the notation and results of Question G.1 and the ideas and result of Question G.3.

(i) Look at the formula of part (iv) of Question G.1. Suppose that, for some given $\alpha>0$, $1\leqslant r\leqslant 2^{m-1}$ and $m\geqslant 1$, we know that $|X_{r,m}|>\alpha$. Write $I=[(r-1)2^{-m+1}, r2^{-m+1}]$, $a=(r-1)2^{-m+1}$, $b=(r-2^{-1})2^{-m+1}$, $c=r2^{-m+1}$. Show that, if $x\in I$, then

$$\sup_{y\in I, y\neq x}\left|\frac{Z(y)-Z(x)}{y-x}\right|\geqslant \max_{y\in\{a,b,c\}, y\neq x}\left|\frac{Z(y)-Z(x)}{y-x}\right|\geqslant 2^{(m-1)/2}\alpha.$$

(ii) Choose an $\alpha>0$ such that $(1/\sqrt{(2\pi)})\int_0^\alpha \exp(-x^2/2)\,dx<1/4$. Fix an integer $N\geqslant 1$ and consider the interval $[0, 2^{1-N}]$. If $|X_{1,N}|\geqslant\alpha$, colour the interval red. If $|X_{1,N}|<\alpha$, colour it blue. Now consider the two daughter intervals $[0, 2^{-N}]$ and $[2^{-N}, 2^{1-N}]$. If the mother interval $[0, 2^{1-N}]$ is red, colour the daughter intervals red. If the mother interval $[0, 2^{1-N}]$ is blue, colour the daughter interval $[0, 2^{-N}]$ red if $|X_{1,N+1}|\geqslant\alpha$, blue otherwise, and colour the daughter interval $[2^{-N}, 2^{1-N}]$ red if $|X_{2,N+1}|\geqslant\alpha$, blue otherwise. Now proceed inductively so that the interval $[(s-1)2^{-N-k}, s2^{-N-k}]$ is coloured red if either the mother interval $[[s/2]2^{-N-k+1}, ([s/2]+1)2^{-N-k+1}]$ is coloured red, or if $|X_{sN+1+k}|\geqslant\alpha$, and coloured blue otherwise. By using Question G.3, show that, with probability 1, eventually all the intervals will be coloured red, i.e. there exists an M (a random variable depending on the random variables $X_{r,m}$) such that, for each $x\in[0, 2^{1-N}]$, there exits a $0\leqslant k\leqslant M$ and an s such that $x\in[s2^{-N-k}, (s+1)2^{-N-k}]$ and $|X_{s,N+1+k}|\geqslant\alpha$.

(iii) Combining (i) and (ii) show that, with probability 1, all $x\in[0, 2^{-N}]$ have the property that there exists a $y\in[0, 2^{-N}]$ with $y\neq x$ and

$$\left|\frac{Z(y)-Z(x)}{y-x}\right|\geqslant 2^{(N-1)/2}\alpha.$$

By applying the same argument to 2^N intervals, show that, with probability 1,

$$\sup_{y\in[0,1], y\neq x}\left|\frac{Z(y)-Z(x)}{y-x}\right|\geqslant 2^{(N-1)/2}\alpha \quad \text{for all } x\in[0, 1].$$

(iv) If events $E_1, E_2, \ldots$ have zero probability, so does the event $\bigcup_{n=1}^\infty E_n$. Conclude that, with probability 1,

$$\sup_{y\in[0,1], y\neq x}\left|\frac{Z(y)-Z(x)}{y-x}\right|=\infty \quad \text{for all } x\in[0, 1],$$

i.e. Z is nowhere differentiable.

G.6 As we said in the first paragraph of Question G.1, since $Z\colon[0,1]\to\mathbb{R}$ is continuous, it must have a Fourier expansion

$$Z(t)\sim A_0+\sum_{m=1}^\infty A_m\cos 2\pi mt+\sum_{m=1}^\infty B_m\sin 2\pi mt.$$

We do not have enough mathematical equipment to determine the coefficients rigorously, but we can make a good guess as to what they must be.

(i) Show that $A_m = 2\int_0^1 Z(t)\cos 2\pi mt\,dt$ $[m \geqslant 1]$. Let us seek an approximation to A_m of the form

$$\alpha_{m,n} = 2n^{-1}\sum_{r=1}^{n} Z(r/n)\cos(2\pi rm/n).$$

Explain why

$$(\sin(2\pi(r+1)m/n) - \sin(2\pi rm/n)) \approx 2\pi(m/n)\cos(2\pi rm/n)$$

and we may replace the approximation $\alpha_{m,n}$ by the new approximation

$$A_{m,n} = (\pi m)^{-1}\sum_{r=1}^{n} Z(r/n)(\sin(2\pi(r+1)m/N) - \sin(2\pi rm/n)).$$

By partial summation, obtain

$$A_{m,n} = -(\pi m)^{-1}\sum_{r=1}^{n}(Z(r/n) - Z((r-1)/n))\sin(2\pi rm/n).$$

Show that $A_{m,n}$ is a normally distributed random variable with mean 0 and variance $(\pi m)^{-2}n^{-1}\sum_{r=1}^{n}(\sin 2\pi rm/n)^2$. Show also that $A_{1,n}, A_{2,n}, \ldots, A_{n-1,n}$ are independent. If we now allow $n \to \infty$, it is reasonable to suppose that the $A_1, A_2, \ldots$ will be independent normally distributed random variables such that A_m has mean 0 and variance $(\pi m)^{-2}\int_0^1(\sin 2\pi t)^2\,dt = 2^{-1}(\pi m)^{-2}$.

(ii) Using similar arguments, show that it is reasonable to approximate B_m by

$$B_{m,n} = (\pi m)^{-1}\left(-Z(1) + \sum_{r=1}^{n}(Z(r/n) - Z((r-1)/n))\cos(2\pi rm/n)\right).$$

Looking at the additional term $-(\pi m)^{-1}Z(1)$ and reflecting on the discontinuity at 1 of the periodic extension of Z, we change our mode of approach slightly. Show that

$$t = -\sum_{m=1}^{\infty}(\pi m)^{-1}\sin(2\pi mt) \quad \text{for } 0 \leqslant t < 1$$

and write

$$Z(t) \sim Z(1)t + A_0 + \sum_{m=1}^{\infty} A_m\cos(2\pi mt) + \sum_{m=1}^{\infty} C_m\sin(2\pi mt).$$

Show that it is reasonable to approximate C_m by

$$C_{m,n} = (\pi m)^{-1}\sum_{r=1}^{n}(Z(r/n) - Z((r-1)/n))\cos(2\pi rm/n).$$

Show that $Z(1)$, $A_{1,n}$, $A_{2,n}$, ..., $A_{n-1,n}$, $C_{1,n}$, $C_{2,n}$, ..., $C_{n-1,n}$ are independent and that it is reasonable to suppose that $Z(1)$, A_1, A_2, ..., C_1, C_2, ... are independent normally distributed random variables. Write down their means and variances.

(iii) All that remains is to look at A_0. Suppose for the moment that our representation

$$Z(t) \sim Z(1)t + A_0/2 + \sum_{m=1}^{\infty} A_m \cos(2\pi mt) + \sum_{m=1}^{\infty} B_m \sin(2\pi mt)$$

actually gives rise to an equality at 0 so that

$$Z(0) = Z(1)0 + A_0 + \sum_{m=1}^{\infty} A_m \cos(2\pi m0) + \sum_{m=1}^{\infty} B_m \sin(2\pi m0)$$

and $0 = A_0 + \sum_{m=1}^{\infty} A_m$. Using this guess, we are lead to our final formula

$$Z(t) \sim X_0 t + \sum_{m=1}^{\infty} (2^{1/2}\pi m)^{-1} X_m((\cos 2\pi mt) - 1)$$

$$+ \sum_{m=1}^{\infty} (2^{1/2}\pi m)^{-1} Y_m \sin(2\pi mt)$$

where X_0, X_1, X_2, ..., Y_1, Y_2, ... are independent normally distributed random variables with mean 0 and variance 1.

(If we now try to follow the line of Question G.2 and *define* Z by the formula

$$Z(t) = X_0 t + \sum_{m=1}^{\infty} (2^{1/2}\pi m)^{-1} X_m((\cos 2\pi mt) - 1)$$

$$+ \sum_{m=1}^{\infty} (2^{1/2}\pi m)^{-1} Y_m \sin(2\pi mt),$$

we run into the immediate difficulty that the Fourier coefficients will take values of size m^{-1} or larger for most values of m. Since $\sum_{m=1}^{\infty} m^{-1}$ diverges, this means that the Fourier series is not absolutely convergent and, as we emphasised in Chapter 9 and elsewhere, much harder to deal with. The problem can be resolved by paying correct attention to the underlying *probabilistic* process and using an inequality of Kolmogorov concerning the maximum and final values of certain random processes. The book *Some Random Series of Functions* by Kahane is written for a substantially more advanced audience than the present book, but, if the reader browses through it, she will glimpse many beautiful and useful results on random Fourier series like the one above.)

G.7 The object of the next two questions is to explore the theorem of Lévy referred to in Appendix G. This question contains various preliminaries.

(i) Check, in as much detail as you require, that the proofs in Chapters 70 and 71 give the following theorem. Suppose that for each $n \geqslant 1$ we have independent identically distributed random variables $Y(1, n)$, $Y(2, n)$, ..., $Y(n, n)$ with mean μ_n and variance σ_n^2. Then

$$Pr\left(\frac{\sum_{r=1}^{n} Y(r, n) - n\mu_n}{n^{1/2}\sigma_n} \in [a, b]\right) \to \frac{1}{\sqrt{(2\pi)}} \int_a^b \exp(-t^2/2)\, dt$$

as $n \to \infty$.

(ii) Suppose $Z_1, Z_2, \ldots$ are random variables, and $\mu_1, \mu_2, \ldots \in \mathbb{R}$ and $\sigma_1, \sigma_2, \ldots \in \mathbb{R}$ with $\sigma_n > 0$ for each $n \geqslant 1$. Suppose

$$Pr\left(\frac{Z_n - \mu_n}{\sigma_n} \in [a, b]\right) \to \frac{1}{\sqrt{(2\pi)}} \int_a^b \exp(-t^2/2)\, dt$$

for each $a < b$ with $a, b \in \mathbb{R}$. Show that, if $n(1) < n(2) < \ldots$ and $\sigma_{n(j)} \to \infty$, then

$$Pr(Z_{n(j)} \in [c, d]) \to 0 \quad \text{as } j \to \infty$$

for each $c < d$ with $c, d \in \mathbb{R}$.

Show also that, if $m(1) < m(2) < \ldots$ and $\sigma_{m(j)}^2 \leqslant K$ for some $K > 0$ and all $j \geqslant 1$ but $|\mu_{m(j)}| \to \infty$ as $j \to \infty$, then

$$Pr(Z_{m(j)} \in [c, d]) \to 0 \quad \text{as } j \to \infty$$

for each $c < d$ with $c, d \in \mathbb{R}$.

(iii) Let Z_j, μ_j and σ_j be as in part (ii), but suppose further that there exists a random variable W such that

$$Pr(|Z_n - W| > \varepsilon) \to 0 \quad \text{as } n \to \infty$$

for each $\varepsilon > 0$. Show, from (ii), that we can find a K such that $|\mu_n|$, $\sigma_n^2 \leqslant K$. Hence deduce that we can find $n(1) < n(2) < \ldots$ together with $\mu, \sigma \in \mathbb{R}$ such that $\sigma \geqslant 0$ and $\mu_{n(j)} \to \mu$, $\sigma_{n(j)} \to \sigma$. If $\sigma = 0$, show that $W = \mu$ with probability 1 (i.e. $\Pr(W = \mu) = 1$) and, if $\sigma > 0$, show that W is normally distributed with mean μ and variance σ^2.

G.8 Suppose $X\colon [0, \infty) \to \mathbb{R}$ is a random function such that

(1) for each $0 \leqslant t \leqslant s$, $X(s) - X(t)$ is independent of the behaviour of $X(y)$ for $0 \leqslant y \leqslant t$ and is distributed like the random variable $W(s - t)$, say;

(2) $X\colon [0, \infty) \to \mathbb{R}$ is continuous.

What can we say about X? (Note that we do not assume $W(s - t)$ has a variance or even that $W(s - t)$ has a mean. We do not assume that $W(s - t)$ has any symmetries.)

Observe first that (2) tells us that, if $0 \leq u < v$, then X is continuous on $[u, v]$ and so uniformly continuous. Thus, for each X and each $\varepsilon > 0$, we can find a $\delta(\varepsilon, X) > 0$ such that

$$|X(s) - X(t)| < \varepsilon \quad \text{for all } s, t \in [u, v] \text{ with } |s - t| < \delta(\varepsilon, X).$$

Unfortunately $\delta(\varepsilon, X)$ depends, not merely on ε, but also on X. However, it seems plausible that the statement just made implies

(2)′ $Pr(|X(s) - X(t)| < \varepsilon$ for all $s, t \in [u, v]$ with $|s - t| < \delta) \to 1$ as $\delta \to 0+$.

This result can be proved rigorously using modern axiomatic probability, but, if the reader is unhappy, she may make (2)′ a hypothesis.

(i) Set

$$Y(r, n) = X(r/n) - X((r-1)/n) \quad \text{if } |X(r/n) - X((r-1)/n)| \leq 1,$$
$$= 0 \qquad \text{otherwise,}$$

and set $Z_n = \sum_{r=1}^{n} Y(r, n)$.

Show using (2)′ that

$$\Pr(Z_n \neq X(1) - X(0)) \to 0 \quad \text{as } n \to \infty.$$

(ii) Show that $Y(1, n), Y(2, n), \ldots, Y(n, n)$ are independent identically distributed random variables. Explain why $Y(1, n)$ has a well-defined mean and variance, $\tilde{\mu}_n$ and $\tilde{\sigma}_n^2$, say. Using part (i) of Question G.7, show that, if $\tilde{\sigma}_n \neq 0$, then

$$Pr\left(\frac{Z_n - n\tilde{\mu}_n}{n^{1/2}\tilde{\sigma}_n} \in [a, b]\right) \to \frac{1}{\sqrt{(2\pi)}} \int_a^b \exp(-t^2/2)\, dt$$

for each $a < b$.

(iii) By setting $W = X(1) - X(0)$, $\mu_n = n\tilde{\mu}_n$ and $\sigma_n = n^{1/2}\tilde{\sigma}_n$ use Question G.7 (iii) to show that *either*

(a) there exists a μ such that $\Pr(X(1) - X(0) = \mu) = 1$, *or*
(b) there exists a μ and a $\sigma > 0$ such that $X(1) - X(0)$ is normally distributed with mean μ and variance σ^2.

(iv) More generally show that, for each $y > 0$, *either*

(a) there exists a $\mu(y)$ such that $\Pr(W(y) = \mu(y)) = 1$, *or*
(b) there exists a $\mu(y)$ and a $\sigma(y) > 0$ such that $W(y)$ is normally distributed with mean $\mu(y)$ and variance $\sigma(y)$.

(v) By observing, once more, that

$$X(1) - X(0) = \textstyle\sum_{r=1}^{n} X(r/n) - X((r-1)/n)$$

show that *either*

(a) there exists a μ such that $Pr(W(n^{-1}) = \mu n^{-1}) = 1$ for all integers $n \geqslant 1$, *or*
(b) there exists a μ and a $\sigma > 0$ such that $W(n^{-1})$ is normally distributed with mean μn^{-1} and variance $\sigma^2 n^{-1}$.

(vi) Extend the result of (v) to cover $W(r/n)$ with r, n integers such that r, $n \geqslant 1$. Now use (2)′ to show that *either*

(a) there exists a μ such that $Pr(W(t) = \mu t) = 1$ for all $t > 0$, *or*
(b) there exists a μ and a $\sigma > 0$ such that $W(t)$ is normally distributed with mean μt and variance $\sigma^2 t$ for all $t > 0$.

Thus $X(t) = a + \mu t + \sigma X_0(t)$ where $X_0(t)$ is a standard Brownian motion and we have vindicated Bachelier's intuition. It is, of course, easy to point to problems with Bachelier's assumptions about the nature of stock exchanges. For example, since buying and selling shares costs money, very small changes in the true value of shares might not be reflected in price changes. However, a lightly modified version of Bachelier's model is now used by large investment houses in valuing stock options.

REFERENCES

Abramowitz, M. and Stegun, I. A. *Handbook of Mathematical Functions*. Dover, 1965. Reprint of 10th Government Printing.

Acton, F. S. *Numerical Methods That Work*. Harper and Row, 1970.

Artin, E. *The Gamma Function*. Holt, Rinehart and Winston, New York, 1985. English translation by M. Butler.

Auckland, G. and Coates, B. *Take Nobody's Word For It*. BBC Books, 1989.

Beardon, A. F. *Iteration of Rational Functions*. Springer, 1991.

Bell, E. T. *Men of Mathematics (2 vols)*. Simon and Schuster, 1937.

Bellman, R. *A brief Introduction to Theta Functions*. Holt, New York, 1961.

Beukers, F. A note on the irrationality of $\zeta(2)$ and $\zeta(3)$. *Bulletin of the London Mathematical Society*, 11: 268–72, 1979.

Bollobás, B. *Graph Theory*. Springer, 1979.

Cheney, E. W. *Introduction to Approximation Theory*. McGraw-Hill, 1966.

Davenport, J. H., Suet, Y. and Tournier, E. *Computer Algebra*. Academic Press, 1988.

Doyle, P. and Snell, J. *Random Walks and Electric Networks*. MAA, 1984. Carus Monographs 22.

Dublins, L. E. and Savage, L. J. *How to Gamble If You Must*. McGraw-Hill, 1965.

Dym, H. and McKean, H. P. *Fourier Series and Integrals*. Academic Press, 1972.

Edwards, H. M. *Riemann's Zeta Function*. Academic Press, 1974.

Feller, W. A direct proof of Stirling's formula. *American Mathematical Monthly,* 74: 1223–5, December 1967.

Feller, W. *An Introduction to Probability Theory and Its Applications (Vol. 1)*, 3rd edition. Wiley, 1968.

Feynman, R. P. *Surely You're Joking, Mr Feynman!* W. W. Norton, New York, 1985.

Feynman, R. P., Leighton R. B. and Sands, M. *The Feynman Lectures on Physics (3 vols)*. Adison-Wesley, 1963.

Gelfond, I. M. and Fomin, S. *Calculus of Variations.* Prentice Hall, 1963. English translation by R. A. Silverman.

Gröchenig, K. Reconstruction algorithms in irregular sampling. *Mathematics of Computation,* 59: 181–94, 1992.

Hardy, G. H. *A Course of Pure Mathematics*. Cambridge University Press, 1914.

Hardy, G. H. *Collected Works (7 vols)*. Oxford University Press, 1979.

Hardy, G. H., Littlewood, J. E. and Pólya, G. *Inequalities*. Cambridge University Press, 1934.

Hardy, G. H. and Wright, E. M. *An Introduction to the Theory of Numbers*. Oxford University Press, 1938.

Helson, H. *Harmonic Analysis*. Adison-Wesley, 1983.

Higgins, J. R. Five short stories about the cardinal series. *Bulletin of the American Mathematical Society*, 12: 45–89, 1985.

Hirsch, M. W. and Smale, S. *Differential Equations, Dynamical Systems and Linear Algebra*. Academic Press, 1974.

Jackson, D. *The Theory of Approximation*. American Mathematical Society Colloquium Publications, Vol. II.

Jones, G. A. and Singerman, D. *Complex Functions.* Cambridge University Press, 1987.

Kahane, J.-P. *Some Random Series of Functions*. Cambridge University Press, 2nd edition, 1985.

Kaku, M. *Introduction to Superstrings*. Springer, 1988.

Katznelson, Y. *An Introduction to Harmonic Analysis*. Wiley, 1983.

Kendall, D. G. Branching processes since 1873. *Journal of the London Mathematical Society*, 41: 385–406, 1966.

Kendall, D. G. The genealogy of genealogy .... *Bulletin of the London Mathematical Society*, X: 225–53, 1975.

Klein, F. *Development of Mathematics in the 19th Century.* Mathematical Science Press, 1979. Translated by M. Ackerman.

Körner, T. W. *Fourier Analysis*. Cambridge University Press, 1988.

Littlewood, J. E. *A Mathematician's Miscellany, 2nd edition, ed. B. Bollobás*. Cambridge University Press, 1969.

Markushevich, A. I. *Theory of Functions of a Complex Variable (3 vols)*. Prentice Hall, 1967. English translation by R. A. Silverman.

Marwick, B. *et al.* Discussion of the Soal-Goldney experiment. *Proceedings of the Society for Psychical Research,* Vol. 56, 1966. Entire issue.

Molk, J. and Tannery, J. *Eléments de la Théorie des Fonctions Elliptiques (4 vols)*. Gauthier Villars, Paris, 1902.

Meyer, Y. *Wavelets and Operators.* Cambridge University Press, to appear. Translated by D. Sallinger.

Natanson, I. P. *Constructive Function Theory (3 vols)*. Fredrick Ungar, New York, 1965. English translation by A. N. Oblensky and J. R. Schulenberger.

Newman, D. J. A simplified version of the fast algorithms of Brent and Salamin. *Mathematics of Computation*, 44: 207–10, 1985.

Patterson, S. J. *An Introduction to the Theory of the Riemann Zeta-Function*. Cambridge University Press, 1988.

Pòlya, G. *Mathematics and Plausible Reasoning (2 vols)*. Princeton University Press, 1954.

Pólya, G. *How to Solve It*, 2nd edition. Princeton University Press, 1957.

Pólya, G. *Collected Works (4 vols)*. MIT, 1984.

Press, W. H., Flannery, B. P., Teukolsky, S. A. and Vetterling, W. T. *Numerical Recipes*. Cambridge University Press, 1986.

Reader, J. *Missing Links*, 2nd edition. Penguin, 1988.

Rees, E. G. *Notes on Geometry*. Springer, 1983.

Richardson, L. F. *Arms and Insecurity*. Boxwood, Pittsburg, 1960.

Richardson, L. F. *Statistics of Deadly Quarrels*. Boxwood, Pittsburg, 1960.

Rivlin, T. J. *The Chebychev Polynomials*. Wiley, 1974.

Rogosinski, W. *Fourier Series*. Chelsea, 1950.

Salem, R. *Algebraic Numbers and Fourier Analysis*. D. C. Heath, 1963.

Shapiro, H. S. *Smoothing and Approximation of Functions*. Van Nostrand, 1969.

Shapiro, H. S. *Topics in Approximation Theory*. Springer, 1971. Lecture Notes In Mathematics 187.

Stewart, I. N. and Tall, D. O. *Algebraic Number Theory*. Chapman and Hall, 1979.

Stromberg, K. R. *An Introduction to Classical Real Analysis*. Wadsworth International, 1981.

van der Poorten, A. J. A proof that Euler missed .... *Mathematical Intelligencer*, 1: 195–203, 1979.

Walker J. *The Flying Circus of Physics*. Wiley, 1977.

Walker, J. S. *Fourier Analysis*. Oxford University Press, 1988.

Wen, L. A space filling curve. *American Mathematical Monthly*, 90: 283, April 1983.

Whittaker, E. T. and Watson, G. N. *Modern Analysis*. Cambridge University Press, 1950.

Wiener, N. *Ex Prodigy*. MIT, 1953.

Wilf, H. S. *Generatingfunctionology*. Academic Press, 1990.

Woods, F. S. *Advanced Calculus*. Ginn, Boston, 1954.

Young, N. *An Introduction to Hilbert Space*. Cambridge University Press, 1988.

Zeeman, M. L. Hopf bifurcations in competitive three-dimensional Lotka–Volterra systems. Preprint 662, Institute For Mathematics And Its Applications, University of Minnesota, 1990.

Zygmund, A. *Trigonometric Series*, 2nd edition. Cambridge University Press, 1959.

INDEX

Printed in the United States
By Bookmasters